T0133973

THE ROLE OF PHYTOCONSTITUTENTS IN HEALTH CARE

Biocompounds in Medicinal Plants

Innovations in Plant Science for Better Health: From Soil to Fork

THE ROLE OF PHYTOCONSTITUTENTS IN HEALTH CARE

Biocompounds in Medicinal Plants

Edited by

Megh R. Goyal, PhD, PE
Hafiz Ansar Rasul Suleria, PhD
Ramasamy Harikrishnan, PhD

APPLE
ACADEMIC
PRESS

Apple Academic Press Inc.
4164 Lakeshore Road
Burlington ON L7L 1A4
Canada

Apple Academic Press Inc.
1265 Goldenrod Circle NE
Palm Bay, Florida 32905
USA

© 2020 by Apple Academic Press, Inc.

First issued in paperback 2021

No claim to original U.S. Government works

ISBN 13: 978-1-77463-512-4 (pbk)
ISBN 13: 978-1-77188-820-2 (hbk)

All rights reserved. No part of this work may be reprinted or reproduced or utilized in any form or by any electronic, mechanical or other means, now known or hereafter invented, including photocopying and recording, or in any information storage or retrieval system, without permission in writing from the publisher or its distributor, except in the case of brief excerpts or quotations for use in reviews or critical articles.

This book contains information obtained from authentic and highly regarded sources. Reprinted material is quoted with permission and sources are indicated. Copyright for individual articles remains with the authors as indicated. A wide variety of references are listed. Reasonable efforts have been made to publish reliable data and information, but the authors, editors, and the publisher cannot assume responsibility for the validity of all materials or the consequences of their use. The authors, editors, and the publisher have attempted to trace the copyright holders of all material reproduced in this publication and apologize to copyright holders if permission to publish in this form has not been obtained. If any copyright material has not been acknowledged, please write and let us know so we may rectify in any future reprint.

Trademark Notice: Registered trademark of products or corporate names are used only for explanation and identification without intent to infringe.

Library and Archives Canada Cataloguing in Publication

Title: The role of phytoconstitutents in health care : biocompounds in medicinal plants / edited by Megh R. Goyal, PhD, Hafiz Ansar Rasul Suleria, PhD, Ramasamy Harikrishnan, PhD.

Names: Goyal, Megh Raj, editor. | Suleria, Hafiz, editor. | Harikrishnan, Ramasamy, editor.

Series: Innovations in plant science for better health.

Description: Series statement: Innovations in plant science for better health : from soil to fork | Includes bibliographical references and index.

Identifiers: Canadiana (print) 20190226250 | Canadiana (ebook) 20190226285 | ISBN 9781771888202 (hardcover) | ISBN 9780429284267 (ebook)

Subjects: LCSH: Herbs—Therapeutic use. | LCSH: Medicinal plants.

Classification: LCC RM666.H33 R618 2020 | DDC 615.3/21—dc23

CIP data on file with US Library of Congress

Apple Academic Press also publishes its books in a variety of electronic formats. Some content that appears in print may not be available in electronic format. For information about Apple Academic Press products, visit our website at **www.appleacademicpress.com** and the CRC Press website at **www.crcpress.com**

OTHER BOOKS ON PLANT SCIENCE FOR BETTER HEALTH FROM APPLE ACADEMIC PRESS, INC.

Book Series: *Innovations in Plant Science for Better Health: From Soil to Fork*
Editor-in-Chief: Hafiz Ansar Rasul Suleria, PhD

Bioactive Compounds of Medicinal Plants: Properties and Potential for Human Health
Editors: Megh R. Goyal, PhD, and Ademola O. Ayeleso

Plant- and Marine-Based Phytochemicals for Human Health: Attributes, Potential, and Use
Editors: Megh R. Goyal, PhD, and Durgesh Nandini Chauhan, MPharm

Human Health Benefits of Plant Bioactive Compounds: Potentials and Prospects
Editors: Megh R. Goyal, PhD, and Hafiz Ansar Rasul Suleria, PhD

Plant Secondary Metabolites for Human Health: Extraction of Bioactive Compounds
Editors: Megh R. Goyal, PhD, P. P. Joy, PhD, and Hafiz Ansar Rasul Suleria, PhD

Bioactive Compounds from Plant Origin: Extraction, Applications, and Potential Health Claims
Editors: Hafiz Ansar Rasul Suleria, PhD, and Colin Barrow, PhD

Phytochemicals from Medicinal Plants: Scope, Applications, and Potential Health Claims
Editors: Hafiz Ansar Rasul Suleria, PhD, Megh R. Goyal, PhD, and Masood Sadiq Butt, PhD

The Therapeutic Properties of Medicinal Plants: Health-Rejuvenating Bioactive Compounds of Native Flora
Editors: Megh R. Goyal, PhD, PE, Hafiz Ansar Rasul Suleria, PhD, Ademola Olabode Ayeleso, PhD, T. Jesse Joel, and Sujogya Kumar Panda

The Role of Phytoconstitutents in Health Care: Biocompounds in Medicinal Plants
Editors: Megh R. Goyal, PhD, Hafiz Ansar Rasul Suleria, PhD, and Ramasamy Harikrishnan, PhD

Assessment of Medicinal Plants for Human Health: Phytochemistry, Disease Management, and Novel Applications
Editors: Megh R. Goyal, PhD, and Durgesh Nandini Chauhan, MPharm

Health Benefits of Secondary Phytocompounds from Plant and Marine Sources
Editors: Hafiz Ansar Rasul Suleria, PhD, and Megh R. Goyal, PhD

ABOUT THE SENIOR EDITOR-IN-CHIEF

Megh R. Goyal, PhD, PE

Retired Professor in Agricultural and Biomedical Engineering, University of Puerto Rico, Mayaguez Campus; Senior Acquisitions Editor, Biomedical Engineering and Agricultural Science, Apple Academic Press, Inc.

Megh R. Goyal, PhD, PE, is a Retired Professor in Agricultural and Biomedical Engineering from the General Engineering Department in the College of Engineering at the University of Puerto Rico–Mayaguez Campus; and Senior Acquisitions Editor and Senior Technical Editor-in-Chief in Agriculture and Biomedical Engineering for Apple Academic Press, Inc. He has worked as a Soil Conservation Inspector and as a Research Assistant at Haryana Agricultural University and Ohio State University.

During his professional career of 51 years, Dr. Goyal has received many prestigious awards and honors. He was the first agricultural engineer to receive the professional license in Agricultural Engineering in 1986 from the College of Engineers and Surveyors of Puerto Rico. In 2005, he was proclaimed as "Father of Irrigation Engineering in Puerto Rico for the Twentieth Century" by the American Society of Agricultural and Biological Engineers (ASABE), Puerto Rico Section, for his pioneering work on micro irrigation, evapotranspiration, agroclimatology, and soil and water engineering. The Water Technology Centre of Tamil Nadu Agricultural University in Coimbatore, India, recognized Dr. Goyal as one of the experts "who rendered meritorious service for the development of micro irrigation sector in India" by bestowing the Award of Outstanding Contribution in Micro Irrigation. This award was presented to Dr. Goyal during the inaugural session of the National Congress on "New Challenges and Advances in Sustainable Micro Irrigation" held at Tamil Nadu Agricultural University. Dr. Goyal received the Netafim Award for Advancements in

Microirrigation: 2018 from the American Society of Agricultural Engineers at the ASABE International Meeting in August 2018.

A prolific author and editor, he has written more than 200 journal articles and textbooks and has edited over 72 books. He is the editor of three book series published by Apple Academic Press: Innovations in Agricultural & Biological Engineering, Innovations and Challenges in Micro Irrigation, and Research Advances in Sustainable Micro Irrigation. He is also instrumental in the development of the book series Innovations in Plant Science for Better Health: From Soil to Fork.

Dr. Goyal received his BSc degree in engineering from Punjab Agricultural University, Ludhiana, India; his MSc and PhD degrees from Ohio State University, Columbus; and his Master of Divinity degree from Puerto Rico Evangelical Seminary, Hato Rey, Puerto Rico, USA.

ABOUT THE EDITOR-IN-CHIEF

Hafiz Ansar Rasul Suleria, PhD

Research Fellow, School of Agric. & Food Facculty of Veterinary & Agric. Sciences The Univ of Melbourne Parkville 3010, Victoria, Au

Hafiz Anasr Rasul Suleria, PhD, is currently working as the Alfred Deakin Research Fellow at Deakin University, Melbourne, Australia. He is also an Honorary Fellow at the Diamantina Institute, Faculty of Medicine, The University of Queensland, Australia.

Recently he worked as a postdoc research fellow in the Department of Food, Nutrition, Dietetic and Health at Kansas State University, USA.

Previously, he has been awarded an International Postgraduate Research Scholarship (IPRS) and an Australian Postgraduate Award (APA) for his PhD research at the University of Queens School of Medicine, the Translational Research Institute (TRI), in collaboration with the Commonwealth and Scientific and Industrial Research Organization (CSIRO, Australia).

Before joining the University of Queens, he worked as a lecturer in the Department of Food Sciences, Government College University Faisalabad, Pakistan. He also worked as a research associate in the PAK-US Joint Project funded by the Higher Education Commission, Pakistan, and the Department of State, USA, with the collaboration of the University of Massachusetts, USA, and National Institute of Food Science and Technology, University of Agriculture Faisalabad, Pakistan.

He has a significant research focus on food nutrition, particularly in the screening of bioactive molecules—isolation, purification, and characterization using various cutting-edge techniques from different plant, marine, and animal sources; and *in vitro*, *in vivo* bioactivities; cell culture; and animal modeling. He has also done a reasonable amount of work on functional foods and nutraceuticals, food and function, and alternative medicine.

Dr. Suleria has published more than 50 peer-reviewed scientific papers in different reputed/impacted journals. He is also in collaboration with more than ten universities where he is working as a co-supervisor/special member for PhD and postgraduate students and is also involved in joint publications, projects, and grants. He is Editor-in-Chief for the book series Innovations in Plant Science for Better Health: From Soil to Fork, published by AAP.

Readers may contact him at: hafiz.suleria@uqconnect.edu.au.

ABOUT THE EDITOR

Ramasamy Harikrishnan, PhD

Assistant Professor, Department of Zoology, Pachaiyappa's College for Men (affiliated with the University of Madras), Kanchipuram, Tamil Nadu, India

Ramasamy Harikrishnan, PhD, is an Assistant Professor in the Department of Zoology at Pachaiyappa's College for Men (affiliated with the University of Madras), Kanchipuram, Tamil Nadu, India. He formerly worked as an Assistant Professor in the Department of Biotechnology at Bharath College of Science and Management, Thanjavur, Tamil Nadu, India; as Research Associate at the Council of Scientific & Industrial Research (CSIR), Government of India; as a postdoctoral fellowship at the Korea Science and Engineering Foundation (KOSEF), South Korea; and as a Research Professor at the College of Ocean Science, Jeju National University, South Korea. He has guided MPhil students and PhD research scholars at Bharathidasan University, Tamil Nadu, India. He has been an invited speaker at the University of South Korea. He has been involved in several projects as co-principal investigator at Jeju National University, South Korea. He has visited several countries for oral presentations, including Malaysia, Thailand, Indonesia, China, and Japan. He also organized several seminars and symposia. He prepared the first cDNA library in a marine kelp grouper, *Epinephelus bruneus,* and identified 2000 mRNA sequences, and some of which have been submitted to the National Center for Biotechnology Information.

He received his BSc degree in Zoology in 1993, his MSc degree in 1995, MPhil degree in 1998, and PhD degree in 2004 from the Bharathidasan University, Tamil Nadu, India. He has published 113 peer-reviewed international articles; five international review articles and seven national research publications; and 22 research articles in the Proceedings of the 7th National Symposium on Advance Research in Biosciences, 3–4 March, 2014 and

7th edition was edited by him. Dr. Harikrishnan is a reviewer and member of the editorial boards for 15 international journals, and has attended many workshops and training courses and national and international conferences and symposiums. Readers may contact him at: rhari123@yahoo.com.

CONTENTS

CONTRIBUTORS

O. Uche Arunsi

Graduate Assistant, Department of Biochemistry, Faculty of Biological and Physical Sciences, Abia State University, Uturu, Abia State University, NYSC; Department of Biochemistry, Faculty of Life Science, Ahmadu Bello University, Zaria, Kaduna State, Nigeria, Mobile: +234-8160874528, E-mail: venniabia@gmail.com

Faiza Ashfaq

National Institute of Food Science and Technology, Faculty of Food, Nutrition and Home Sciences, University of Agriculture, Faisalabad, Pakistan, E-mail: kahloonfazi@yahoo.com

Kanza Aziz Awan

National Institute of Food Science and Technology, Faculty of Food, Nutrition and Home Sciences, University of Agriculture, Faisalabad, Pakistan, E-mail: kahloonfazi@yahoo.com

Chellam Balasundaram

Professor, Department of Herbal and Environmental Science, Tamil University, Thanjavur–613005, Tamil Nadu, India, Mobile: +91-9443821666, E-mail: bal333.chellam@gmail.com

Ahmad Bilal

Assistant Professor, University Institute of Diet and Nutritional Sciences, University of Lahore, Lahore, Pakistan, E-mail: ahmedbilal1317@yahoo.com

Masood Sadiq Butt

Dean, Faculty of Food, Nutrition, and Home Sciences, University of Agriculture, Faisalabad, Pakistan, E-mail: drmsbutt@yahoo.com

S. Chieme Chukwudoruo

Research Assistant, Department of Biochemistry, Faculty of Biological and Physical Sciences, Abia State University, Uturu, Abia State, Nigeria, Mobile: +234-8069114183, E-mail: cchukwudoruo@gmail.com

Djadouni Fatima

University Professor and Researcher, Biology Department, Faculty of Natural Sciences and Life, Mascara University, Mascara – 29000, Algeria, Tel.: +0021-354 0039761, E-mail: Sdjadouni@gmail.com

Megh R. Goyal

Retired Faculty in Agricultural and Biomedical Engineering from College of Engineering at the University of Puerto Rico–Mayaguez Campus; and Senior Technical Editor-in-Chief in Agricultural and Biomedical Engineering for Apple Academic Press Inc.; PO Box 86, Rincon–PR–006770086, USA, E-mail: goyalmegh@gmail.com

Ramasamy Harikrishnan

Assistant Professor, Department of Zoology, Pachaiyappa's College for Men, Kanchipuram–631501, Tamil Nadu, India, Mobile: +91-8940283621, E-mail: rhari123@yahoo.com

E. Martina Ilondu
Senior Lecturer, Department of Botany, Faculty of Sciences, Delta State University, Abraka, Delta State, Nigeria, Mobile: +234-8036758249, E-mail: ebelemartina@gmail.com; ilondu@delsu.edu.ng

P. C. Jessykutty
Professor and Head, Department of Plantation Crops and Spices, College of Agriculture, Vellayani, Kerala, India, Tel.: 9497484060, E-mail: pcjessy@gmail.com

C. Stanley Okereke
Senior Lecturer, Department of Biochemistry, Faculty of Biological and Physical Sciences; and Director, Examination Center, Abia State University, Uturu, Abia State, Nigeria, Mobile: +234-7036769255, E-mail: okerekecstan@gmail.com

N. S. Sonia
PhD Scholar, Department of Plantation Crops and Spices, College of Agriculture, Vellayani, Kerala, India, Tel.: 8129448004, E-mail: coa2008soniya@gmail.com

Hafiz Ansar Rasul Suleria
School of Agric. & Food, Facculty of Veterinary & Agric. Sciences, The Univ of Melbourne Parkville 3010, Victoria, Au

K. M. Vidya
PhD Scholar, Department of Plantation Crops and Spices, College of Agriculture, Vellayani, Kerala, India, Tel.: 9964498871, E-mail: guddy.vidyagowda@gmail.com

ABBREVIATIONS

AAI	anti-atherogenic index
ACAT	Acyl-CoA: cholesterol acyltransferase
AICD	activation-induced cell death
AIDS	acquired immune deficiency syndrome
ALP	alkaline phosphatase
ALT	alanine transaminase
ART	anti-retroviral treatment
AST	aspartate transaminase
BAP	benzyl amino purine
BMI	body mass index
CAT	catalase
CCl_4	carbon tetrachloride
CK	creatine kinase
CK-MB	creatine kinase-MB
CRP	C-reactive protein
DISC	death-inducing signaling complex
DMAPP	dimethylallyl pyrophosphate
DNA	deoxyribonucleic acid
DPPH	2,2-diphenyl-1-picrylhydrazy
FADD	Fas-associated death domain
FDA	Food and Drug Administration
FID	flame ionization detector
FOSHU	foods for specific health use
FRAP	fluorescence recovery after photobleaching
FSH	follicle-stimulating hormone
FTIR	Fourier-transform infrared spectroscopy
GAE	gallic acid equivalents
GC	gas chromatograph
GC-MS	gas chromatography-mass spectrometry
GLS	glucosinolates
GPx	glutathione peroxidase
GR	glutathione reductase

GSH	glutathione
GST	glutathione-S-transferase
HIV	human immunodeficiency virus
HPLC	high-performance liquid chromatography
IBA	indole-3-butyric acid
IL-6	interleukin-6
IPGRI	International Board for Plant Genetic Resources
IPP	isopentenyl pyrophosphate
IR	infrared
JKT	Jurkat
LAB	lactic acid bacteria
LCAT	lecithin cholesterol acyltransferase
LDH	lactate dehydrogenase
LH	luteinizing hormone
MAC	membrane attack complex
MAC	mitochondrial apoptosis-induced channel
MDA	malondialdehyde
MEP	methylerythritol phosphate
MS	Murashige-Skoog's
MVA	mevalonic acid
NACO	National AIDS Control Organization
NAFLDs	non-alcoholic fatty liver diseases
NBJ	nutrition business journal
NDEA	N-nitrosodiethyamine
NF-κB	nuclear factor-κB
NIST	National Institute of Standard and Technology
NMR	nuclear magnetic resonance
NO	nitric oxide
NSAIDs	non-steroidal anti-inflammatory drugs
PCO	protein carbonyl
PON1	paraoxonase 1
PTZ	pentylenetetrazole
RE	rutin equivalents
RNS	reactive nitrogen species
ROM	reactive oxygen metabolites
ROS	reactive oxygen species
SMAC	second mitochondria-derived activator of caspases
SOD	superoxide dismutase

STZ	streptozotocin
TCM	Traditional Chinese Medicine
THC	tetrahydrocannabinol
TLC	thin layer chromatography
TNFR	TNF receptor
TNF-α	tumor necrosis factor-α
TRADD	TNF-receptor associated death domain
UNAIDS	United Nations Program on HIV/AIDS
WHO	World Health Organization

PREFACE 1

Globally, non-communicable diseases (NCDs) like HIV/AIDS and cancer are emerging as the most predominant challenges in human health. According to the UNAIDS survey of 2017, about 36.9 million people in the world live with HIV/AIDS, including 1.8 million children less than 15 years of age. Similarly, cancer is another leading cause of suffering and globally about 9.6 million people died in 2018, shockingly about 70% of these deaths (about 1 in 6) are in the under-developing and developing countries.

Hence any new tool to tackle the NCDs can supplement/supplant in the disease management scenario, and the available treatments are often symptomatic of the interest in the emerging traditional medicines because of low cost and easy availability. The wide variety of potential plant species are amazing as about 250,000 higher level and 215,000 lower level plant species have been identified. However, only about 6% of these species have been screened for their biological properties while about 15% have been subjected to phytoconstituent analysis.

Phytoconstituents comprise of chemical bioactive compounds that occur naturally in plants, and these are generally called as 'biologically significant chemicals,' but not essentially are considered as nutrients. The plant bioactive compounds (such as flavanones, chromones, benzenoids, inositol, pyrimidines, triterpenoids, steroids, etc.) are well-established secondary metabolites with complex organic molecules and possess several physiological or toxicological effects in humans and animals. In the modern era, a number of novel techniques are available for the extraction of bioactive components and essential oils from the medicinal and aromatic plants. These phytomedicines (phytoconstituents) have been used throughout the world as an integral part of our healthcare system since a long time ago. Although modern drugs are effective in preventing NCDs disorders, yet their use is often limited due to their adverse side effects.

The WHO estimated that more than 80% of the human population living in developing countries still relies on herbal drugs for their primary healthcare. Recent scientific literature has documented the link between certain phytoconstituents and inhibition or protection against NCDs.

It has been recently estimated that natural products have enjoyed a 100 times higher hit rate when compared with synthetic drugs. The philosophy behind these approaches is the long-standing therapeutic experiences and the belief that a complex pathophysiological process can be treated more effectively with the combination of low dose extracts due to synergetic effects and apparently with fewer side-effects than with a single high dosage of an isolated compound.

Several research laboratories are actively involved in phytomedicine research, and new chemical entities are being successfully used over the years, and many are in the clinical trial phase. Correspondingly, the current global trade of herbs has increased to a yearly turnover of US$100 billion. In India and China, the medicinal plant trade has increased from about $2 billion to $5 billion when compared to $1.00 billion annually in Germany.

Under these circumstances, this book volume aims to review the therapy with herbal phytoconstituents with special reference to the current scenario on the basis of: type of disorders, mode of action, and pharmacological screening. This book volume, *The Role of Phytoconstituents in Health Care: Biocompounds in Medicinal Plants*, under the book series on Innovations in Plant Science for Better Health: From Soil to Fork focuses on: Part I: Herbs and Their Extracts: Scope and Role of Bioactive Compounds in Human Health; Part II: Functional Activities of Selected Plants. The information in five chapters in this book volume is essential for herbal drug researchers and scientists in pharmaceutical industries. The collated information will not only provide information on the management or prevention of HIV/AIDS and cancer disorders but also pave the way for future research.

I am beholden to Dr. Megh R. Goyal for inviting me to join his team in publishing this book volume. He is a well-known international scientist and engineer with expertise in agricultural and biological engineering, a rare blend indeed. He is a patron, leader, mentor, and model for budding scientists.

—Ramasamy Harikrishnan, PhD
Editor

PREFACE 2

To be healthy, it is our moral responsibility,
Towards Almighty God, ourselves, our family, and our society;
Eating fruits and vegetables makes us healthy,
Believe and have a faith;
Reduction of food waste can reduce the world hunger and
can make our planet eco-friendly.

—Megh R. Goyal

Medicinal plants contain certain chemicals in leaves, stems, roots, and fruits that can provide therapeutic benefits against different kinds of diseases. These chemicals are often referred to as "phytochemicals." The word "phyto" is a Greek word that means "plant." Phytochemicals are natural non-essential bioactive compounds found in plants and plant foods. In our world today, many commercially available drugs have plant-based origins, with more than 30% of modern medicines directly or indirectly derived from medicinal plants. The use of medicinal plants has largely increased because they are locally accessible, economical, less toxic with fewer side-effects, and vital in promoting health. However, scientific data and information regarding the safety and efficacy of these medicinal plants are inadequate.

The book volume will further encourage the preservation of traditional medical knowledge of medicinal plants. These plant products are drawing the attention of researchers/policymakers because of their demonstrated beneficial effects against diseases with high global burdens. This book volume is a treasure house of information and, provides excellent reference material for researchers, scientists, students, growers, traders, processors, industries, dieticians, medical practitioners, and others. We hope that this compendium will be useful for students and researchers as well as those working in the food, nutraceutical, and herbal industries.

The contribution of all cooperating authors to this book volume has been most valuable in the compilation. Their names are mentioned in each chapter and in the list of contributors. We appreciate you all for having patience with our editorial skills. This book would not have been written without the valuable cooperation of these investigators, many of whom are renowned scientists who have worked in the field of plant science and food science throughout their professional career.

The goal of this book volume is to guide the world science community on how plant-based secondary metabolites can alleviate us from various conditions and diseases.

We thank editorial and production staff, and Ashish Kumar, Publisher and President, at Apple Academic Press, Inc., for making every effort to publish this book when all are concerned with health issues.

We express our admiration to our families and colleagues for their understanding and collaboration during the preparation of this book volume. As an educator, I share with you a piece of advice to one and all in the world: *"Permit that our almighty God, our Creator, provider of all and excellent Teacher, feed our life with Healthy Food Products and His Grace; and Get married to your profession."*

We request the reader to offer your constructive suggestions that may help to improve the next edition.

Megh R. Goyal, PhD, PE
Hafiz Ansar Rasul Suleria, PhD
Editors

PART I

Herbs and Their Extracts: Scope and Role of Bioactive Compounds in Human Health

CHAPTER 1

POTENTIAL OF HERBAL EXTRACTS AND BIOACTIVE COMPOUNDS FOR HUMAN HEALTHCARE

RAMASAMY HARIKRISHNAN and CHELLAM BALASUNDARAM

ABSTRACT

Globally, herbal therapy is still the backbone of about 80% of the population in developing countries as essential healthcare, because the herbals are cheapest, widely available anywhere, and with no side effects. Many of Western medicines have been derived from plant resources, e.g., magnaprin from bark of willow, dioxin from foxglove, qualaquin from cinchona bark, and zomorph from the opium poppy. Scientists, doctors, and pharmaceutical companies have access to information on herbal medicines; however, the main issue is the absence of scientific validation. The mode of action of many plant-derived methodologies can be negatively affected by the isolation technique, concentration, time, and administrative route, lack of uniform formulation within or among manufacturers. In addition, the efficiency of herbal medicine may be due to symbiotic activity of multi-pace bio-compounds and antagonist actions, so that testing of single bio-compound may not be an appropriate evaluation of the efficiency of the "crude" extracts used by patients. Phytochemicals in the form of dietary supplements, herbs, and spices, comprise an unlimited source of molecules available for improving human health. The claimed beneficial values of the described species give a sound basis for the significant role of folk medicine in the development of new drug.

Further insights on the mechanisms of action for different herbals and their bioactive compounds are necessary to discover the molecular targets for human diseases. Such mastery will help to enlighten specialists, permit them to give exceptional guidance to their patients and thus can support in the improvement of innovative pharmaceuticals via a rediscovery of age-old compounds. For example, the biocompounds in the fennel plant can be used for different drug formulations, which play a significant role in our health. Majority of studies were mainly conducted on the anti-oxidant, antimicrobial activities in different animal models. Natural products have traditionally been favored for the exploration of potential targets for drug production, and its function is still clear today. The pharma-industry has been facing a serious threat as the drug discovery practice is becoming exceptionally costly, sensitive, and typically ineffective. Natural materials may serve as the main source of drugs and more than half of the pharma preparations in use today are derived from natural plant resources. However, careful experimental designs using multidisciplinary pathways together with consistency and assessment of natural products are important for the successful production of innovative and promising therapies.

1.1 INTRODUCTION

According to the World Health Organization (WHO), about one-third of all deaths are due to infectious diseases, and nearly 50,000 people die each day globally. Antibiotics continue to be a major therapeutic discovery in the effective control of pathogens. Numerous broad spectrum and pathogen-specific drugs to control microbes are available today. Though the treatment with antibiotics offers significant advantages, yet their use is still arbitrary today, thus resulting in an alarming increase in antibiotics resistance of virulent strains. The indiscriminate or improper usage of antimicrobial drugs has increased drug resistance among microorganisms [971]. The high degree of multidrug resistance is due to the presence of antibiotic efflux systems, which trigger resistance to pathogens [5]. Indeed drug resistance among pathogens poses a serious threat throughout the world [5, 210, 751, 878]. Therefore, the exploration of a new generation of drugs is a continuous pursuit, because the target microorganisms continue to develop new virulent strains that have become further resistant to currently existing antimicrobial drugs [254, 1031]. Also, new families of antimicrobial drugs frequently have short life expectancy [184]. Hence, there is a dire need to explore novel groups of antibacterial agents.

At present, an important challenge is a multidrug resistance, which is frequently not controlled to the specific antibiotic but commonly extends to other related antimicrobial compounds; also, common microbes can also become opportunistic pathogens posing a threat to disease management in humans. The future scope of controlling diseases with the wide use of antimicrobial drugs is vague. Indiscriminate use of antibiotics is alarming in developing and developed countries. Also, the antibiotics are known for bioaccumulation, which adversely affects the immunity status of the host [249]. In this backdrop, there is an urgent need for alternative antimicrobial strategies. In this direction, the emerging focus is now on plant sources since natural products are considered safer than the traditional chemotherapeutics [19]. This has led to the growing interest in ancient phytotherapy [609], since antimicrobials of plant origin have better therapeutic potential with few side effects [133, 323].

WHO reports that medicinal plants are the ideal source to develop novel drugs [13]. The traditional use of plants and their bioproducts began since time immemorial in the form of folk medicine; and over time, its application has also been extended into allopathic medicine [233]. Today, a number of herbs for use as traditional and folk medicine continue to explore further understanding and to document their potentials, protection, and effectiveness with modern tools of research. Indeed since antiquity, the pharmacological bioproperties of various herbs were documented; they possess antimicrobial secondary metabolites, such as: alkaloids, amides, glycosides, essential oils, flavonoids, furostanol, phytosterols, spirostanol, saponins, steroids, tannins, terpenoids, etc. which are utilized against infectious pathogens [69, 79, 288, 309, 386, 433, 487, 723, 970]. Many herbal medicines possess immunomodulatory, antibacterial, and antimicrobial activities [206, 271, 717, 972].

Various advantages of using plant-based antimicrobial compounds include: few side effects, increased patient tolerance, cheaper than allopathy drugs, acceptable for long-term use, being eco-friendly, renewable in nature [321], increased utilization efficiency, and sustained bioavailability [398]. Besides their antimicrobial and immunomodulatory properties, they can be gainfully used in the management of non-communicable diseases. Indeed a number of recent studies on breast, lung, and colon cancer cell lines indicate that herbal extracts have the potential to inhibit the growth of pathogens [859]. The combinations of crude herbal extract and protein inhibitors exert the highest inhibition against infectious agents. Hence, there is a global awareness to make innovative antimicrobial agents as an alternative tool for chemotherapy [26, 737].

Another novel approach is the synergistic effect between antibiotics and bioactive plant extracts, which lead to the development of newer, more effective complexes either by inhibition of cell-wall synthesis or lyzes or leads to the mortality of target microbes. Numerous studies indicated that the effects of such drugs in treatment and control of cancer could circumvent the adverse effects of synthetic drugs on human health [351].

World's population will exceed 7.5 billion after 10 to 15 years in the developing countries, which impel the need to find alternative methods to control multidrug-resistant pathogens; till date about 70 to 80% of world population will be using herbal remedies as essential healthcare due to cultural suitability and compatibility with least significant side effects on human body [392, 790, 959]. In non-industrialized countries, the survival of 60 to 90% of the population relies on medicinal plants for family health care either totally or partially [1040].

Civilizations (that originated from 3000 BC in Egypt, Middle East, India, and China) began using medicinal herbs for health care. The American Indians have used herbal medicines for remedy of malaria, mutism, abnormalities in menstrual irregularities, menopause, kidney problems, sore throat, etc. [92, 98, 497]. Even today, many of these medicinal plants are being used to treat diseases in countries like China, Japan, and Korea; e.g., *Cimicifuga* species can be used to treat fever, pain, and inflammation [821, 823]. Also, many synthetic drugs cause environmental problems since the chemical industries release a large amount of waste by-products [835]. Medicinal plants play a unique role in pharmacological research and drug discovery; constituents of herbal are used not only directly but also pave the way for the biosynthesis of novel drugs or serve as models for pharmacological compounds [1106].

Secondary metabolites of plants (viz. alkaloids, essential oils, glycosides, flavonoids, lignins, phenolics, saponins, sterols, terpenes, tannins, tirpenes, etc. [238, 385, 434, 513]) are considered as essential for metabolism, but generally they are not implicated in the metabolic function of the host; and these are used successfully in traditional and as components in allopathic medicine [233]. They provide many biological remedial benefits, such as, anti-cancer, anti-microbial, anti-oxidant, anti-diarrheal, anti-aging, anti-carcinogen, analgesic, and wound healing compounds and protect from Alzheimer's, brain dysfunctions like Parkinson's, cardiovascular, Huntington's diseases, immune/autoimmune diseases, etc. [511, 906]; traditionally plants have also been used for anthelmintic, antilithic diuretic, stomachic, anti-malarial [214, 1025], and anti-inflammatory [997] conditions.

Therefore, the identification of novel compounds in herbals and their mode of action are imperative to evaluate their potential for clinical use. Since improper stimulation of NF-κB is implicated in various human diseases [146], there is an urgent need for novel therapeutic intervention. Although their innovation and effectiveness aim towards the arrest of disease development, yet they are costly therapeutic regimens. On the other hand, plant-based natural products and therapies have advantages to combat cancer-like diseases; and the search began in the 1950s with the exploration and growth of the vinca alkaloids (like vinblastine and vincristine) and their successful use in cancer management [305].

India is known as the storehouse of herbal plants, out of which many have been incorporated in the Ayurvedic, Unani, and Siddha systems to prevent or cure various diseases, such as: arthritis, cold, cough, chronic fever, convulsions, dysentery, diarrhea, diabetes, malaria, skin diseases, respiratory disorder, trauma, and in treatment of internal organ, hepatic, vessel, and immunologic disorders [43, 663]. The plant phytoconstituents are being exploited in the industrial production of cosmetics, food ingredients, perfumes, and pharmaceuticals [501].

India has a rich biodiversity that facilitates the discovery of its novel bioactive molecules in an effort to discover new drugs [177]. Traditionally, herbal-based medicines have been extensively used by Indian people. Indeed about 7500 plant species have been used as herbal medicines out of which 90% plants are used in formulations; and over 90% of the medicinal herbs are wild species [619]. Isolation and characterization of plant-derived natural drugs provide new avenues of research in the development of drugs to target important enzymes that hinder the onset of life-threatening diseases. The expenditures for alternative treatment averages to $1127/year/patient, when compared with $1148 for traditional treatment [791]. Numerous plant-based natural-product drugs include arteether, galanthamine, and triotopium approved recently by the US Food and Drug Administration (FDA) for clinical trials.

This chapter focuses on the potential of herbal drugs and bioactive compounds for human healthcare.

1.2 HISTORY OF TRADITIONAL HERBAL MEDICINES

Since time immemorial, traditional medicine continues to remain as the most inexpensive and easily available resource in primary healthcare. The typical recent example is the widely recommended oral administration of *Andrographis paniculata* and Papaya leaf extract as an antidote for dengue. The application of herbal dates before middle Paleolithic age for health problems by early man, which is documented in fossil records found in caves where a Neanderthal man was buried 6,000 years ago.

Ayurveda means "the Science of Life," as it offers a rich and holistic approach to a healthy life. According to Western scholars, Ayurveda marked the actual beginning of the ancient Indian medical science from 2500 to 600 B.C.; it comprises of eight divisions that cover various aspects of the science of life and the art of remedial.

At present, the Indian System of medicine has branched into different systems, namely: Ayurveda, Siddha, Unani, Homeopathy, and Tibetan. Folk traditions and knowledge have codified around 8,000 plant species as herbal medicines. However, pollen analysis indicates that the numerous medicinal plants with different remedial values for various illnesses about 4,000 year ago in the Ancient Egyptian civilization have been buried; and these plants constitute a record of wealth of information used as remedies mainly for pain relief, heart, and circulatory disorders preserved 3,500 years ago [1064]. Of these, over 12,000 items are commonly used by folk therapists till date; while around 500 items are rare. Medicinal herbal products are used after several processing methods, such as: shade drying, grinding, frizzle with vinegar or wine.

About 5,000 traditional herbal remedies are available and regularly used mainly in rural areas of China [557]. Writings on papyrus leaves have been found in Ancient China and Egypt thus describing plant-based medicines as early as 3,000 B.C. African tribes and Amerindian cultures document the use of medicinal herbs in the remedial act; and other traditions developed Siddha, Ayurveda, Unani, and traditional Chinese medicines (TCM) [28]. However, these traditional medical systems did not gain popularity in the West with the development of more well documented synthetic drugs that are available commonly. Despite these discoveries from synthetic medicines, the traditional systems were used by the majority of people, such as: Siddha or Ayurveda remedy in India, Kampo medicines in Japan, TCM in China, and Unani remedy Southern Asia [655].

Many traditional therapies found their origin in China toward the Japanese systems of folk therapy in the 9th century [820]. In India, traditional remedy was found in Vedic Sanskrit text dating back to 1600–3500 B.C. It was estimated that about 250,000 higher level and 215,000 lower level plant species were identified [37]; and total species were as high as 500,000 at an upper level. However, only about 6% of these have been screened for biological properties, and only about 15% were subjected to phytochemical research [953].

In India, the ancient physicians recorded the medicinal properties and therapeutic uses of herbs empirically in the indigenous medical system, which constitutes the foundation of ancient medical science [763]. Indeed this information is a rich resource forming the basis for natural drug research and development. Chinese, native Egypt, Hellenes, Romanian, and Arabian texts are well documented dating back to 5,000 years. The Latin Indian texts include Sanskrit and *Charaka Samhita* Ayurveda [993]. In India, still indigenous systems follow the tenets envisaged in Ayurveda, Siddha, Unani, Homeopathy, Yoga, and Naturopathy [259]. In several developing countries, still the majority of people rely on the folk specialist and their techniques of medicinal herbs for health care, due to the historical and cultural reasons along with allopathy [998].

The value of natural products can be accessed by:

- Introductory price of new chemical material of broad structure diversification helps for semi or total synthetic transition;
- Numerous diseases treated with these substances; and
- Their efficacy in the healing of diseases.

In recent years, increased attention has surged in the field of natural products attributed to several factors, such as: unmade remedial use, discovery of diverse chemical structures and their functions of natural secondary metabolites, the efficiency of novel bioactive compounds as biochemical exploration, the improvement of perceptive procedure, isolation, purification, and characterization to existence of natural compounds, etc. Aptly, WHO has constituted design, guidance, and specification importance of folk remedy. It has provided the agro-industrial techniques for the cultivation and processing of medicinal plants and herbal drugs [1038].

There is no doubt that the herbal therapy is popular in rural and urban Indian communities and other parts of the world; because natural products are safe, non-toxic, and have fewer side effects apart from being cost-effective. Therefore, the demand is continuously broadening in developed and developing countries for herbal-based products in healthcare, cosmetics, pharma drugs, food ingredient, etc. [259].

1.3 TRADITIONAL HERBAL MEDICINES IN DEVELOPED COUNTRIES

Plants and their phytochemical constituents have an ancient history of use in certain systems of traditional medicine and in modern (Western) medicine. Application of herbal medicines for a health problem in developed countries has stretched particularly to the latter half of the 20th century. Interest in medicinal plant research has been revived recently [965] in European and American countries, where the herbal substance is being included in alternative, integral, and holistic system of therapy. Such therapy in the Western World has increasing search for primitive interest in these resources; and has been accepted by most of the academic and manufacturing sectors [937].

In pursuance, a number of monographs is available on selected medicinal plants in the European, ESCOP [257], German Commission-E [91] and the WHO [1037], which describes similar indigenous names, diversity, and mode of actions of the plants in macro or microscopic investigation. Methods of purity testing include uses of parts, active principles, dosage, and forms of dosing, pharmacology, contraindications, adverse reactions, and medicine uses [279]. A number of herbal products sold in the European market are still marketed as admixture, tablets, or extracts exhibiting anti-oxidant, antiseptic, diuretic, modulation of the central nervous system, digestive problem, etc. Such products are believed to be safer for a healthy lifestyle than synthetic drugs [847].

The secondary metabolites of plants act as antioxidants; for example, the Phenolic compounds are the largest group with significant antioxidant potential [196]. Among different species of *Thymus* to prevent insect bites, muscle swelling, rheumatism, pains, etc. [673], the essential oil is a mixture of monoterpenes, and it possesses the anti-oxidative, anti-microbial, medicinal drug, anti-tissue, anti-spasmodic, and anti-bacterial properties [614].

1.4 MODERN HERBAL MEDICINES

Medicinal plants play a crucial role in the formulation of innovative medicines. Indeed in modern medicine, plant-derived drugs have a long record of their use as a whole or parts of the plant that is evidenced in folklore or traditional systems. About 100 higher plants based innovative drugs have been included in the USA drug market out which examples are: canescine, moderil, raudixin, velban, and leurocristine approved during 1950–1970. Indeed several new drugs are available in the world market, such as: etopophos, lectinan, eguggulsterone, vumon, cesamet, kelnac, Z-guggulsterone, and ginkgolides between 1971 and 1990. About 2% of the drugs introduced in the market include paclitaxel, toptecan, gomishin, irinotecan, etc., during 1991 to 1995.

Herbal drug serpentine, derived from *Rauwolfia serpentina* root in 1953, is used to reduce hypertension and pressure in blood [914] while vinblastine (derived from *Catharanthus roseus*) is employed to reduce choriocarcinoma, Hodgkins, non-Hodgkins lymphomas, leukemia in children, testicular, and neck cancer [266].

The phytophyllotoxin isolated from *Podophyllum emodi* successfully treated testicular, lung cancer, and lymphomas. Vincristine isolated from *C. roseus* is recommended in the management of childhood lymphocytic leukemia, lung, uterus, and breast cancer [265]. Vumon and Etopophos purified from *Podophyllum* species were able to reduce testis and lung cancer [394, 897]. Taxol isolated from *Taxus brevifolia* is used to treat metastatic ovarian and lung cancer [814]. Number of plant-based drugs has been commonly used to treat diabetes, hypertension, jaundice, mental, and skin problems, TB; and possess anti-bacterial, anti-diabetic, anti-microbial, and anti-cancer [745] properties. Triterpenoid extracted from *Hemidesmus indicus* possesses antiodote or anti-venom property to Russell's viper [141]. A recent study on experimental animals indicated that biocompound extracted from *Strychnus nux-vomica* seed inhibit viper venom through lipid peroxidation [140]. The mode of action of these herbal compounds induces venom neutralization that could be used against snakebite.

1.5 SIGNIFICANCE OF TRADITIONAL HERBAL MEDICINES

Over 80% of peoples in developing countries have been using herbal therapy as primary healthcare, which is safe, effective, cultural-privileged, and least side effects. Ancient text also documents the use of herbal medicines against age-related diseases like diabetic, immune, and liver disorders, memory loss, and osteoporosis. Remarkably, these medicines are made from renewable bio-resource through eco-friendly processes; therefore, these ensure economic importance to the marginalized farmers by growing the raw materials.

1.6   ECONOMIC IMPORTANCE OF TRADITIONAL HERBAL MEDICINES

Herbal drugs and natural medicines offer novel and crucial leads against diseases, e.g., cardiovascular diseases, cancer, malaria, and neurological disorders, etc. [626, 785]. Currently, the global trade of herbs has a yearly turnover of US$100 billion. The medicinal plant trade in India and China is about $2 billion to $5 billion compared to $1.00 billion annually in Germany [882]. Several herbs and herbal extracts are used as prescribed drugs in Germany and France, and its sale in the European Union is about $20 billion today. The plant-based drugs sold in health food stores in the USA were worth about $4 billion in 1996 [795]. In India, the export of herbal drugs is about $80 million [85]. Currently, the sale of herbal medicines as health foods or nutraceuticals is valued at about $80 billion to $250 billion in the USA and Europe [107]. Since the cost of health care in the allopathic system is increasing at an alarming rate, there is a steadily growing market for phytopharmaceuticals in the world.

1.7   SIGNIFICANCE OF DRUG DISCOVERY FROM NATURAL PRODUCTS

WHO stressed the important role of traditional medicine as herbal drugs for therapeutic purposes that continue to be in use till date, thus paving the way for the development of modern medicine [1035]. The traditional medicine is synthesized with the remedial experience of generations through enthusiastic physicians in primitive medicine. The folk measures included herbal products, minerals, organic matter, etc. Synthetic drugs have been effective in curing many diseases. More than 3 million new chemicals developed in pharmaceutical industries have been based on their efforts to produce new drugs. Their victory in developing drugs to treat or cure many diseases has been successful and gave desired results immediately in AIDS, cancer, diabetes, and heart diseases and has been challenging due to their complexity in disease management.

In this context, many societies have their own rich plant pharmacopeias, which are on the revival. In consonance, the search for the pharmacological properties of higher plants has led to the innovation of different novel drugs. Consequently, they still continue to be a potential reservoir to find out useful chemical compounds not only as drugs, but also as templates for synthetic analogs.

1.8   GLOBAL TRADE OF HERBAL MEDICINES

The importance of medicinal herbs as a viable resource of overseas trade in developed and developing countries has turned to be used as exportable raw and processed materials (such as: plants and their parts like bark, flowers, fruit, leaves, roots, seeds, tubers, and wood extract) in the pharmaceutical industry. Herbal medicines constitute the mainstay in the healthcare of over 75% global population, mostly in the developing countries, because of good cultural satisfaction, fewer side effects, and consistency within the human body. With relaxing guidelines of the FDA for the sale of herbal supplements [306], the market is flourishing with herbal products [104].

In general, herbal medicines are readily preferred by consumers, especially due to adverse side effects of modern remedies and the trust in herbal remedies for safe use. It constitutes an excellent alternative to preventive medicine; and herbal medicines confer effective health benefits in certain diseases [779]. WHO has reported that the world market of herbal products valued at about US $83 billion annually [397]. World Bank has reported that the requirement of herbal-related drugs and the raw material is growing at 5 to 15%/year. In European Union, the herbal market has exceeded $20 billion; and is about $3 billion in Germany, $1.6 billion in France, $0.6 billion in Italy, and $1.5 billion in Japan [107]. In Germany and France, the herbal essence is sold as a recommended drug, which are covered by countrywide health insurance. The Global pharma market worth has risen from US $550 billion in 2004 to US$900 billion in 2012 [795]. In India, the herbal-based market is about US$10 billion/year, and exported value is US$1.1 billion with 50% contribution by classical

Ayurvedic preparations. The annual trade in China is about US$48 billion, and export worth US$3.6 billion [435].

India exported 32,600 tons of raw drugs worth US$46 million during 1995 [223]. China exported 120,000 tons of crude drugs that were worth US$264.5 million [789]. The European market value of herbal products was worth $7.0 billion in 1997. Among these, the German market alone was about $3.5 billion compared to $1.8 billion in France, $700 million in Italy; $400 million in the UK, $300 million in Spain, and $100 million in the Netherlands [779].

Medicinal plant-based products are essential for folk medicine formulation in Germany and Russia, from where these are imported by various European countries. The present international herbal market is worth about US250 billion [107]. TCM reached US$400 billion in 2010 [1020]. Presently, the US has the biggest market of Indian herbal products for 50% of total exports. However, Asian countries (like Japan, Hong Kong, Korea, Singapore, etc.) are the top importers of the herbal products making about 66% share from China [779]. The herbal trade is expected to increase @ 6.4% annually [99].

1.9 HERBAL PHARMACEUTICALS AS BUSINESS

Globally, there is a growing need for herbal medicines because of their wide phytochemical properties, along with the least side effects and lower cost compared to synthetic drugs [487]. Medicinal plants have tremendous potential for sustained economic growth. Among these, there is a global demand for nutraceuticals (Health foods), particularly in the USA, Europe, and Japan [281].

Huge nutraceutical trade was increased after US Congress passed the Dietary Supplement Health Education Act of 1994, which allows extraordinary health benefits of herbal foods or as dietary supplements for anticipation and healing of diseases [833]. With such rapid growth for herbal products in perfumery and allied industries, the herbal materials play a significant role in the national economy of India [487]. For instance, in India, approximately 25,000 effective plant relevant drugs were used by over 1.5 million specialists in the folk system in rural communities for their preventive, promotional, and remedial effects. More than 7, 800 herbal-based industries in India devour over 2,000 tons/annum [789]. The sale of Ayurvedic medicines in the market grows about 20% annually, while the sale of medicinal plants registered a growth of nearly 25% from 1987 to 1996 [487].

1.10 CURRENT STATUS OF HERBAL MEDICINE MARKET

The widely use of herbal pharma is not barred to developing countries, e.g., in France and Germany, about 70% of all physicians commonly recommend herbal medicines [104]. The ambition of the patient for herbal therapy is growing rapidly [194]. The pharma herbal market is booming after US-FDA relaxed guidelines for the trade of herbal-added products [237]. The herbal medicine market in the European Union is valued at $20 billion compared to $6 billion in 1991 [342]. In 1999, the international herbal-added trade rose to US$15 billion, of which it was US$7 billion in Europe, US$2.4 billion in Japan, US$3 billion in Northern America, and US$2.7 billion in Asia. In the USA, herbal-added dietary food products were worth $14 billion in 1996 [843].

Europe alone imported 400,000 tons of herbal products/annum from Africa and Asia, with an average market value of US$1 billion. Germany spent US$2.2 billion/annum in herbal products, followed by £116 million in France and by £88 million in the UK [616].

In the last few decades, the curious emphasis has been on herbal products [435]. During 1983–1994, it is estimated that about 39% of all 520 innovative approved drugs were of natural-based materials; and out of which 60–80% were antibacterial and anti-cancer drugs [551, 660]. Herbal drug is also the essential ingredient of most modern preparations for reducing hypertension and mental illness.

1.11 FUTURE PROSPECTS OF TRADITIONAL HERBAL MEDICINES

Globally, almost three-fourths of the widely used herbal drugs have their roots from traditional knowledge. About 25% of modern medicines are from synthetic analogs prototype compounds isolated from herbs. The medicinal plants have a dual role for healthcare products and income source; and supplying botanicals and herbals to the drug industry is a booming business. The scientists, doctors, and pharmaceutical companies will continue their search for such new herbal supplements in countries like China and India, etc.

1.12   SOURCES OF NEW HERBAL MEDICINE FOR FUTURE RESEARCH

Ethnomedical treatment has been commonly used by tribes in most of the countries, e.g., treatment of asthma, aging, and cancer, cut wounds, diabetes, eczema, gastric ulcer, jaundice, mental illness, skin infection, swelling, scabies, venereal diseases, and snakebite [745, 774]. However, this information on herbals has not been documented since the message is transferred from generation to generation by vocally [222]. In this regard, WHO mainly aims to document the available information on the methods of herbal usage by the tribes across the world [430, 1033]. In this direction, several developing countries have begun to document the ethnomedical data through field research to document scientific claims by tribal healers. Once these primary ethnomedical methods are scientifically accepted and dispersed properly, the herbal effectiveness will be improving health [608].

1.13   ROLE OF BIOACTIVE COMPOUNDS IN HUMAN HEALTHCARE

Recently much interest has been focused on herbal bioactive compounds because of their ability to promote our health, with reference to the amelioration and better management of deleterious cancer, diabetes, and cardiovascular disorders, etc. [188, 420, 466]. The associations of free radicals by tumor or proto-oncogenes are involved in the progress of cancers [328, 329, 587]. Phenolics plant compounds have recently gained great popularity due to their significant bioactives and positive effects on human health [46, 63]. It is important to emphasize the need to increase the consumption of foods rich in Phenolic compounds aimed at preventive action [63]. The chlorophyll compound has a significant antioxidant effect [804]. Parthenolide and some other metabolites have been determined as an inhibitor of human blood platelet function [359].

Plant bioactive compounds are secondary metabolites that exhibit physiological or toxicological effects in humans and animals [73, 296]. Flavanones, chromones, benzenoids, inositol, pyrimidines, triterpenoids, and steroids isolated from *Viscum coloratum* significantly inhibited the production of superoxide anion by neutrophils into fMLP (formyl-L-methionyl-L-leucyl-L-phenylalanine) [553]. Different parts of the plant are still used in local systems of medicine to cure human ailments like bowels, cholera, ulcers, and wounds [828] and also in folk therapy to cure diarrhea, fevers, and as a tonic in psychiatry cases [705]. However, the identification of the plant(s) and its availability, their specific chemical composition, dose, potential use for sickness, the mechanism of action, unpredictable toxicity, and cost are major constraints that hinder the use and widespread use of herbal preparations.

1.13.1 BIOACTIVE PROPERTIES OF PLANT-BASED NATURAL PRODUCTS

Numerous medicinal plants have shown bioactivities of aqueous or active solvent extracts of different parts (listed in Appendix 1.1), such as:

- Anti-apoptotic;
- Anti-bacterial;
- Anti-cancer;
- Anti-clastogenic;
- Anti-diarrheal;
- Anti-hyper ammonic activities;
- Anti-inflammatory;
- Anti-malarial;
- Anti-helmintic;
- Anti-neoplastic;
- Anti-neuropathic;
- Anti-obesity;
- Anti-oxidant;
- Anti-platelet aggregation;
- Anti-ulcer;

- Anti-viral;
- Cardio protective;
- Cellular pathways;
- Cytotoxicity;
- Decrease peroxidation;
- Free radicals;
- Gene expression;
- Immunostimlant;
- Inhibits kinase activity;
- Modulate enzyme;
- Nerve impulse transmission;
- Restoration of synaptic;
- Rheumatic activity;
- RNA synthesis;

Plant-based bioactive compounds have been used to treat or cure diseases such as (Appendix 1.1):

- Abdominal pain;
- Abortifacient;
- Acne;
- Acrid;
- Acute viral hepatitis;
- Allergic rhinitis;
- Allergies;
- Amenorrhea;
- Analgesic;
- Anemia;
- Anodyne;
- Anthelmintic;
- Anti-aging;
- Anti-arthritic;
- Anti-atherosclerotic;
- Anti-fertility;
- Anti-histamine;
- Anti-pyretic;
- Anti-septic;
- Anti-spasmodic;
- Anti-stress;
- Antiperiodic;
- Anxiolytic;
- Aphrodisiac;
- Aphtha;
- Arthritis;
- Asthma;
- Astringent;
- Atherosclerosis;
- Backache;
- Bilious fever;
- Bitter;
- Bleeding;
- Body ache;
- Brady cardiac;
- Burses and dropsy;
- Bursitis;
- Cardiovascular conditions;
- Cerebrovascular disorders;
- Chest pain;
- Cholagogue;
- Cholera;
- Choleretic;
- Chronic pain;
- Cirrhosis;
- Cold;
- Colic;
- Colitis;
- Common cold;
- Conjunctivitis;
- Constipation;
- Convulsions;
- Coronary heart disease;
- Cough;
- Demulcent;
- Dermopathy;
- Destroys worms;
- Diabetes mellitus;
- Diaphoretic;
- Diarrhea;
- Diuretic and expectorant;
- Dizziness,
- Dysentery;
- Dyspepsia;
- Eczema;
- Emetic;
- Enhancing neuronal synthesis;
- Enteritis;
- Epilepsy;
- Estrogenic;
- Eye infections;
- Febrifuge;
- Flatulence;
- Flu;
- Gastric pain;
- Gastro protective;

- Gleet;
- Gonorrhea;
- Gout;
- Hair loss;
- Headache;
- Helminthiasis;
- Hemodynamic;
- Hemorrhoids;
- Hepatoprotective laxative;
- Herpes simplex type II infection;
- Herpes;
- HIV/AIDS;
- Hypertension;
- Hypolipidemic;
- Inflammations;
- Ingestion problems;
- Jaundice;
- Joint pain;
- Lactagogue;
- Laxation and cleansing of bowels;
- Laxative;
- Leprosy;
- Leucoderma;
- Leucorrhea;
- Liver and cardio tonic;
- Longevity;
- Mastitis;
- Menstrual disorders;
- Menstruation;
- Migraine;
- Motion sickness;
- Nausea;
- Nephritis;
- Ophthalmic;
- Oral thrush;
- Paralysis, peptic ulcers;
- Piles;
- Postoperative vomiting;
- Pulmonary heart disease;
- Purgative;
- Rejuvenation;
- Repair of damaged neurons;
- Respiratory disorders;
- Ringworm;
- Sinus congestion;
- Skin infections;
- Skin-diseases;
- Small pox;
- Snakebite;
- Sore mouths;
- Sore throat;
- Spasms;
- Spleen;
- Spleenopathy;
- Stomachic;
- Stomatitis;
- Stop bleeding;
- Strangury;
- Stroke;
- Styptic;
- Sudorific;
- Thermogenic;
- Thrombopoeitic;
- Trauma;
- Tuberculosis;
- Tumors;
- Typhoid fever;
- Ulcer colitis;
- Ulcers;
- Upper-respiratory tract infections;
- Urethrorrhea;
- Urticaria;
- Uterine stimulant;
- Venereal diseases;
- Vermifuge;
- Vomiting in pregnancy;
- Vomiting;
- Whooping cough.
- Wounds;

1.14 HERBAL EXTRACTS FOR HUMAN IMMUNE SYSTEM

Different solvent extracts from herbal parts elicit an increase in the human PBMCs, Neutrophils, T-lymphocytes, Jurkat (JKT) cells leading to enhanced production of antibodies (Appendix 1.2). Both the innate and acquired immunities are extremely important to ensure good health. The immune system influences the suppression of infectious pathogens or exogenous antigen. A small defect in the innate or acquired immunity can lead to a serious problem in the prevention of infections. The adaptive immune response comprises of two key features: (1) specificity and memory, which include lymphocytes, antibodies, and other molecules; (2) while innate/non-specific defense immunity affords first-line protection during infection.

Variety of components is present in innate immunity before the start of the disease, which evokes a set of disease resistances against a specific pathogen; they comprise cellular and molecular components that recognize the pathogens. The ability of the innate immune system facilitates the elimination of some pathogens; evolutionary innovations have resulted in the development of adaptive immune responses [113]. The dysfunction of the immune defense is important for the onset of infectious diseases, such as: allergy, airways, cancer, etc. Therefore, enhancing immune function is vital to control/suppress epidemic diseases.

As long as the phagocytes work efficiently, there is no need to trigger acquired immunity. For example, in innate immunity, phagocytes (such as macrophages and neutrophils) produce different types of antimicrobial compounds that play an important role in the immune system. The main function of the phagocytes is to remove or eliminate the microorganisms or other dead bodies. Neutrophils possess phagocytosis, which have an important role in host defense against infectious agents. The innate immunity and the phagocytosis of macrophages provide the first barrier against infection and in acquired immunity, which contributes to regulate both humoral and cellular immune functions [438]. The phagocytosis has a burst through oxidative metabolism that results in the generation of ROS, which can be determined by NBT assay to confirm the intracellular killing activity of phagocytosis macrophage [697]. Phagocytosis is the primary defense mechanism against any foreign body entering the body, which is offered by neutrophils and macrophages.

Lymphocytes and splenocytes are the most important components of acquired immunity; and are directly or indirectly involved in the production of antibodies. They contain acid phosphatases that are useful in the destruction of microbes and their antigens [314]. Lymphocytes are type of white blood cells that origin from bone marrow through hematopoiesis and flow into blood and lymphatic streams to bind the receptors on the cell surface. Lymphocytes produce immune specificity, multiplicity, memory as well as recognition, by self or nonself. The lymphocytes are two major groups: B-lymphocytes and T lymphocytes [301]. The B-lymphocytes and plasma cells produce a different type of antibody molecules (IgG, IgM, etc.) that play an important role in humoral immune functions. The IgG and IgM are immunoglobulin proteins that regulate complement pathways, including opsonization and neutralization of microorganisms and its toxins [633]. Cell-mediated protection involved as effector systems are accomplished by T lymphocytes, and its secretion of lymphokines is responsible for the protection of microorganisms, cancer, and type IV hypersensitivity reactions [633]. The B-lymphocytes contain surface receptors that bind to antigens manifesting into an endosomal anatomy and development present in MHC class II to the TH2 cells. Further, it is triggered to produce a large number of clone daughter cells; some of these cells serve as memory cells, and other cells differentiate into plasma cells to produce a large amount of specific antibodies. The IgM is secreted initially and often increasing the proportion of IgG [203, 301] that extends a secondary antibody response against the same antigen for the production of antibodies.

Macrophages play important role in specific and nonspecific and defense functions. These are associated with neutrophils, eosinophils, and natural killer cells that constitute the first line of defense to identify and eliminate the microorganisms or toxic macromolecules. Macrophages are a pivotal key in the maintenance of tissue homeostasis and are responsible for detecting, engulfing, and destroying pathogens [674]. The states of macrophages activation and T- and B-lymphocytes play an important role in the pathogenesis [801, 810]. In response to injury, macrophages bind to invading pathogens and deliver these to other components of adaptive immunity, thus is constituted by antibody to produce cell-mediated immune responses performed by B- and T-lymphocytes [4, 674, 675]. During macrophage activation, several compounds are released, such as: cytokines, ROS, NO, and lipid inflammatory mediators, which are implicated in the inflammatory response [1, 971].

An immunomodulator is a biological or synthetic substance, which suppresses or modulates any component of the immune system, and the effects are known as immunostimulation and immunosuppression. The concept of immunomodulation is closely related to the non-specific immune system, because it stimulates the function and efficacy of macrophages, granulocytes, complement. The NK cells and lymphocytes (which are activate) produce various effector molecules. Most of the drugs trigger only one receptor; however, immunomodulator activates at the same time both immunosuppressant and immunostimulant on different immune compounds within the immune system. Immunomodulation is a process to regulate the immune system of an organism by interfering with its functions either through immunostimulation or immunosuppression leading to enhance immune reactions, which is commonly known as immunostimulative drug. These stimulate primarily the non-specific system that includes granulocytes, macrophages, complement, T-lymphocytes, and various effector substances. Immunosuppression can reduce immune resistance to infections. Both immunostimulation and immunosuppression are required to tackle normal immunological functioning. Modulation of the immune system is important for disease management, and the concept of 'Rasayana' is based on the principle of immunomodulatory activities in Ayurveda. Therefore, both immunostimulation and immunosuppression agents are important for searching better immuno agents [738, 960].

Several research studies have reported that plant substances possess immunomodulating properties to stimulate non-specific innate immunity [44]. Immunomodulation by medicinal plants can afford an alternative to conventional chemotherapy in a variety of diseases [324, 989]. Both the immunostimulation and immunosuppression are important to regulate normal immunological performance and its adaptation with pathophysiological processes; hence, these are called immunomodulatory agents. In this regard, the medicinal plants have rich source of immunomodulating substances (especially granulocytes, macrophages, natural killer cells) to induce immunity and complement functions [411, 819].

1.15 HERBS AND THEIR BIOACTIVE COMPOUNDS: EFFECTS ON HIV/AIDS AND VIRAL DISEASES

Human immunodeficiency virus (HIV) is an etiological agent of acquired immune deficiency syndrome (AIDS) [65, 285]. About 87% of reported AIDS cases are due to heterosexual sex, which is the major transmission route (sexually transmitted disease) worldwide, according to the National AIDS Control Organization (NACO) [445, 676]. In addition, HIV transmission by blood and blood products, by infected mother to infants is of a major concern. Once HIV virus enters the body, then it multiplies in CD^{4+} cells or macrophages through CD^4 receptor and CCR5 or CXCR4 [72, 965]. HIV-1 steadily multiplies in the lymphoid tissues resulting in progressive deterioration of immunity, which leads to AIDS. Finally, CD^{4+} helper cell count begins a steady decline (below 200 cells/mm^3) accompanied by an increase in plasma HIV-RNA concentration, thus facilitating possible risk of opportunistic infections that more often leads to neoplasia. The decline of CD^{4+} T-cells initiated by activation-induced cell death (AICD) and apoptosis plays a critical role on non-infected cells [71, 90, 632].

Recent treatment of HIV disease is highly active antiretroviral treatment (ART) using nucleoside or non-nucleoside analogs of reverse transcriptase; and protease inhibitor drugs can modulate hyperimmune response beneficial for HIV suffered patients. However, these drugs have prolonged limited success because of their toxicity [273, 315, 744].

The efficiency of HAART is control of HIV-1 multiplications, but the drug-resistant virus patients need to continue the search for new inhibitors for HIV-1 [991, 931]. Viruses are genetically highly variable RNA genome. Herbal plants are effective and affordable therapeutic agents. It would be logical to assume that these plant active compounds have anti-retroviral properties [383, 529, 629, 844, 918, 927, 980, 1094, 1097]. The traditional use of herbs to treat HIV is limited [171, 307, 453]. The herbal therapy is considered as complementary medicine for HIV patients in European countries [711].

Several medicinal plants inhibit virus replication and boost the immune system against opportunistic infections due to HIV/AIDS [514, 701]. Traditional healers have been practicing some medicinal herbs to control HIV [432, 635, 639, 644]. Medicinal herbal products offer an attractive alternative medicine to HIV or anti-HIV drug target for the mode of attachment, entry, and influence replication [111, 530, 598, 703, 899, 1096]. The anti-HIV plant active compounds have diverse chemical structures, e.g., the efficacy of glycyrrhizin from *Glycyrrhiza uralensis* has been studied in patients with AIDS [344, 404]. An earlier study revealed that *Moringa oleifera* powder supplementation strengthens the immune system for patients suffering from HIV infection [116]. Number of plant-based bioactive compounds (such as: tannins, flavones, alkaloids, polysaccha-rides lignans, coumarins, and terpenes) have shown anti-viral activity [213, 278, 382, 617, 682, 845, 887, 963]. Therefore, such plants can serve as useful sources leading to the discovery of novel anti-HIV compounds (Appendix 1.3).

1.16   GLOBAL STATUS OF CANCER

The potential manifestation of cancer includes abnormal bleeding, swelling, longer cough, weight loss, etc. [215]. They form a subset of neoplasm that usually undergoes an unregulated growth of tissue but does not always form a mass or lump, which is commonly referred to as a tumor [86, 190]. Around 13% of people (about 12.7 million people) have died due to cancers in 2008 [415]. About 5.8 million people died in 1990 compared to 7.98 million in 2010 [590]. Nearly 90.5 million persons were afflicted by cancer in 2015 [293]. In humans, the most common cause for mortality has been lung (1.4 million), stomach (740,000), liver (700,000), colorectal (610,000), and breast (460,000) cancers [123]. The cancer of lung, prostate, colorectal, and stomach is common in males [1042], whereas cancer of breast, colorectal, cervical, and lung is common in females [1042]. In children, the common cancers are acute lymphoblastic leukemia and brain tumors [1041]. In developed countries, invasive cancer is a major cause of death than that in developing countries [415].

Invasive cancer may strike at any age, but generally, most patients over the age of 65 are more vulnerable [185]; and at this age, the probability of cancer affliction is high [421]; the immunosenescence has the relationship between aging and cancer [740]. DNA errors can occur once in a lifetime [20] to cause age-related endocrine changes [31]. In humans, over 100 types of cancers have been identified. Among these, the mortality due to tobacco cancer is about 22% [121]; and 10% is due to obesity, deficiency of diet, less physical work, more alcohol, radiation effect, and environmental contamination [29, 121, 413, 699]. Hepatitis B and C, and papillomavirus can cause over 20% mortality [121], while 5–10% of cancers are inherited [355]. In children, cancer cases are attributed to 34% of leukemia, 12% of lymphomas, and 3% brain tumors [428, 1022].

The characteristic features of all tumor cells are continuous improper signals of cell growth or cell multiplication, programed cell death inter-ruption, an endless number of cell growth, developing excessive blood vessel network, tissue invasion, and metastasis [332]. Carcinoma cancers are formed from epithelial cells that are most common among prostate, breast, pancreas, colon, and lung cancer. The sarcoma cancers (formed from cartilage, bone, fat, and nerve tissues) develop mesenchymal cells on the outside of bone marrow. However, the lymphoma or leukemia is formed from hematopoietic cells on bone marrow that mature in parts of lymph and blood vessels. The germ-cell tumor is formed from embryonic stem cells that are mostly present in the testis or ovary, particularly called seminoma and dysgerminoma. The blastoma cancers are also formed from immature embryonic stem cells.

The number of cancers shows signs of recovery with the stopping of smoke, maintaining good health, low alcohol consumption, and red meat intake, eating fresh fruits and whole grains, plenty of clean vegetables, vaccination against certain infectious diseases, and avoiding exposure to high sunlight [509, 731]. Approximately 70% to 90% of common cancers are due to environmental hazards and hence are potentially preventable. Over 30% of cancer deaths are controlled by risk factors, including air contamination, obesity, unbalanced diet, lack of physical work, over alcohol, sexually transmitted diseases, and tobacco intake [1036].

1.17 CANCER THERAPY AND TREATMENT

Cancer is treated synergistically with a combined therapy of radiotherapy, chemotherapy, target therapy, and ablation [121, 808, 936]. Breast cancer can be reduced after taking daily tamoxifen or raloxifene medication [946]. Vitamin D_3 intake or exposure to morning or twilight exposure to sunlight may lower the risk of death from cancer [89]. Vaccines have been developed and used to control carcinogens [122], papillomavirus [122], and hepatitis B virus [122]. Chemotherapy is the effective treatment of cancer with single or combination with chemotherapeutic drugs that kill cancer cells rapidly.

Targeted therapy is another form of chemotherapy, which targets cancer cells or normal cells based on molecular differences. It generally blocks the estrogen receptor molecules that inhibit the growth of cancer cells [677]. However, this chemotherapy may be effective based on the types and stages of cancer. The combination therapy with ablation and chemotherapy has been found useful in the management of breast, rectal, pancreatic, osteosarcoma, testis, ovary, multiple myeloma, lymphoma, prostate, and certain lung cancers [363, 679].

Radiotherapy involves treating or curing of cancer by DNA damage in cancerous tissues. The surgical method is the removal of solid cancerous tissues for palliation, which helps to prolong survival [363]. Palliative care includes the reduction of physical, emotional, spiritual, and psycho-social distress to enhance the quality of life. Immunotherapy stimulates or helps to boost the immune system to fight cancer. This approach has been practiced since 1997, and includes the use of antibodies, checkpoint therapy, and adoptive cell transference [1003]. Complementary medicine uses the substance with conventional medicine for treating cancers, whilst the alternative medicine uses some compounds as a substitute of conventional medicine [1034].

1.18   APOPTOSIS

Apoptosis is a process of programmed cell death that appeared in eukaryotic by several biochemical changes resulting in shrinking and morphology changes of cell, spherical bubbles, nuclear or DNA damage, and chromatin compression. In this process, they produce cell fragments known as apoptotic bodies, which are engulfed and rapidly removed by phagocytic cells before the apoptotic contents leak out to damage the neighboring cells [22]. However, the process of apoptosis cannot stop once it has begun. Every day about 50 to 70 billion cells die in a normal human adult [21] compared to death of 20 to 30 billion cells in child [424].

The apoptosis can be initiated by *an intrinsic pathway* that activates the stressed cells by intracellular signals and release proteins from mitochondrial intermembrane [21], while the *extrinsic pathway* is activated the cell-surface death receptors through binding with extracellular ligands leads to formation of death-inducing signaling complex (DISC) [21]. In *intrinsic pathway*, an apoptotic protein target mitochondria is affected by swelling in the formation of membrane pores or it may increase the mitochondrial membrane mobility that leaks out the apoptotic bodies [192, 303]. The *intrinsic pathway* tumors occur frequently than the *extrinsic pathway* being very sensitive [647]. There is evidence that nitric oxide (NO) has ability to promote apoptosis through mitochondria membrane potential making it more permeable [108]; it also implicates in initiation and inhibition of apoptosis [109]. Mitochondria releases cytochrome-c by mitochondrial apoptosis-induced channel (MAC) that involved morphological changes as related apoptosis [192]. The cytochrome-c once binds to apoptotic protease activating factor-1 (Apaf-1), ATP, and pro-caspase-9 to form an apoptosome protein complex, which cleaves the pro-caspase-9 into caspase-3. MAC (Membrane Attack Complex) formed by complement activation regulated by various proteins, such as anti-apoptotic genes like mammalian encoded Bcl-2 family of protein are capable of inhibiting apoptosis through MAC [216, 589]. The Bax makes the pore, while Bcl-2, Bcl-xL or Mcl-1 inhibits the pore formation.

In mammals, the *extrinsic pathway* begins with the direct initiation of apoptosis through TNF (tumor necrosis factor) induction or Fas-Fas ligand-mediated induction that involved with the TNF receptor (TNFR) family [1002]. Once macrophages activate to release TNF-α cytokine, which is a main extrinsic mediator of apoptosis. The majority of human cells are present TNF-α receptors, namely TNFR1 and TNFR2. Once TNF-α binds to TNFR1, initiate the caspase activation pathways through TNF-receptor associated death domain (TRADD) and Fas-associated death domain (FADD) proteins; however, the cIAP1/2 may probably inhibit TNF-α signals once bind to TRAF2. The FLIP can also inhibit the activation of caspase-8 [147]. TNFR1 signal pathway may promote apoptosis through a caspase-independent manner [156]. The TNF-α binds with apoptosis gives rich TNF-α cytokine production that plays to produced several autoimmune diseases. The *Fas* receptor is the first signal of apoptosis known as Apo-1 or CD95, which is a TNF family transmembrane protein that possibly binds to Fas ligand (*Fas*L) [1002]. The link between Fas and *Fas*L leads to form DISC that contains FADD and caspase-8, -10. In some cells, type I is prepared caspase-8 directly and activates other caspase family members results, which triggers the strange of apoptosis while other type II cells, that combination with Fas-DISC, starts increase to release of proapoptotic factors in mitochondria after activation of caspase-8 [1001].

TNFR1 and *Fas* are activation in mammalian cells, a balance between pro-apoptotic and anti-apoptotic Bcl-2 family members [659]. This balance formed from pro-apoptotic homodimers on mitochondrion outer-membrane required to make permeable for the release of caspase activators (cytochrome-c and second mitochondria-derived activator of caspases (SMAC)). The caspases play a major role of ER apoptotic signals transduction that remarkably conserved, cysteine-dependent aspartate-specific proteases. The caspases are divided into: i) initiator caspases, caspase-2, -8, -9, -10, -11, -12; and ii) effector caspases, caspase-3, -6, -7. The caspases activation is initiated by binding with a specific oligomeric activator protein; then, the caspases act as active effector caspases by proteolytic cleavage. Further, it was proteolytically degrade on a host of intracellular proteins to execute apoptosis through caspase-independent apoptotic pathway [910]. However, the destruction of cellular organelles and mRNA decay are started very early, and this mechanism is not yet fully understood [945]. Cell shrinkage and morphological changes occur by retraction lamellipodia, and the breakdown of the cell occurs by the proteinaceous cytoskeleton through caspases [95].

The chromatin undergoes condensation into compact patches against a perinuclear envelope process called pyknosis. This process becomes discontinuous, and the DNA becomes damage as the process known as karyorrhexis [603, 909]. Then the nucleus becomes breakdown into different distinct chromatin bodies [666]. Both of these pathways induce apoptosis through the activation of caspases by proteases or enzymes. These pathways initiate or executioner caspases and then kill the cell by proteolysis indiscriminately. The cell activates intracellular apoptotic signals through binding with nuclear receptors by glucocorticoids, heat, radiation, nutrient deprivation, viral infection, hypoxia, etc. that paves the way for cell suicide by damaging the membrane triggering signals pathways.

Research on apoptosis has been initiated since early 1990s because of its implication in biological activity due to the implications of defective apoptotic processes in numerous diseases. Excessive cell death is formed atrophy, while the inadequate amount of apoptosis results in the manifes-tation of uncontrolled cell growth like cancer. The apoptosis is usually promoted by *Fas* receptors and caspases though it can be inhibited in a number of Bcl-2 family of proteins. Apoptosis can be regulated by poly ADP ribose polymerase known as cellular components [168].

The phagocytic cells are eliminated of dead cells in an orderly fashion without elicit any inflammatory process known as efferocytosis, which marks the terminal stage of apoptosis [498] and display phosphatidylserine on phagocytic cell surface [558]. The phosphatidylserine normally found on the inner leaflet surface of the plasma membrane and redistributed during apoptosis to the extracellular surface by a protein known as scram-blase [1017]. These molecules imprint the cell for phagocytosis through appropriate receptors, such as macrophages [840]. During the apoptosis, the cellular RNA and DNA are disassociated from each other into various apoptotic bodies [327].

1.19   HERBAL BIOACTIVE COMPOUNDS

The secondary metabolites (including antibiotics, mycotoxins, alkaloids, pigments, and Phenolics groups) are the most common bioactive compounds [362, 689–691]. More than 30,000 functional bioactive terpenoid compounds have been identified in various plants [112]. Each terpene may be easily classified since they possess a unifying feature that basically contains a repeating 5-carbon isoprene unit; it is a distinctive group of hydrocarbon-based natural products formed structure theoretically from isoprene, and it may be divided further into isopentane units. Flavonoids consist of another most rich class of plant polyphenols having over 6000 structures [338]. It was participating and joined a carbon skeleton of diphenyl propanes, and two benzene rings (A and B) formed a linear three-carbon chain ($C_6C_3C_6$). Among these, the central chain is typically formed by a pyran ring (C) and a benzene ring. Based on the difference, the flavonoids are classified into six subclasses namely: anthocyanidins, flavones, isoflavonoids, flavanones, flavanols, and flavonols.

Phenolics acids are the most common non-flavonoids that contain two distinct carbon skeletons: the hxydroxycinnamic (C_6C_3) and hydroxybenzoic (C_6C_1) [197, 358]. The most common hydroxybenzoic acids are gallic, phydroxybenzoic, protocatechuic, vanillic, syringic acids, and the most common hydroxycinnamic acids are caffeic, ferulic, p-coumaric, and sinapic acids [103]. Very recently, an attention has focused on the bioactive compounds because of their capacity to improve or maintained good health in humans, e.g., diabetes, cancer, cardiovascular problems [189, 330, 419, 496], anti-oxidant, anti-mutagenic, anti-allergenic, anti-inflammatory, anti-microbial properties [58, 331, 732]. Hence, these innumerable beneficial effects on human health, researches have been conducted to find out new novel bioactive compounds from fruits, vegetables, plants, agri-, agro-sources. For example, the pomegranate wastes contain a rich source of Phenolics compounds, including anthocyanins, hydrolyzable tannins, and lignans derivatives [3, 97, 199] that have potential anti-oxidant, anti-mutagenic, anti-inflammatory, anti-cancer properties [299, 678, 680]. *Larrea tridentata* has also been used as a rich source of lignan (nordihydroguaiaretic acid) [389], and that possess anti-cancer and anti-viral activity [200, 388, 988].

Bioactive flavonoids are used as immunostimulators to enhance the growth, development, and immunity [856]. Therefore, various bioactive compounds (like alkaloids, glucans, isoflavonoids, phytosterols, polysaccharides, sesquiterpenes, tannins, and vitamins) can modulate the immune systems [166, 920]. Flavonoids and hydroxylated phenols can respond to infection [226]. Saponins compounds are responsible to decrease paw edema, probably through increased release of serine proteases and cytokines [877], and it also activated the macrophages that remove foreign agents and gradually reduced the edema [877]. Flavonoid bioactive compounds are generally non-toxic and possess different favorable biological activity. Epidemiological analysis reveals that dietary intake of flavonoids with fruits and vegetables can reduce prevalence of human cancers through inactivation of carcinogens, anti-proliferation, apoptosis, angiogenesis, and anti-oxidation [379]. On the other hand, saponins generally are toxic, though its consumption may be beneficial in reducing heart disease by humans by binding saponins with plasma membrane and cholesterol [772].

1.20 BIOACTIVE COMPOUNDS IN HERBAL EXTRACTS FOR CANCER TREATMENT

Aqueous and methanolic extracts of *Agastache rugosa* roots exhibit significant activity against HIV integrase [469]. Methanol extract of *Agrimonia pilosa* demonstrates anti-HIV-1 activity [637]. Aqueous extract of *A. pilosa* roots exhibits hepatoprotective activity against chemically induced cytotoxicity in human hepatic cells [725]. *A. officinarum* rhizome has demonstrated anti-inflammatory activity *in vitro* in human cells [1072]. *Ebenus boissieri* hydroalcoholic extract at doses up to 1000 μg/mL inhibits the proliferation of 293T cells and increases caspase-3, -9 activity in a time-dependent manner and also significantly enhances the levels of TNF-α or IFN-γ in MDA-MB231 cells indicating its potential anti-cancer activity against breast cancer cells [395]. *Withania somnifera* can inhibit the growth of human breast, lung, and colon cancer [858]. Seham and Moustafa et al., [850] have tested 200 herbal methanol extracts for cytotoxicity against human cancer cell lines. Three cycloartane triterpenoids isolated from the aerial parts of *Actaea dahurica* possess potential antitumor activity [950]. *Adina rubella* possesses anti-cancer activity [377], and the methanolic extract of *Aegle marmelos* inhibits proliferation of human tumor cells *in vitro* [516, 519]. Number of other medicinal herbal extracts possesses anti-cancer activity against human cell lines (Appendix 1.4).

Phenolics and flavonoids possess cytotoxic and anti-tumor properties [714]. Triterpenes of *Ganoderma lucidum* exhibit significant pharmacological and therapeutic potentials for cancer [862, 1099]. Many subtypes of the triterpene are directly involved apoptosis in cancer cell lines [1099]. Ganodermanontriol is a triterpene extract from *G. lucidum* that reduces the proliferation of colon cancer cells (HCT116 and HT-29) [414]. Further, triterpenes used as a therapeutic anti-cancer agent, arrest the cell cycle in hepatocellular carcinoma (HuH-7), whereas it did not wield any effect in normal human liver cell line [422]. Alkaloids such as Vinca alkaloids, paclitaxel, or camptothecin are used as chemotherapeutics in cancer therapy [239, 240, 275, 648, 987]. Camptothecin produced from *Camptotheca acuminata*, an inhibitor of DNA topoisomerase, is used in cancer therapy [239, 240, 275, 938, 939, 985, 1000, 1047, 1049]. Alkaloids such as berberine and sanguinarine isolated from *Chelidonium* extracts inhibit ABC (ATP-binding cassette) transporters of several cancer cells [713]. Saponins demonstrate the hypocholesterolemic effect and anti-cancer activity [861]. Quercetin shows anti-cancer activity and inhibits cell proliferation of human liver HepG2 and breast MCF-7 breast cancer cells [169, 347, 348, 523]. A vast number of other plant active compounds inhibit cytotoxicity and apoptosis in a variety of human cell lines (Appendix 1.5).

1.21 SUMMARY

This chapter collates the information of herbs and its products documenting their potential to arrest or decrease the growth of progressive and often fatal diseases like cancer. Herbs can also be gainfully used as an adjuvant with other regular treatments like chemotherapy to ameliorate side effects for cancer patients. The medicinal herbs have exhibited curative properties in the treatment of leukemia, skin cancer, and sarcomas in animal models. The clinical effectiveness and toxicity of various herbal medicines are yet to be documented and hence remain uncertain. The available investigations on the effect of herbs have played an important role in disease management in humans. However, still there is not enough of knowledge on the anti-cancer effect of number of plants, while still there is a high percentage of worldwide mortality due to cancer. Therefore, there is a dire necessity to boost future research on the effect of chosen herbs on cancer cell lines. In this regard, India has been the storehouse of medicinal plants most of which are yet to be documented. The growing environmental issues and health concerns of society are also demanding green technology, which indicates the potential of herb-based drugs in future. In this backdrop, this review collates the information that is less focused on role, function, and future prospects in search of novel herbal remedies with least side-effects to augment human health, paving the way for researchers to explore, supplement, and supplant the search of new herbal-based chemical entities based on scientific research and indicate its vast potential for pharma sector.

KEYWORDS

- **bioactive compounds**
- **Food and Drug Administration (FDA)**
- **herbal medicine**
- **human immunodeficiency virus (HIV)**
- **metabolites**
- **World Health Organization (WHO)**

REFERENCES

1. Abbas, A. K., Williams, M. E., Burstein, H. J., Chang, T. L., Bossu, P., & Lichtman, A. H. Activation and functions of CD4+ T-cell subsets. *Immunological Reviews,* **1991,** *123,* 5–22.
2. Abubakar, B. A., Aliyu, M. M., Mikhail, S. A., Hamisu, I., & Adebayo, O. O. Phytochemical screening and antibacterial activities of *Vernonia ambigua, Vernonia blumeoides* and *Vernonia oocephala* (Asteraceae). *Acta Poloniae Pharmaceutica,* **2011,** *68,* 67–73.

3. Adams, L. S., Seeram, N. P., Aggarwal, B. B., Takada, Y. S., & Heber, D. Pomegranate juice, total pomegranate ellagitannins, and punicalagin suppress inflammatory cell signaling in colon cancer cells. *Journal of Agricultural and Food Chemistry*, **2006**, *54*, 980–985.

4. Aderem, A. Role of toll-like receptors in inflammatory response in macrophages. *Crit. Care Med.*, **2001**, *29*, 16–18.

5. Adwan, G. M., Abu-shanab, B. A., & Adwan, K. M. *In vitro* activity of certain drugs in combination with plant extracts against *Staphylococcus aureus* infections. *African Journal of Biotechnology*, **2009**, *8*, 4239–4241.

6. Afolayan, A. J. Extracts from the shoots of *Arctotis arctotoides* inhibit the growth of bacteria and fungi. *Pharmaceutical Biology*, **2003**, *41*, 22–25.

7. Agarwal, C., Sharma, Y., & Agarwal, R. Anticarcinogenic effect of a polyphenolic fraction isolated from grape seeds in human prostate carcinoma DU145 cells: Modulation of mitogenic signaling and cell-cycle regulators and induction of G1 arrest and apoptosis. *Molecular Carcinogenesis*, **2000**, *28*, 129–138.

8. Agrawal, R. R. C. Evaluation of anticarcinogenic and antimutagenic effects of triphala extract. *International Journal of Scientific Research*, **2012**, *2*, 1–6.

9. Agullo, G., Gamet, L., Besson, C., Demigne, C., & Remesy, C. Quercetin exerts a preferential cytotoxic effect on active dividing colon carcinoma HT29 and caco-2 cells. *Cancer Letters*, **1994**, *87*, 55–63.

10. Ahirwal, L., Mehta, A., Mehta, P., John, J., & Singh, S. Anthelmintic potential of *Gymnema sylvestre* and *Swertia chirata*. *Journal of Ethanopharmacol*, **2010**, *7*, 9–10.

11. Ahirwal, L., Singh, S., Mehta, A., & Rajoria, A. Evaluation of antimicrobial potential of *Gymnema Sylvestre* leaves extracts. *International Journal of Pharmaceutical Sciences Review and Research*, **2012**, *16*, 43–46.

12. Ahmad, A., & Khan, A. U. Prevalence of Candida species and potential risk factors for *Vulvovaginal candidiasis* in Aligarh, India. *The European Journal of Obstetrics & Gynecology and Reproductive Biology*, **2009**, *44*, 68–71.

13. Ahmad, I., & Beg, A. Z. Antimicrobial and phytochemical studies on 45 Indian medicinal plants inst multiple drug resistant human pathogens. *Journal of Ethnopharmacology*, **2001**, *74*, 113–123.

14. Ahmad, N., Adhami, V. M., Afaq, M., Feyes, D. K., & Mukhtar, H. Resveratrol causes WAF-1/p21-mediated G1-phase arrest of cell cycle and induction of apoptosis in human epidermoid carcinoma A431 cells. *Clinical Cancer Research*, **2001**, *7*, 1466–1473.

15. Ahmad, N., Feyes, D. K., Nieminen, A. L., Agarwal, R., & Mukhtar, H. Green tea constituent epigallocatechin-3-gallate and induction of apoptosis and cell cycle arrest in human carcinoma cells. *Journal of the National Cancer Institute*, **1997**, *89*, 1881–1886.

16. Ahn, M. J. Inhibition of HIV-1 integrase by galloyl glucoses from *Terminalia chebula* and flavonol glucoside gallates from *Euphorbia pekinensis*. *Planta Medica*, **2002**, *68*, 454–457.

17. Ahn, M. J., Kim, C. Y., Lee, J. S., Kim, T. G., Kim, S. H., Lee, C. K., Lee, B. B., Shin, C. G., Huh, H., & Kim, J. Inhibition of HIV-1 integrase by galloyl glucoses from *Terminalia chebula* and flavonol glycoside gallates from *Euphorbia pekinensis*. *Planta Medica*, **2002**, *68*, 457–459.

18. Aiyar, S. E., Park, H., Aldo, P. B., Mor, G., Gildea, J. J., Miller, A. L., Thompson, E. B., Castle, J. D., Kim, S., & Santen, R. J. TMS, a chemically modified herbal derivative of resveratrol, induces cell death by targeting Bax. *Breast Cancer Research and Treatment*, **2010**, *124*, 265–277.

19. Alagesaboopathi, C. Antimicrobial potential and phytochemical screening of *Andrographis affinis* Nees an endemic medicinal plant from India. *International Journal of Pharmacy and Pharmaceutical Sciences*, **2011**, *3*, 157–159.

20. Alberts, B., Johnson, A., Lewis, J., Raff, M., Roberts, K., & Walter, P. Studying gene expression and function. In: *Molecular Biology of the Cell* (4th edn., pp. 1616). Garland PublishingInc. New York, USA, **2002**.

21. Alberts, B., Johnson, A., Lewis, J., Morgan, D., Raff, M., Roberts, K., & Walter, P. The cell cycle. In: *Molecular Biology of the Cell* (6th edn., pp. 1616). Garland Publishing, Inc. New York, USA, **2015**.

22. Alberts, B., Johnson, A., Lewis, J., Raff, M., Roberts, K., & Walter, P. Apoptosis: Programed cell death eliminates unwanted cells. Chapter 18. In: *Molecular Biology of the Cell* (6th edn., pp. 1616). Garland Publishing, Inc. New York, USA, **2008**.

23. Al-Madhagi, W., Albarai, I., Alariqi, W., Alammari, M., Noman, N. M., & Mohamed, K. Extraction, formulation and evaluation of yemeni *Argemone Mexicana linn* as antimicrobial and wounds healing agent. *European Journal of Biomedical and Pharmaceutical Sciences*, **2016**, *3*, 627–633.

24. Al-Rehaily, A. J., Al-Howiriny, T. A., Al-Sohaibani, M. O., & Rafatullah, S. Gastro-protective effects of Amla, *Emblica officinalis* on *in vivo* test models in rats. *Phyto-medicine*, **2002**, *9*, 515–522.

25. Altaf, R., Asmawi, M. Z., Dewa, A., Sadikun, A., & Umar, M. I. Phytochemistry and medicinal properties of *Phaleria macrocarpa* (Scheff.) Boerl extracts. *Pharmacog-nosy Review*, **2013**, *7*, 73–80.

26. Alviano, D. S., & Alviano, C. S. Plant extracts: Search for new alternatives to treat microbial diseases. *Current Pharmaceutical Biotechnology*, **2009**, *10*, 106–112.

27. Amic, D., Davidovic-Amic, D., Beslo, D., & Trinajstic, N. Structure-radical scavenging activity relationships of flavonoids. *Croatica Chemica Acta*, **2003**, *76*, 55–61.

28. Ampofo, A. J., Andoh, A., Tetteh, W., & Bello, M. Microbiological profile of some Ghanaian herbal preparations-safety issues and implications for the health professions. *Open Journal of Medical Microbiology*, **2012**, *2*, 121–130.

29. Anand, P., Kunnumakkara, A. B., Kunnumakara, A. B., Sundaram, C., Harikumar, K. B., Tharakan, S. T., Lai, O. S., Sung, B., & Aggarwal, B. B. Cancer is a prevent-disease that requires major lifestyle changes. *Pharmaceutical Research*, **2008**, *25*, 2097–2116.

30. Anand, T., Kumar, G. P., Pandareesh, M. D., Swamy, M. S. L., Farhath, K., & Bawa, A. S. Effect of bacoside extract from *Bacopa monniera* on physical fatigue induced by forced swimming. *Phytotherapy Research*, **2012**, *26*, 587–593.

31. Anisimov, V. N., Sikora, E., & Pawelec, G. Relationships between cancer and aging: A multilevel approach. *Biogerontology*, **2009**, *10*, 323–338.

32. Annan, K., & Dickson, R. Evaluation of wound healing actions of *Hoslundia opposita* VAHL, *anthocleista nobilis* G. DON. and *Balanites aegyptiaca* L. *The Journal of Science and Technology*, **2008**, *28*, 26–35.

33. Antosiewicz, J., Herman-Antosiewicz, A., Marynowski, S. W., & Singh, S. V. c-Jun NH(2)-terminal kinase signaling axis regulates diallyl trisulfide-induced generation of reactive oxygen species and cell cycle arrest in human prostate cancer cells. *Cancer Research*, **2006**, *66*, 5379–5386.

34. Anuchapreeda, S., Thanarattanakorn, P., Sittipreechacharn, S., Chanarat, P., & Limtrakul, P. Curcumin inhibits WT1 gene expression in human leukemic K562 cells. *Acta Pharmacologica Sinica*, **2006**, *27*, 360–366.

35. Anuchapreeda, S., Tima, S., Duangrat, C., & Limtrakul, P. Effect of pure curcumin, demethoxycurcumin, and bisdemethoxycurcumin on WT1 gene expression in leukemic cell lines. *Cancer Chemotherapy and Pharmacology*, **2008**, *62*, 585–594.

36. Anuya, A. R., Ramakrishna, Y. A., & Ranjana, A. D. *In vitro* testing of anti-HIV activity of some medicinal plants. *Indian Journal of Natural Products and Resources*, **2010**, *1*, 193–199.

37. Arash, R., Koshy, P., & Sekaran, M. Antioxidant potential and phenol content of ethanol extract of selected Malaysian plants. *Research Journal Biotechnology*, **2000**, *5*, 16–19.

38. Arena, A., Bisifnano, G., Pavone, B., Tomaino, A., Bonina, F. P., Saija, A., Cristani, M., D'Arrigo, M., & Trombetta, D. Antiviral and immunomodulatory effect of a lyophilized extract of *Cappris spinosa* L. buds. *Phytotherapy Research*, **2008**, *22*, 313–317.

39. Armania, N., Yazan, L. S., Ismail, I. S., Foo, J. B., Tor, Y. S., Ishak, N., Ismail, N., & Ismail, M. *Dillenia suffruticosa* extract inhibits proliferation of human breast cancer cell lines (MCF-7 and MDA-MB-231) via induction of G2/M arrest and apoptosis. *Molecules*, **2013**, *18*, 13320–13339.

40. Aslan, M., Deliorman, O. D., Orhan, N., Sezik, E., & Yesilada, E. *In vivo* antidiabetic and antioxidant potential of *Helichrysum plicatum* ssp. *plicatum* capitulums in streptozotocin-induced-diabetic rats. *Journal of Ethnopharmacology*, **2007**, *109*, 54–59.

41. Asres, K., Seyoum, A., Veeresham, C., Bucar, F., & Gibbons, S. Naturally derived anti-HIV agents. *Phytotherapy Research*, **2005**, *19*, 557–581.

42. Astorg, P. Food carotenoids and cancer prevention: An overview of current research. *Trends in Food Science & Technology*, **1997**, *8*, 406–413.

43. Atal, C. K., & Kapoor, B. M. Cultivation and utilization of medicinal plants (Eds PID CSIR). *Asian Pacific Journal of Tropical Biomedicine*, **1989**, *3*, 780–784.

44. Atal, C. K., Sharma, M. L., Kaul, A., & Khajuria, A. Immunomodulating agents of plant origin. I: preliminary screening. *Journal of Ethanopharmacology*, **1986**, *18*, 133–141.

45. Attard, E., & Cuschieri, A. *In vitro* immunomodulatory activity of various extracts of Maltese plants from the *Asteraceae* family. *Journal of Medicinal Plants Research*, **2009**, *3*, 457–461.

46. Auclair, S., Silberberg, M., Gueux, E., Morand, C., Mazur, A., Milenkovic, D., & Scalbert, A. Apple polyphenols and fibers attenuate atherosclerosis in apolipoprotein E-deficient mice. *Journal of Agricultural and Food Chemistry*, **2008**, *56*, 5558–5563.

47. Aviello, G., Rowland, I., Gill, C. I., Acquaviva, A. M., Capasso, F., McCann, M., Capasso, R., Izzo, A. A., & Borrelli, F. Anti-proliferative effect of rhein, an

anthraquinone isolated from *Cassia* species, on Caco-2 human adenocarcinoma cells. *Journal of Cellular and Molecular Medicine,* **2010**, *14*, 2006–2014.

48. Avila, M. A., Velasco, J. A., Cansado, J., & Notarlo, V. Quercetin mediates the down-regulation of mutant p53 in the human breast cancer cell line MDA-MB468. *Cancer Research*, **1994**, *54*, 2424–2428.

49. Awah, F. M., Uzoegwu, P. N., Oyugi, J. O., & Rutherford, J. Free radical scavenging activity and immunomodulatory effect of *Stachytarpheta angustifolia* leaf extract. *Food Chemistry*, **2010**, *119*, 1409–1416.

50. Awah, F. M., Uzoegwu1, P. N., & Ifeonu, P. *In vitro* anti-HIV and immunomodulatory potentials of *Azadirachta indica* (Meliaceae) leaf extract. *African Journal of Pharmacy and Pharmacology*, **2011**, *5*, 1353–1359.

51. Aydın, A. A., Zerbes, V., Parlar, H., & Letzel, T. The medical plant butterbur (*Petasites*): analytical and physiological (re)view. *Journal of Pharmaceutical and Biomedical Analysis*, **2013**, *5*, 220–229.

52. Ayisi, N. K., & Nyadedzor, C. Comparative *in vitro* effects of AZT and extracts of *Ocimum gratissimum, Ficus polita, Clausena anisata, Alchornea cordifolia*, and *Elaeophorbia drupifera* against HIV-1 and HIV-2 infections. *Antiviral Research*, **2003**, *58*, 25–33.

53. Bach, J. P., Deuster, O., Balzer-Geldsetzer, M., Meyer, B., Dodel, R., & Bacher, M. The role of macrophage inhibitory factor in tumorigenesis and central nervous system tumors. *Cancer*, **2009**, *115*, 2031–2040.

54. Badisa, R. B., Badisa, V. L., Walker, E. H., & Latinwo, L. M. Potent cytotoxic activity of *Saururus cernuus* on human colon and breast carcinoma cultures under normoxic conditions. *Anticancer Research*, **2007**, *27*, 189–193.

55. Badria, F. A., Mikhaeil, B. R., Maatooq, G. T. M. A., & Amer, M. Immunomodulatory triterpenoids from the Oleogum Resin of *Boswellia carterii* Birdwood. *Zeitschrift für Naturforschung*, **2003**, *58c*, 505–516.

56. Bag, A., Bhattacharyya, S. K., Pal, K. K., & Chattopadhyay, R. R. *In vitro* anti-microbial potential of *Terminalia chebula* fruit extracts against multidrug-resistant uropathogens. *Asian Pacific Journal of Tropical Biomedicine*, **2012**, *2*, S1883–S1887.

57. Bailly, F., Mbemba, G., & Cotelle, P. Synthesis and anti HIV integrase activities of caffeic acid dimmers derived from *Salvia officinalis. Bioorganic & Medicinal Chemistry Letters*, **2005**, *15*, 5053–5056.

58. Balasundram, N., Sundram, K., & Samman, S. Phenolic compounds in plants and agriindustrial by-products: Antioxidant activity, occurrence, and potential uses. *Food Chemistry*, **2006**, *99*, 191–203.

59. Baliga, M. S., Meera, S., Mathai, B., Rai, M. P., & Pawar, V. Scientific validation of the ethnomedicinal properties of the ayurvedic drug Triphala: A review. *Chinese Journal of Integrative Medicine*, **2012**, *18*, 946–954.

60. Ballal, M. Screening of medicinal plants used in rural folk medicine for treatment of diarrhea, 2005. https://www.pharmainfo.in (Accessed on 30 September 2019).

61. Balunas, M. J., & Kinghorn, A. D. Drug discovery from medicinal plants. *Life Sciences*, **2005**, *78*, 431–441.

62. Bandopadhyaya, S., Mani, R., Ramesh, P. T., & Yogisha, S. *In-vitro* evaluation of plant extracts against colorectal cancer using HCT 116 cell line. *International Journal of Plant Science and Ecology*, **2015**, *1*, 107–112.

63. Barba, F. J., Esteve, M. J., & Frígola, A. Bioactive components from leaf: Appendix-products. *Studies in Natural Products Chemistry*, **2014**, *41*, 321–346.

64. Barik, B. R., Bhowmik, T., Dey, A. K., Patra, A., Chatterjee, A., & Joy, S. S. Prem-nazole an isoxazole alkaloid of *Premna integrifolia* and *Gmelina arborea* with anti-inflammatory activity. *Fitoterapia*, **1992**, *63*, 295–299.

65. Barre-sinoussi, F., Chermann, J. C., Rey, F., Nugeyre, M. T., Chamaret, S., Gruest, C., Dauguest, C., Axler-blin, C., Vezinet-Brun, F., Rouzioux, C., Rozenbaum, W., & Montagnier, L. Isolation of a T Lymphotrophic retrovirus from a patient at risk for acquired immune deficiency syndrome (AIDS). *Science*, **1983**, *220*, 868–871.

66. Barton, B. E., Karras, J. G., Murphy, T. F., Barton, A. B., & Huang, H. F. Signal transducer and activator of transcription 3 (STAT3) activation in prostate cancer: Direct STAT3 inhibition induces apoptosis in prostate cancer lines. *Molecular Cancer Therapeutics*, **2004**, *3*, 11–20.

67. Basma, A. A., Zakaria, Z., Latha, L. Y., & Sasidharan, S. Antioxidant activity and phytochemical screening of the methanol extracts of *Euphorbia hirta* L. *Asian Pacific Journal of Tropical Medicine*, **2011**, *4*, 386–390.

68. Basu, N. K., & Dandiya, P. C. Chemical investigation of *Premna integrifolia* Linn. *Journal of the American Pharmaceutical Association*, **1947**, *36*, 389–391.

69. Baytop, T. Turk eczacilik tarihi arastirmalari. In: *Researches on the History of Turkish Pharmacy* (p. 97). Sinangin Matbaasi, Istanbul, **2000**.

70. Baytop, T. *Turkiye'de Bitkilerle Tedavi* (Treatment of plants in Turkey). *Istanbul University Yay.*, *No. 2355*, Istanbul, **1984**, p. 73.

71. Bentwich, Z., Kalinkovich, A., Weisman, Z., & Grossman, Z. Immune activation in the context of HIV infection. *Clinical & Experimental Immunology*, **1998**, *111*, 1–2.

72. Berger, E. A., Murphy, P. M., & Farber, J. M. Chemokine receptors as HIV-1 co-receptors: Roles in viral entry, tropism and disease. *Annual Review of Immunology*, **1999**, *17*, 657–700.

73. Bernhoft, A. A brief review on bioactive compounds in plants. In: *Proceedings from a Symposium* (pp. 11–17). The Norwegian Academy of Science and Letters, Oslo-Norway, **2010**.

74. Berry, M. I. Feverfew faces the future. *Pharmaceutical Journal*, **1984**, *232*, 611–614.

75. Bessong, P. O., Obi, C. L., Andreola, M. L., Rojas, L. B., Pouyzegu, L., Igumbor, E., Meyer, J. J., Quideau, S., & Litvak, S. Evaluation of selected South African medicinal plants for inhibitory properties against human immunodeficiency virus type 1 reverse transcriptase and integrase. *Journal of Ethnopharmacology*, **2005**, *99*, 83–91.

76. Bessong, P. O., Obi, C. L., Igumbor, E., Andreola, M. L., & Litvak, S. *In vitro* activity of three selected South African medicinal plants against human immunodeficiency virus type 1 reverse transcriptase. *The African Journal of Biotechnology*, **2004**, *3*, 555–559.

77. Bhalodia, N. R., Nariya, P. B., Acharya, R. N., & Shukla, V. J. Evaluation of *in vitro* antioxidant activity of flowers of *Cassia fistula* Linn. *International Journal of Pharmtech. Research*, **2011**, *3*, 589–599.

78. Bhandary, M. J., Chandrash, S. K. R., & Kaveriappa, K. M. Medical ethnobotany of the Siddis of Uttara Kannada District, Karnataka, India. *Journal of Ethnopharmacology*, **1995**, *47*, 149–158.

79. Bhattacharya, A., Chatterjee, A., Ghosal, S., & Bhattacharya, S. K. Antioxidant activity of active tannoid principles of *Emblica officinalis* (amla). *Indian Journal of Experimental Biology*, **1999**, *37*, 676–680.

80. Bhattacharya, A., Ghosal, S., & Bhattacharya, S. K. Antioxidant activity of tannoid principles of *Emblica officinalis* (Amla) in chronic stress induced changes in rat brain. *Indian Journal of Experimental Biology*, **2000**, *38*, 877–880.

81. Bhattarai, G., Lee, Y. H., Lee, N. H., Lee, I. K., Yun, B. S., Hwang, P. H., & Yi, H. K. Fomitoside-K from *Fomitopsis nigra* induces apoptosis of human oral squamous cell carcinomas (YD-10B) via mitochondrial signaling pathway. *Biological and Pharmaceutical Bulletin*, **2012**, *35*, 1711–1719.

82. Bhaumik, S., Anjum, R., Rangaraj, N., Pardhasaradhi, B. V., & Khar, A. Curcumin mediated apoptosis in AK-5 tumor cells involves the production of reactive oxygen intermediates. *FEBS Letters*, **1999**, *456*, 311–314.

83. Bhutkar, M. A., & Bhise, S. B. Comparison of antioxidant activity of some antidiabetic plants. *Journal of Pharmacy Technology*, **2011**, *4*, 415–420.

84. Bicchi, C., Rubiolo, P., Ballero, M., & Sanna, C. HIV-1 inhibitory activity of the essential oil of *Ridolfia segetum* and *Oenanthe crocata*. *Planta Medica*, **2009**, *75*, 1331–1335.

85. Anonymous. Sectoral study on Indian medicinal plants. In: *Status, Perspective and Strategy for Growth* (p. 230). Biotech Consortium India Ltd., New Delhi, **1996**.

86. Birbrair, A., Zhang, T., Wang, Z. M., Messi, M. L., Olson, J. D., Mintz, A., & Delbono, O. Type-2 pericytes participate in normal and tumoral angiogenesis. *American Journal of Physiology-Cell Physiology*, **2014**, *307*, C25–C38.

87. Birt, D. F., Widrlechner, M. P., Hammer, K. D. P., Hillwig, M. L., Wei, J., Kraus, G. A., Murphy, P. A., Mc Coy, J., Wurtele, E. S., Neighbors, J. D., Wiemer, D. F., Maury, W. J., & Price, J. P. Hypericum in infection: Identification of antiviral and anti-inflammatory constituents. *Pharmaceutical Biology*, **2009**, *47*, 774–782.

88. Bisi-Johnson, M. A., Obi, C. L., Kambizi, L., & Nkomo, M. A survey of indigenous herbal diarrheal remedies of O.R. Tambo district, Eastern Cape Province. *The African Journal of Biotechnology*, **2010**, *9*, 1245–1254.

89. Bjelakovic, G., Gluud, L. L., Nikolova, D., Whitfield, K., Wetterslev, J., Simonetti, R. G., & Gluud, C. Vitamin D supplementation for prevention of mortality in adults. *The Cochrane Database of Systematic*, **2014**, *7*, CD-007470, E-article.

90. Blankson, J. N., Persaud, D., & Siliciano, R. The challenge of viral reservoir in HIV-1 infection. *Annual Review of Medicine*, **2002**, *53*, 557–593.

91. Blumenthal, M., Busse, W. R., Goldberg, A., Gruenwald, J., Hall, T., & Riggins, C. W. The complete german Commission E monographs, therapeutic guide to herbal medicines. In: *American Botanical Council, Boston: Integrative Medicine Communications* (p. 685). Austin, Texas, **1998**.

92. Blumenthal, M., Goldberg, A., & Brinkman, J. Black cohosh root. In: Blumenthal, M., (ed.), *Herbal Medicine: The Expanded Commission E Monographs* (pp. 22–27). American Botanical Council: Newton, MA, **2000**.

93. Boateng, J., & Verghese, M. Protective effects of the phenolic extracts of fruits against oxidative stress in human lung cells. *International Journal of Pharmacology*, **2012**, *8*, 52–160.

94. Bodiwala, H. S., Sabde, S., & Bhutani, K. K. Anti HIV diterpenoids from *Coleus forskohlii*. *Natural Product Communications*, **2009**, *4*, 1173–1175.

95. Böhm, I. Disruption of the cytoskeleton after apoptosis induction by autoantibodies. *Autoimmunity*, **2003**, *36*, 183–189.

96. Bokesch, H. R., Pannell, L. K., & Cochran, P. K. A novel anti-HIV macrocyclic peptide from *Palicourea condensata*. *Journal of Natural Products*, **2001**, *64*, 249–250.

97. Bonzanini, F., Bruni, R., Palla, G., Serlataite, N., & Caligiani, A. Identification and distribution of lignans in *Punica granatum* L. fruit endocarp, pulp, seeds, wood knots and commercial juices by GC-MS. *Food Chemistry*, **2009**, *117*, 745–749.

98. Boon, H., & Smith, M. The botanical pharmacy. In: *The Pharmacology of 47 Common Herbs* (pp. 41–45). Quarry Health Books, Kingston, Ontario, Canada, **1999**.

99. Borris, J. Natural products research perspectives from a major pharmaceutical company. Merck Research Laboratories. *Journal of Ethnopharmcology*, **1996**, *51*, 29.

100. Bowen-Forbes, C. S., Zhang, Y., & Nair, M. G. Anthocyanin content, antioxidant, anti-inflammatory and anticancer properties of blackberry and raspberry fruits. *Journal of Food Composition and Analysis*, **2010**, *23*, 554–560.

101. Bown, D. *Encyclopaedia of Herbs and Their Uses* (p. 336). Dorling Kindersley, London, UK, **1995**.

102. Boyd, M. R., Hallock, Y. F., Cardellina, II J. H., Manfredi, K. P., Blunt, J. W., McMahon, J. B., Buckheit, Jr. R. W., Bringmann, G., & Schaeffer, M. Anti-HIV michellamines from *Ancistrocladus korupensis*. *Journal of Medicinal Chemistry*, **1994**, *37*, 1740–1745.

103. Bravo, L. Polyphenols: Chemistry, dietary sources, metabolism, and nutritional significance. *Nutrition Reviews*, **1998**, *56*, 317–333.

104. Brevoort, P. The booming US botanical market: A new over view. *Herbal Gram*, **1998**, *44*, 33–44.

105. Brien, S., Lewith, G. T., & McGregor, G. Devil's claw (*Harpagophytum procumbens*) as a treatment for osteoarthritis: A review of efficacy and safety. *Journal of Alternative and Complementary Medicine*, **2006**, *12*, 981–993.

106. Brindha, D., Saroja, S., & Jeyanthi, G. P. Protective potential (correction of potencial) of *Euphorbia hirta* against cytotoxicity induced in hepatocytes and a HepG2 cell line. *Journal of Basic and Clinical Physiology and Pharmacology*, **2010**, *21*, 401–413.

107. Brower, V. Nutraceuticals: Poised for a healthy slice of the healthcare market? *Nature Biotechnology*, **1998**, *16*, 728–731.

108. Brune, B. Nitric oxide: NO apoptosis or turning it ON? *Cell Death and Differentiation*, **2003**, *10*, 864–869.

109. Brune, B., Von Knethen, A., & Sandau, K. B. Nitric oxide (NO): An effector of apoptosis. *Cell Death and Differentiation*, **1999**, *6*, 969–975.

110. Bruno, M., Rosselli, S., Pibiri, I., Kilgore, N., & Lee, K. H. Anti-HIV agents from the ent-kaurane diterpenoid linearol. *Journal of Natural Products*, **2002**, *65*, 1594–1597.

111. Buckheit, R. W., Russell, J., Xu, Z. Q., & Flavin, M. Anti-HIV-1 activity of cala-nolides used in combination with other mechanistically diverse inhibitors of HIV-1 replication. *Antiviral Chemistry and Chemotherapy*, **2000**, *11*, 321–327.

112. Buckingham, J. *Dictionary of Natural Products* (p. 8584). Chapman & Hall, London, **1998**.

113. Budhia, S., Haringm, L. F., McConnellm, I., & Blacklawsm, B. A. Quantitation of ovine cytokine mRNA by real-time RT-PCR. *The Journal of Immunological Methods*, **2006**, *309*, 160–172.

114. Buhagiar, J. A., Podesta, M. T., Wilson, A. P., Micallef, M. J., & Ali, S. The induction of apoptosis in human melanoma, breast and ovarian cancer cell lines using an essential oil extract from the conifer *Tetraclinis articulata*. *Anticancer Research*, **1999**, *19*, 5435–5443.

115. Bunyapraphatsara, N. *Medicine Plants* (pp. 477–842). Prachachon Publishing Ltd, Bangkok, **2000**.

116. Burger, D. J., Fuglie, L., & Herzig, J. W. The possible role of *Moringa oleifera* in HIV/AIDS supportive treatment. *Int. Conf. The International AIDS Conference*, **2002**, *12*, 14.

117. Burkill, H. M. The useful plants of west tropical Africa families MFT. *Royal Botanic Garden Kew*, **1994**, *4*, 605–610.

118. Buzzini, P., Arapitsas, P., Goretti, M., Branda, E., Turchetti, B., Pinelli, P., Ieri, F., & Romani, A. Antimicrobial and antiviral activity of hydrolysable tannins, mini-review. *Mini-Reviews in Medicinal Chemistry*, **2008**, *8*, 1179–1187.

119. Caceres, D. D., Hancke, J. L., Burgos, R. A., Sandberg, F., & Wikman, G. K. Use of visual analog scale measurements (VAS) to assess the effectiveness of standardized *Andrographis paniculata* extract SHA-10 in reducing the symptoms of common cold. A randomized double blind-placebo study. *Phytomedicine*, **1999**, *6*, 217–223.

120. Calabrese, C. A phase I trial of andrographolide in HIV positive patients and normal volunteers. *Phytotherapy Research*, **2002**, *14*, 333–338.

121. Cancer Fact Sheet N°297. World Health Organization, February **2014**. http://www.who.int/news-room/fact-sheets/detail/cancer (Accessed on 30 September 2019).

122. Cancer Vaccine Fact Sheet. NCI, **2006**. https://www.cancer.gov/about-cancer/causes-prevention/vaccines-fact-sheet (Accessed on 30 September 2019).

123. Cancer. World Health Organization, **2010**. www.who.int/news-room/fact-sheets/detail/cancer (Accessed on 30 September 2019).

124. Candan, F., Unlu, M., Tepe, B., Daferera, D., Polissiou, M., Sokmen, A., & Akpulat, H. A. Antioxidant and antimicrobial activity of the essential oil and methanol extracts of *Achillea millefolium* subsp. *millefolium* Afan. (Asteraceae). *Journal of Ethnopharmacology*, **2003**, *87*, 215–220.

125. Cao, W., Li, X. Q., Liu, L., Wang, M., Fan, H. T., Li, C., Lv, Z., Wang, X., & Mei, Q. Structural analysis of water-soluble glucans from the root of *Angelica sinensis* (Oliv.) diels. *Carbohydrate Research*, **2006**, *341*, 1870–1877.

126. Carter, B. Z., Mak, D. H., Schober, W. D., McQueen, T., Harris, D., Estrov, Z., Evans, R. L., & Andreeff, M. Triptolide induces caspase-dependent cell death mediated via the mitochondrial pathway in leukemic cells. *Blood*, **2006**, *108*, 630–637.

127. Cazarolli, L. H., Zanatta, L., Jorge, A. P., De Sousa, E., Horst, H., Woehl, V. M., Pizzolatti, M. G., Szpoganicz, B., & Silva, F. R. Follow-up studies on glycosylated

flavonoids and their complexes with vanadium: Their anti-hyperglycemic potential role in diabetes. *Chemico-Biological Interactions*, **2006**, *163*, 177–191.

128. Cerda, B., Tomas-Barberan, F. A., & Espin, J. C. Metabolism of antioxidant and chemopreventive ellagitannins from strawberries, raspberries, walnuts, and oak-aged wine in humans: Identification of biomarkers and individual variability. *Journal of Agricultural and Food Chemistry*, **2005**, *53*, 227–235.

129. César, G. Z., Alfonso, M. G., Marius, M. M., Elizabeth, E. M., Angel, C. B., Maira, H. R., Guadalupe, C. L., Manuel, J. E., & Ricardo, R. C. Inhibition of HIV-1 reverse transcriptase, toxicological and chemical profile of *Calophyllum brasiliense* extracts from Chiapas, Mexico. *Fitoterapia*, **2011**, *82*, 1027–1034.

130. César, G. Z., Alfonso, M. G., Marius, M. M., Elizabeth, E. M., Angel, C. B., Maira, H. R., Guadalupe, C. L., Manuel, J. E., & Ricardo, R. C. Curcumin and curcumin derivatives inhibit Tat mediated transactivation of type-1 human immunodeficiency virus long terminal repeat. *Research in Virology*, **1998**, *149*, 43–52.

131. Chan, E. W., Soh, E. Y., Tie, P. P., & Law, Y. P. Antioxidant and antibacterial properties of green, black, and herbal teas of *Camellia sinensis*. *Pharmacognosy Research*, **2011**, *3*, 266–272.

132. Chan, W. H., Wu, H. J., & Hsuuw, Y. D. Curcumin inhibits ROS formation and apoptosis in methylglyoxal-treated human hepatoma G2 cells. *Annals of the New York Academy of Sciences*, **2005**, *1042*, 372–378.

133. Chanda, S., Dudhatra, S., & Kaneria, M. Antioxidative and antibacterial effects of seeds and fruit rind of nutraceutical plants belonging to the family Fabaceae family. *Food & Function*, **2010**, *1*, 308–315.

134. Chandra-Kuntal, K., & Singh, S. V. Diallyl trisulfide inhibits activation of signal transducer and activator of transcription 3 in prostate cancer cells in culture and *in vivo*. *Cancer Prevention Research (Phila)*, **2010**, *3*, 1473–1483.

135. Chang, C. W., Linn, M. T., Lee, S. S., Liu, K. C. S. C., Hsu, F. L., & Lin, J. Y. Differential inhibition of reverse transcriptase and cellular DNA polymerase-alpha activities by lignans isolated from Chinese herbs, *Phyllanthus myrtifolius* Moon, and tannins from *Lonicera japonica* Thunb and *Castanopsis hystrix*. *Antiviral Research*, **1995**, *27*, 367–374.

136. Chang, I. M. Liver-protective activities of aucubin derived from traditional oriental medicine. *Research Communications in Molecular Pathology and Pharmacology*, **1998**, *102*, 189–204.

137. Chang, R. S., & Young, H. W. Inhibition of growth of human immunodeficiency virus *in vitro* by crude extracts of Chinese medicinal herbs. *Antiviral Research*, **1988**, *9*, 163–176.

138. Chang, R. S., Ding, L., Chen, G. Q., Pan, Q. C., Zhao, Z. L., & Smith, K. M. Dehydroandrographolide succinic acid monoester as an inhibitor against the human immunodeficiency virus. *Proceedings of the Society for Experimental Biology and Medicine*, **1991**, *197*, 59–66.

139. Chang, W. T., Kang, J. J., Lee, K. Y., Wei, K., Anderson, E., Gotmare, S., Ross, J. A., & Rosen, G. D. TPL and chemotherapy cooperate in tumor cell apoptosis: A role for the p53 pathway. *The Journal of Biological* Chemistry, **2001**, *276*, 2221–2227.

140. Chatterjee, I., Chakravarty, A. K., & Gomes, A. Antisnake venom activity of ethanolic seed extract of *Stychnos nux vomica* Linn. *Indian Journal of Experimental Biology*, **2004**, *42*, 468–475.

141. Chatterjee, I., Chakravarty, A. K., & Gomesa, A. *Daboia russellii* and *Naja kaouthia* venom neutralization by lupeol acetate isolated from the root extract of Indian sarsaparilla *Hemidesmus indicus* R. Br. *Journal of Ethnopharmacology*, **2006**, *106*, 38–43.

142. Chaturvedi, A. N. *Firewood Farming on the Degraded Lands of the Gangetic Plain, U.P* (Vol. 1, p. 286). Forest Bulletin No. 50, India Government of India Press, Lucknow, India, **1985**.

143. Chauhan, D., & Chauhan, J. S. Flavonoid glycosides from *Pongamia pinnata*. *Pharmaceutical Biology*, **2002**, *40*, 171–174.

144. Chemler, J. A., Lock, L. T., Koffas, M. A., & Tzanakakis, E. S. Standardized biosynthesis of flavan-3-ols with effects on pancreatic beta-cell insulin secretion. *Applied Microbiology and Biotechnology*, **2007**, *77*, 797–807.

145. Chen, C. Y., Yang, W. L., & Kuo, S. Y. Cytotoxic activity and cell cycle analysis of hexahydrocurcumin on SW 480 human colorectal cancer cells. *Natural Product Communications*, **2011**, *6*, 1671–1672.

146. Chen, F., Castranova, V., Shi, X., & Demers, L. M. New insights into the role of nuclear factor-kappa-B, a ubiquitous transcription factor in the initiation of diseases. *Clinical Chemistry*, **1999**, *45*, 7–17.

147. Chen, G., & Goeddel, D. V. TNF-R1 signaling: A beautiful pathway. *Science*, **2002**, *296*, 1634–1635.

148. Chen, H. W., Hsu, M. J., Chien, C. T., & Huang, H. C. Effect of alisol B acetate, a plant triterpene, on apoptosis in vascular smooth muscle cells and lymphocytes. *European Journal of Pharmacology*, **2001**, *419*, 127–138.

149. Chen, J. C., Chang, N. W., Chung, J. G., & Chen, K. C. Saikosaponin-A induces apoptotic mechanism in human breast MDA-MB-231 and MCF-7 cancer cells. *The American Journal of Chinese Medicine*, **2003**, *31*, 363–377.

150. Chen, J. C., Lu, K. W., Lee, J. H., Yeh, C. C., & Chung, J. G. Gypenosides induced apoptosis in human colon cancer cells through the mitochondria-dependent pathways and activation of caspase-3. *Anticancer Research*, **2006**, *26*, 4313–4326.

151. Chen, J. C., Lu, K. W., Tsai, M. L., Hsu, S. C., Kuo, C. L., Yang, J. S., Hsia, T. C., Yu, C. S., Chou, S. T., Kao, M. C., Chung, J. G., & Wood, W. G. Gypenosides induced G0/G1 arrest via CHk2 and apoptosis through endoplasmic reticulum stress and mitochondria-dependent pathways in human tongue cancer SCC-4 cells. *Oral Oncology*, **2009**, *45*, 273–283.

152. Chen, N. H., Liu, J. W., & Zhong, J. J. Ganoderic acid Me inhibits tumor invasion through down-regulating matrix metalloproteinases 2/9 gene expression. *Journal of Pharmacological Sciences*, **2008**, *108*, 212–216.

153. Chen, P. N., Chu, S. C., Chiou, H. L., Chiang, C. L., Yang, S. F., & Hsieh, Y. S. Cyanidin 3-glucoside and peonidin 3-glucoside inhibit tumor cell growth and induce apoptosis *in vitro* and suppress tumor growth *in vivo*. *Nutrition and Cancer*, **2005**, *53*, 232–243.

154. Chen, P. S., Shih, Y. W., & Huang, H. C. H. W. Diosgenin, a steroidal saponin, inhibits migration and invasion of human prostate cancer PC-3 cells by reducing matrix metalloproteinases expression. *PLoS One*, **2011**, *6*, e20164.

155. Chen, S. X., Wan, M., & Loh, B. N. Active constituents against HIV-1 protease from *Garcinia mangostana*. *Planta Medica*, **1996**, *62*, 381–382.

156. Chen, W., Li, N., Chen, T., Han, Y., Li, C., Wang, Y., He, W., Zhang, L., Wan, T., & Cao, X. The lysosome-associated apoptosis-inducing protein containing the pleckstrin homology (PH) and FYVE domains (LAPF), representative of a novel family of PH and FYVE domain-containing proteins, induces caspase-independent apoptosis via the lysosomal-mitochondrial pathway. *The Journal of Biological Chemistry*, **2005**, *280*, 40985–40995.

157. Chen, Y., Gong, Z., Chen, X., Tang, L., Zhao, X., Yuan, Q., & Cai, G. *Tripterygium wilfordii* Hook F (a traditional Chinese medicine) for primary nephrotic syndrome. *The Cochrane Database of Systematic Reviews*, **2013**, *8*, CD-008568.

158. Chen, Y. F., Tsai, H. Y., & Wu, T. S. Anti-inflammatory and analgesic activities from roots of *Angelica pubescens*. *Planta Medica*, **1995**, *61*, 2–8.

159. Cheng, C. R., Yue, Q. X., Wu, Z. Y., Song, X. Y., Tao, S. J., Wu, X. H., Xu, P. P., Liu, X., Guan, S. H., & Guo, D. A. Cytotoxic triterpenoids from *Ganoderma lucidum*. *Phytochemistry*, **2010**, *71*, 1579–1585.

160. Cheng, S., Huang, C., Chen, Y., Yu, J., Chen, W., & Chang, S. Chemical compositions and larvicidal activities of leaf essential oils from two eucalyptus species. *Bioresource Technology*, **2009**, *100*, 452–456.

161. Cheng, Y. L., Chang, W. L., Lee, S. C., Liu, Y. G., Chen, C. J., & Lin, S. Z. E. A. Acetone extract of *Angelica sinensis* inhibits proliferation of human cancer cells via inducing cell cycle arrest and apoptosis. *Life Sciences*, **2004**, *75*, 1579–1594.

162. Chevallier, A. *The Encyclopedia of Medicinal Plants* (p. 423). Dorling Kindersley, London, UK, **1996**.

163. Chhabra, S. C., Mahunnah, R. L., & Mshiu, E. N. Plants used in traditional medicine in Eastern Tanzania. I. pteridophytes and angiosperms (Acanthaceae to Canellaceae). *Journal of Ethnopharmacology*, **1987**, *21*, 253–277.

164. Chiang, C. T., Way, T. D., Tsai, S. J., & Lin, J. K. Diosgenin, a naturally occurring steroid, suppresses fatty acid synthase expression in HER2-overexpressing breast cancer cells through modulating Akt, mTOR and JNK phosphorylation. *FEBS Letter*, **2007**, *581*, 5735–5742.

165. Chiang, J. H., Yang, J. S., Ma, C. Y., Yang, M. D., Huang, H. Y., Hsia, T. C., Kuo, H. M., Wu, P. P., Lee, T. H., & Chung, J. G. Danthron, an anthraquinone derivative, induces DNA damage and caspase cascades-mediated apoptosis in SNU-1 human gastric cancer cells through mitochondrial permeability transition pores and Bax-triggered pathways. *Chemical Research in Toxicology*, **2011**, *24*, 20–29.

166. Chiang, L. C., Ng, L. T., Chiang, W., Chang, M. Y., & Lin, C. C. Immunomodulatory activities of flavonoids, monoterpenoids, triterpenoids, iridoid glycosides and phenolic compounds of *Plantago* species. *Planta Medica*, **2003**, *69*, 600–604.

167. Chiang, L. C., Ng, L. T., Lin, I. C., Kuo, P. L., & Lin, C. C. Anti-proliferative effect of apigenin and its apoptotic induction in human Hep G2 cells. *Cancer Letters*, **2006**, *237*, 207–214.

168. Chiarugi, A., & Moskowitz, M. A. PARP-1-a perpetrator of apoptotic cell death? *Science*, **2002**, *297*, 259–263.

169. Chinnici, F., Bendini, A., Gaiani, A., & Riponi, C. Radical scavenging activities of pees and pulps from cv. golden delicious apples as related to their phenolic composition. *Journal of Agricultural and Food Chemistry*, **2004**, *52*, 4684–4689.

170. Chinnici, F., Bendini, A., Gaiani, A., & Riponi, C. Radical scavenging activities of peels and pulps from cv. golden delicious apples as related to their phenolic composition. *Journal of Agricultural and Food Chemistry*, **2004**, *52*, 4684–4689.

171. Chinsembu, K. C., & Mutirua, T. *Validation of Traditional Medicines for HIV/AIDS Treatment in Namibia. A Report of the Study Visit to Zambia and South Africa* (p. 112). Windhoek, Namibia, **2008**.

172. Chiu, Y. J., Hour, M. J., Lu, C. C., Chung, J. G., Kuo, S. C., Huang, W. W., Chen, H. J., Jin, Y. A., & Yan, J. S. Novel quinazoline HMJ-30 induces U-2 OS human osteogenic sarcoma cell apoptosis through induction of oxidative stress and up-regulation of ATM/p53 signaling pathway. *Journal of Orthopaedic Research*, **2011**, *29*, 1448–1456.

173. Cho, J. Y., Yoo, E. S., Baik, K. U., Park, M. H., & Han, B. H. *In vitro* inhibitory effect of protopanaxadiol ginsenosides on tumor necrosis factor (TNF)-alpha production and its modulation by known TNF-alpha antagonists. *Planta Medica*, **2001**, *67*, 213–218.

174. Choi, D. K., Koppula, S., & Suk, K. Inhibitors of microglial neurotoxicity: Focus on natural products. *Molecules*, **2011**, *16*, 1021–1043.

175. Chopra, R. N., & Ghosh, S. Some observation on the pharmacological action and therapeutic properties of *Adhatoda vasica*. *The Indian Journal of Medical Research*, **1925**, *13*, 205–210.

176. Chou, C. C., Pan, S. L., Teng, C. M., & Guh, J. H. Pharmacological evaluation of several major ingredients of Chinese herbal medicines in human hepatoma Hep3B cells. *European Journal of Pharmaceutical Sciences*, **2003**, *19*, 403–412.

177. Chu, W. L., & Radhakrishnan, A. K. Research on Bioactive molecules: Achievements and the way forward. *International e-Journal of Science, Medicine & Education*, **2008**, *2*(1), S21–S24.

178. Chuang, J. Y., Huang, Y. F., Lu, H. F., Ho, H. C., Yang, J. S., Li, T. M., Chang, N. W., & Chung, J. G. Coumarin induces cell cycle arrest and apoptosis in human cervical cancer HeLa cells through a mitochondria- and caspase-3 dependent mechanism and NF-kappaB down-regulation. *In Vivo*, **2007**, *21*, 1003–1009.

179. Chun, H. S., Kim, H. J., & Choi, E. H. Modulation of cytochrome P4501-mediated bioactivation of benzo[a]pyrene by volatile allyl sulfides in human hepatoma cells. *Biosci. Biotechnol. Biochem.*, **2001**, *65*, 2205–2212.

180. Cichewicz, R. H., Zhang, Y., Seeram, N. P., & Nair, M. G. Inhibition of human tumor cell proliferation by novel anthraquinones from daylilies. *Life Sciences*, **2004**, *74*, 171–179.

181. Circosta, C., Occhiuto, F., Ragusa, S., Trovato, A., Tumino, G., Briguglio, F., & De Pasquale, A. A drug used in traditional medicine: *Harpagophytum procumbens* DC, II: Cardiovascular activity. *Journal of Ethnopharmacology*, **1984**, *11*, 259–274.

182. Claeson, U. P., Malmfors, T., Wikman, G., & Bruhn, J. G. *Adhatoda vasica*: A critical review of ethnopharmacological and toxicological data. *Journal of Ethnopharmacology*, **2000**, *72*, 1–20.

183. Clement, F., Pramod, S. N., & Venkatesh, Y. P. Identity of the immunomodulatory proteins from garlic (*Allium sativum*) with the major garlic lectins or agglutinins. *International Immunopharmacology*, **2010**, *10*, 316–324.

184. Coates, A., Hu, Y. M., Bax, R., & Page, C. The future challenges facing the development of new antimicrobial drugs. *Nature Reviews Drug Discovery*, **2002**, *1*, 895–910.

185. Coleman, W. B., & Rubinas, T. C. In molecular pathology: In: Tsongalis, G. J., & Coleman, W. L., (eds.), *The Molecular Basis of Human Disease* (p. 66). Elsevier Academic Press, Amsterdam, **2009**.

186. Collins, J. R., Burt, S. K., & Erickson, J. W. Flap opening in HIV-l protease simulated by 'activated' molecular dynamics. *Nature Structural Biology*, **1995**, *2*, 334–338.

187. Collins, R. A., Ng, T. B., Foung, W. P., Wan, C. C., & Yeung, H. W. A comparison of human immunodeficiency virus type 1 inhibition by partially purified aqueous extracts of Chinese medicinal herbs. *Life Sciences*, **1997**, *60*, 345–351.

188. Conforti, F., Sosa, S., Marrelli, M., Menichini, F., Statti, G. A., Uzunov, D., Tubaro, A., & Menichini, F. The protective ability of Mediterranean dietary plants against the oxidative damage: The role of radical oxygen species in inflammation and the polyphenol, flavonoid and sterol contents. *Food Chemistry*, **2009**, *112*, 587–594.

189. Conforti, F., Menichini, F., Formisano, C., Rigano, D., Senatore, F., Arnold, N. A., & Franco, P. Comparative chemical composition, free radical-scavenging and cytotoxic properties of essential oils of six *Stachys* species from different regions of the Mediterranean Area. *Food Chemistry*, **2009**, *116*, 898–905.

190. Cooper, G. M. *Elements of Human Cancer* (p. 116). Jones and Bartlett Publisher, Boston, **1992**.

191. Corbiere, C., Liagre, B., Bianchi, A., Bordji, K., Dauca, M., Netter, P., & Beneytout, J. L. Different contribution of apoptosis to the antiproliferative effects of diosgenin and other plant steroids, hecogenin and tigogenin, on human 1547 osteosarcoma cells. *International Journal of Oncology*, **2003**, *22*, 899–905.

192. Cotran, R. S., & Kumar, C. *Robbins Pathologic Basis of Disease* (p. 213). W.B Saunders Company, Philadelphia – PA, **1998**.

193. Coyle, T., Levante, S., Shetler, M., & Winfield, J. *In vitro* and *in vivo* cytotoxicity of gossypol against central nervous system tumor cell lines. *Journal of Neuro-Oncology*, **1994**, *19*, 25.

194. Cragg, G. M., Newman, D. J., & Snader, K. M. Natural products in drug discovery and development. *Journal of Natural Products*, **1997**, *60*, 52–60.

195. Creagh, T., Ruckle, J. L., Tolbert, D. T., Giltner, J., Eiznhamer, D. A., Dutta, B., Flavin, M. T., & Xu, Z. Q. Safety and pharmacokinetics of single doses of (+)-calanolide a, a novel, naturally occurring nonnucleoside reverse transcriptase inhibitor, in healthy, human immunodeficiency virus-negative human subjects. *Antimicrobial Agents and Chemotherapy*, **2000**, *45*, 1379–1386.

196. Cronquist, A. *The Evolution and Classification of Flowering Plants* (p. 98). Botanical Garden, Bronx-New York, **1988**.

197. Croteau, R. The discovery of terpenes. *Discoveries in Plant Biology*, **1998**, *1*, 329–343.

198. Csupor-Löffler, B., Hajdú, Z., Zupkó, I., Réthy, B., Falkay, G., Forgo, P., & Hohmann, J. Antiproliferative effect of flavonoids and sesquiterpenoids from

Achillea millefolium s.l. on cultured human tumor cell lines. *Phytotherapy Research*, **2009**, *23*, 672–676.

199. Cuccioloni, M., Mozzicafreddo, M., Sparapani, L., Spina, M., Eleuteri, A. M., Fioretti, E., & Angeletti, M. Pomegranate fruit components modulate human thrombin. *Fitoterapia*, **2009**, *80*, 301–305.

200. Cui, Y., Lu, C., Liu, L., Sun, D., Yao, N., Tan, S., Bai, S., & Ma, X. Reactivation of methylation-silenced tumor suppressor gene p161NK4a by nordihydroguaiaretic acid and its implication in G1 cell cycle arrest. *Life Sciences*, **2008**, *82*, 247–255.

201. Cuong, N. D., & Quynh, N. H. *Pseuderanthemum Palatiferum (Nees) Radlk* (p. 714). Pharmaceutical encyclopedia, Encyclopedia Publisher, Hanoi, **1999**.

202. Da Rocha, M. D., Viegas, F. P., Campos, H. C., Nicastro, P. C., Fossaluzza, P. C., Fraga, C. A., Barreiro, E. J., & Viegas, C. Jr. The role of natural products in the discovery of new drug candidates for the treatment of neurodegenerative disorders II: Alzheimer's disease. *CNS & Neurological Disorders – Drug Targets*, **2011**, *10*, 251–270.

203. Dale, M. M., & Formanj, C. *Text-Book of Immunopharmacology* (2nd edn., p. 244). Blackwell Scientific Publication, Oxford, UK, **1989**.

204. Dalvi, S. Sutherlandia for immune support. *Positive Health*, **2003**, 23–25.

205. Danbara, N., Yuri, T., Tsujita-Kyutoku, M., Tsukamoto, R., Uehara, N., & Tsubura, A. Enterolactone induces apoptosis and inhibits growth of colon 201 human colon cancer cells both *in vitro* and *in vivo*. *Anticancer Research*, **2005**, *25*, 2269–2276.

206. Darwish, R. M., & Aburjai, T. A. Effect of ethnomedicinal plants used in folklore medicine in Jordan as antibiotic resistant inhibitors on *Escherichia coli*. *BMC Complementary and Alternative Medicine*, **2010**, *10*, 9–16.

207. Das, N. D., Jung, K. H., Park, J. H., Mondol, M. A., & Shin, H. J. *Terminalia chebula* extract acts as a potential NF-κB inhibitor in human lymphoblastic T cells. *Phytotherapy Research*, **2011**, *25*, 927–934.

208. Das, S. K., Hashimoto, T., & Kanazawa, K. Growth inhibition of human hepatic carcinoma HepG2 cells by fucoxanthin is associated with down-regulation of cyclin D. *Biochimica et Biophysica Acta*, **2008**, *1780*, 743–749.

209. Das, S. K., Hashimoto, T., Shimizu, K., Yoshida, T., Sakai, T., Sowa, Y., Komoto, A., & Kanazawa, K. 903 Fucoxanthin induces cell cycle arrest at G0/G1 phase in human colon carcinoma cells through up-regulation 904 of p21WAF1/Cip1. *Biochimica et Biophysica Acta*, **2005**, *1726*, 328–335.

210. Davis, J. Inactivation of antibiotic and the dissemination of resistance genes. *Science*, **1994**, *264*, 375–382.

211. D'cruz, O. J., Waurzyniak, B., & Uckun, F. M. Mucosa toxicity studies of a gel formulation of native pokeweed antiviral protein. *Toxicologic Pathology*, **2004**, *32*, 212–221.

212. De Araújo, J. R. F., De Souza, T. P., Pires, J. G., Soares, L. A., De Araújo, A. A., Petrovick, P. R., Mâcedo, H. D., De Sá Leitão, O. A. L., & Guerra, G. C. A dry extract of *Phyllanthus niruri* protects normal cells and induces apoptosis in human liver carcinoma cells. *Experimental Biology and Medicine*, **2012**, *237*, 1281–1288.

213. De Rodriguez, D. J., Chula, J., Simons, C., Armoros, M., Veriohe, A. M., & Girre, L. Search for *in vitro* antiviral activity of a new isoflavone glycoside from *Vlex europeus*. *Planta Medica*, **1990**, *50*, 59–62.

214. De, W. W. P., & Jansen, P. C. M. Selaginella. In: De Winter, W. P., & Amoroso, V. B., (eds.), *Plant Resources of South-East Asia – Cryptogams: Ferns and Fern Allies* (pp. 178–184). Backhuys Publishers, Netherlands, **2003**.
215. Defining Cancer. National Cancer Institute. Archived from the original on 25 June **2014**. https://www.cancer.gov/about-cancer/understanding/what-is-cancer (Accessed on 30 September 2019).
216. Dejean, L. M., Martinez-Caballero, S., Manon, S., & Kinnally, K. W. Regulation of the mitochondrial apoptosis-induced channel, MAC, by BCL-2 family proteins. *Biochimica et Biophysica Acta*, **2006**, *1762*, 191–201.
217. De Luca, P., Rossetti, R. G., Alavian, C., Karim, P., & Zurier, R. B. Effects of gammalinolenic acid on interleukin-1b and tumor necrosis-a secretion by stimulated human peripheral blood monocytes: studies *in vitro* and *in vivo*. *Journal of Investigative Medicine*, **1999**, *47*, 246–250.
218. Demissie, A. G., & Lele, S. S. Bioactivity-directed isolation and identification of novel alkaloid from *Jatropha curcas* (Linn.). *Research Journal of Chemical and Environmental Sciences*, **2013**, *1*, 22–28.
219. Deore, S. L., Khadabadi, S. S., Kamdi, K. S., Ingle, V. P., Kawalkar, N. G., Sawarkar, P. S., Patil, U. A., & Vyas, A. J. *In vitro* Anthelmintic activity of *Cassia tora*. *International Journal of Chem. Tech. Research*, **2009**, *2*, 177–179.
220. Derwich, E., Benzian, Z., & Boukir, A. Antibacterial activity and chemical composition of the essential oil from flowers of *Nerium oleander*. *Electronic Journal of Environmental, Agricultural and Food Chemistry*, **2010**, *9*, 1074–1084.
221. Deters, A. M., Schröder, K. R., & Hensel, A. Kiwi fruit (*Actinidia chinensis* L.) polysaccharides exert stimulating effects on cell proliferation via enhanced growth factor receptors, energy production, and collagen synthesis of human keratinocytes, fibroblasts, and skin equivalents. *Journal of Cellular Physiology*, **2005**, *202*, 717–722.
222. Dhar, M. L., Dhar, M. M., Dhawan, B. N., Mehrotra, B. N., & Ray, C. Screening of Indian plants for biological activity: Part I. *Indian Journal of Experimental Biology*, **1968**, *7*, 232–247.
223. Dhar, U., Manjkhola, S., Joshi, M., Bhatt, A., & Joshi, M. Current status and future strategy for development of medicinal plants sector in Uttranchal, India. *Current Science*, **2002**, *83*, 956–964.
224. Dharmaratne, H. R. W., Tan, G. T., Marasinghe, G. P. K., & Pezzuto, J. M. Inhibition of HIV-1 reverse transcriptase and HIV-1 replication by *Callophyllum coumarins* and xanthones. *Planta Medica*, **2002**, *68*, 86–87.
225. Dihal, A. A., Woutersen, R. A., Van Ommen, B., Rietjens, I. M., & Stierum, R. H. Modulatory effects of quercetin on proliferation and differentiation of the human colorectal cell line Caco-2. *Cancer Letters*, **2006**, *238*, 248–259.
226. Dixon, R. A., Dey, P. M., & Lamb, C. Phytoalexins: Enzymology and molecular biology. *Advances in Enzymology and Related Areas of Molecular Biology*, **1983**, *55*, 1–136.
227. Djuric, Z., Severson, R. K., & Kato, I. Association of dietary quercetin with reduced risk of proximal colon cancer. *Nutrition and Cancer*, **2012**, *64*, 351–360.
228. Dong, X., & Shivendra, V. S. Diallyl trisulfide, a constituent of processed garlic, inactivates Akt to trigger mitochondrial translocation of BAD and caspase-mediated apoptosis in human prostate cancer cells. *Carcinogenesis*, **2006**, *27*, 533–540.

229. Dong, X., Yan, Z., & Shivendra, V. S. Diallyl Trisulfide-induced apoptosis in human cancer cells is linked to checkpoint kinase 1-mediated mitotic Arrest. *Molecular Carcinogenesis*, **2009**, *48*, 1018–1029.

230. Doss, A., & Rangasamy, D. Preliminary phytochemical screening and antibacterial studies of leaf extract of *Solanum trilobatum* Linn. *Ethnobotanical Leaflets*, **2008**, *12*, 638–642.

231. Dreger, M., Stanisławska, M., Krajewska-Patan, A., Mielcarek, S., Mikołajczak, P. Ł., & Buchwald, W. Pyrrolizidine alkaloids-chemistry, biosynthesis, pathway, toxicity, safety and perspectives of medicinal usage. *Herba Polonica Journal*, **2009**, *55*, 127–147.

232. Duan, H., Takaishi, Y., Imakura, Y., Jia, Y., Li, D., Cosentino, L. M., & Lee, K. H. Sesquiterpene alkaloids from *Tripterigium hypoglaucum* and *Tripterygium wilfordii*: A new class of potent anti-HIV agents. *Journal of Natural Products*, **2000**, *63*, 357–361.

233. Dubey, R., Dubey, K., Sridhar, C., & Jayaveera, K. N. Human vaginal pathogen inhibition studies on aqueous methanolic and saponins extracts of stem barks of *Ziziphus mauritiana*. *International Journal of Pharmaceutical Sciences and Research*, **2011**, *2*, 659–663.

234. Dudhgaonkar, S., Thyagarajan, A., & Sliva, D. Suppression of the inflammatory response by triterpenes isolated from the mushroom *Ganoderma lucidum*. *International Immunopharmacology*, **2009**, *9*, 1272–1280.

235. Duh, C. Y., Wang, S. K., Chu, M. J., & Sheu, J. H. Cytotoxic sterols from the soft coral *Nephthea erecta*. *Journal of Natural Products*, **1998**, *61*, 1022–1024.

236. Duke, J. A., Bogenschutz-Godwin, M. J., Ducelliar, J., & Duke, P. A. K. *Handbook of Medicinal Herbs* (2nd edn., pp. 70, 71). CRC Press, Boca Raton – FL, **2002**.

237. Dwyer, J., & Rattray, D. Plant, people and medicine. In: *Magic and Medicine of Plant* (pp. 48–73). Reader's Digest – General Book, New York, **1993**.

238. Edeoga1, H. O., Okwu, D. E., & Mbaebie, B. O. Phytochemical constituents of some Nigerian medicinal plants. *African Journal of Biotechnology*, **2005**, *4*, 685–688.

239. Efferth, T., Fu, Y., Zu, Y., Schwarz, G., Newman, D., & Wink, M. Molecular target-guided tumor therapy with natural products derived from Traditional Chinese Medicine. *Current Medicinal Chemistry*, **2007**, *14*, 2024–2032.

240. Efferth, T., & Wink, M. Chemical-biology of natural products from medicinal plants for cancer therapy. In: Alaoui-Jamali, M., (ed.), *Alternative and Complementary Therapies for Cancer* (pp. 557–582). Springer, New York, USA, **2010**.

241. Ekta, S., & Sheel, S. Phytochemistry, traditional uses and cancer chemopreventive activity of Amla (*Phyllanthus emblica*): The sustainer. *Journal of Applied Pharmaceutical Science*, **2011**, *2*, 176–183.

242. El Bouzidi, L., Larhsini, M., Markouk, M., Abbad, A., Hassani, L., & Bekkouche, K. Antioxidant and antimicrobial activities of *Withania frutescens*. *Natural Product Communications*, **2011**, *6*, 1447–1450.

243. El Dine, R. S., El Halawany, A. M., Ma, C. M., & Hattori, M. Anti-HIV-1 protease activity of lanostane triterpenes from the vietnamese mushroom *Ganoderma colossum*. *Journal of Natural Products*, **2008**, *71*, 1022–1026.

244. El-Azab, M., Hishe, H., Moustafa, Y., & El-Awady, S. Anti-angiogenic effect of resveratrol or curcumin in Ehrlich ascites carcinoma-bearing mice. *European Journal of Pharmacology*, **2011**, *652*, 7–14.

245. Elizabeth, K. M. Antimicrobial activity of *Terminalia bellerica*. *Indian Journal of Clinical Biochemistry*, **2005**, *20*, 150–153.

246. Elizabeth, M. W. Major herbs of Ayurveda. Churchill Livingstone, New York Kapil. *Journal of Ethnopharmacology*, **2002**, *58*, 89–91.

247. El Mekkawy, S., Meselhy, M. R., Kusumoto, I. T., Kadota, S., Hattori, M., & Namba, T. Inhibitory effects of Egyptian folk medicines on human immunodeficiency virus (HIV) reverse transcriptase. *Chemical and Pharmaceutical Bulletin*, **1995**, *43*, 641–648.

248. El-Readi, Z. M., Hamdan, D., Nawal, M., Farrag, A., El-Shazly, A., & Wink, M. Inhibition of β-glycoprotein by limonin and other secondary metabolites from *Citrus* species in human colon and leukemia cell lines. *European Journal of Pharmacology*, **2010**, *626*, 139–145.

249. Elsje, P., Nicholas, A. C., Lin, P. L., Véronique, D., Joshua, T. M., Russ, B., JoAnne, L. F., Denise, E. K., & Jennifer, J. L. A computational tool integrating host immunity with antibiotic dynamics to study tuberculosis treatment. *Journal of Theoretical Biology*, **2015**, *367*, 166–179.

250. Emmanuel, S., Ignacimuthu, S., Perumalsamy, R., & Amalraj, T. Anti-inflammatory activity of *Solanum trilobatum*. *Fitoterapia*, **2006**, *77*, 611–612.

251. Encalada, M. A., Hoyos, K. M., Rehecho, S., Berasategi, I., García-Íñiguez de Ciriano, M., Ansorena, D., Astiasarán, I., Navarro-Blasco, I., Cavero, R. Y., & Calvo, M. I. Anti-proliferative effect of *Melissa officinalis* on human colon cancer cell line. *Plant Foods for Human Nutrition*, **2011**, *66*, 328–334.

252. Endrini, S., Rahmat, A., Ismail, P., & Taufiq-Yap, Y. H. Comparing of the ctotoxicity properties and mechanism of *Lawsonia inermis* and *Strobilanrhes crispus* extract against several cancer cell lines. *Journal of Medical Sciences*, **2007**, *7*, 1098–1102.

253. Engelbrecht, A. M., Mattheyse, M., Ellis, B., Loos, B., Thomas, M., Smith, R., Peters, S., Smith, C., & Myburgh, K. Proanthocyanidin from grape seeds inactivates the PI3-kinase/PKB pathway and induces apoptosis in a colon cancer cell line. *Cancer Letters*, **2007**, *258*, 144–153.

254. Enne, V. I., Livermore, D. M., Stephens, P., & Hal, L. M. C. Persistence of sulphonamide resistance in *Escherichia coli* in the UK despite national prescribing restriction. *The Lancet*, **2001**, *28*, 1325–1328.

255. Eo, H. J., Park, J. H., Park, G. H., Lee, M. H., Lee, J. R., Koo, J. S., & Jeong, J. B. Anti-inflammatory and anticancer activity of mulberry (*Morus alba* L.) root bark. *BMC Complementary and Alternative Medicine*, **2014**, *14*, 200–208.

256. Ermakova, S., Sokolova, R., Kim, S. M., Um, B. H., Isakov, V., & Zvyagintseva, T. Fucoidans from brown seaweeds *Sargassum hornery, Eclonia cava, Costaria costata*: Structural characteristics and anticancer activity. *Applied Biochemistry and Biotechnology*, **2011**, *164*, 841–850.

257. ESCOP (European Scientific Cooperative on Phytotherapy). *Monographs on the Medicinal Uses of Plant Drugs* (p. 435). ESCOP, Exeter, UK, **1999**, Part 5.

258. Eswani, N., Abd, K. K., Nazre, M., Awang, N. A. G., & Ali, M. Medicinal plant diversity and vegetation analysis of logged over hill forest of Tekai Tembeling forest reserve, Jerantut, Pahang. *The Journal of Agricultural Science*, **2010**, *2*, 189–210.

259. Evans, M. *A Guide to Herbal Remedies* (pp. 24–30). Orient Paperbacks, New Delhi, **1994**.

260. Fagelman, E., & Lowe, F. C. Herbal medications in the treatment of benign prostatic hyperplasia (BPH). *Urologic Clinics of North America*, **2002**, *29*, 23–29.

261. Falkiewicz, B., & Lukasiak, J. Vilcacora *Uncaria tomentosa* (Willd.) DC. and *Uncaria guianensis* (Aublet) Gmell.-a review of published scientific literature. *Case Reports Journal | Clinical Case Reports and Reviews*, **2001**, *2*, 305–316.

262. Faried, A., Faried, L. S., Kimura, H., Sohda, M., Nakajima, M., Miyazaki, T., Kato, H., Kanuma, T., & Kuwano, H. Differential sensitivity of paclitaxel-induced apoptosis in human esophageal squamous cell carcinoma cell lines. *Cancer Chemotherapy and Pharmacology*, **2006**, *57*, 301–308.

263. Faried, A., Kurnia, D., Faried, L. S., Usman, N., Miyazaki, T., & Kato, H. Anticancer effects of gallic acid isolated from Indonesian herbal medicine, *Phaleria macrocarpa* (Scheff.) Boerl, on human cancer cell lines. *International Journal of Oncology*, **2007**, *30*, 605–613.

264. Fariza, I. N., Fadzureena, J., Zunoliza, A., Chuah, A. L., Pin, K. Y., & Adawiah, I. Anti-inflammatory activity of the major compound from methanol extract of *Phaleria macrocarpa* leaves. *Journal of Applied Sciences*, **2012**, *12*, 1195–1198.

265. Farnsworth, N. R., & Bingel, A. S. Problems and prospects of discovery new drugs from higher plants by pharmacological screening. In: Wagner, H., & Wolff, P., (eds.), *New Natural Products and Plant Drugs with Pharmacological, Biological and Therapeutical Activity* (pp. 1–22). Springer Verlag, Berlin, **1977**.

266. Farnsworth, N. R., Blowster, R. N., Darmratoski, D., Meer, W. A., & Cammarato, L. V. Studies on *Catharanthus* alkaloids IV Evaluation by means of TLC and ceric ammonium sulfate spray reagent. *Lloydia*, **1967**, *27*, 302–314.

267. Fassina, G., Buffa, A., Benelli, R., Varnier, O. E., Noonan, D. M., & Albini, A. Polyphenolic antioxidant (-)-epigallocatechin-3-gallate from green tea as a candidate anti-HIV agent. *AIDS*, **2002**, *16*, 939–941.

268. Ferenci, P., Dragosics, B., Dittrich, H., Frank, H., Benda, L., Lochs, H., Meryn, S., Base, W., & Schneider, B. Randomized controlled trial of silymarin treatment in patients with cirrhosis of the liver. *Journal of Hepatology*, **1989**, *9*, 105–113.

269. Fernández-Herrera, M. A., López-Muñoz, H., Hernández-Vázquez, J. M., López-Dávila, M., Escobar-Sánchez, M. L., Sánchez-Sánchez, L., Pinto, B. M., & Sandoval-Ramírez, J. Synthesis of 26-hydroxy-22-oxocholestanic frameworks from diosgenin and hecogenin and their *in vitro* antiproliferative and apoptotic activity on human cervical cancer CaSki cells. *Bioorganic & Medicinal Chemistry*, **2010**, *18*, 2474–2484.

270. Finkel, T., & Holbrook, N. J. Oxidants, oxidative stress and the biology of aging. *Nature*, **2000**, *408*, 239–247.

271. Fisgin, N. T., Cayci, Y. T., Coban, A. Y., Ozatli, D., Tanyel, E., Durupinar, B., & Tulek, N. Antimicrobial activity of plant extract Ankaferd Blood Stopper®. *Fitoterapia*, **2009**, *80*, 48–50.

272. Foster, S. *Herbs for Your Health* (p. 98). Interweave Press, Loveland, CO, **1996**.

273. Francois, C., & Hance, A. J. HIV drug resistance. *The New England Journal of Medicine*, **2004**, *350*, 1023–1035.

274. Frederico, P., Rafael, C. D., Dalton, D. J., Miriam, T. P. L., & Nadia, R. B. Antioxidant and cytotoxic activities of *Centell asiatica*. *International Journal of Molecular Sciences*, **2009**, *10*, 3713–3721.

275. Fu, Y., Li, S., Zu, Y., Yang, G., Yang, Z., Luo, M., Jiang, S., Wink, M., & Efferth, T. Medicinal chemistry of paclitaxel and its analogs. *Current Medicinal Chemistry*, **2009**, *16*, 3966–3985.

276. Fu, Z., Zhen, W., Yuskavage, J., & Liu, D. Epigallocatechin gallate delays the onset of type 1 diabetes in spontaneous non-obese diabetic mice. *British Journal of Nutrition*, **2011**, *105*, 1218–1225.

277. Fujioka, T. L., & Kashiwada, Y. Anti-AIDS agents 11. Betulinic acid and platonic acid as anti-HIV primciples from *Syzigium calviflorum* and the anti HIV activity of structurally related triterpenoids. *Journal of Natural Products*, **1994**, *57*, 243–247.

278. Fukuchi, K., Sakagarmi, H., Okuda, T., Hatano, T., Tanuma, S., Kitajima, K., Inoue, Y., Inoue, S., Ichikawa, S., Nonoyama, M., & Konno, K. Inhibition of herpes simplex virus infection by tannis and related compounds. *Antiviral Research*, **1989**, *11*, 285–297.

279. Fumonisin, B. *Some Traditional Herbal Medicines, Some Mycotoxins, Naphthalene and Styrene*. IARC monographs on the evaluation of carcinogenic risks to humans, No. 82, **2002**.

280. Furse, R. K., Rossetti, R. G., & Zurier, R. B. Gammalinolenic acid, an unsaturated fatty acid with anti-inflammatory properties, blocks amplification of IL-1 beta production by human monocytes. *The Journal of Immunology*, **2001**, *167*, 490–496.

281. Gadre, A. Y., Uchi, D. A., Rege, N. N., & Daha, S. A. Nuclear variations in HPTLC fingerprint patterns of marketed oil formulations of *Celastrus paniculates*. *Indian Journal of Pharmacology*, **2001**, *33*, 124–145.

282. Gaherwal, S., Shiv, G., & Wast, N. Antifungal activity of *Solanum xanthocarpum* (kantkari) leaf extract. *World Journal of Zoology*, **2014**, *9*, 111–114.

283. Gajja, S., Hugh, J. M., & Verwijs, J. The Bax-alpha: Bcl–2 ratio modulates the response to dexamethasone in leukaemic cells and is highly variable in childhood acute leukemia. *International Journal of Cancer*, **1997**, *71*, 959–965.

284. Gali, K., Ramakrishnan, G., Kothai, R., & Jaykar, B. *In-vitro* anti-cancer activity of methanolic extract of leaves of *Argemone mexicana* Linn. *International Journal of Pharm. Tech. Research*, **2011**, *3*, 1329–1333.

285. Gallo, R. C., Sarin, P. S., Gelman, E. P., Robert-Guroff, M., Richardson, E., Kalyanaraman, V. S., Mann, D., Sidhu, G. D., Stahl, R. E., Zolla-Pazner, S., Leibowitch, J., & Popovic, M. Isolation of human T cell leukemia virus in acquired immune deficiency syndrome (AIDS). *Science*, **1983**, *220*, 865–867.

286. Galve-Roperh, I., Sanchez, C., Cortes, M. L., Pulgar, T. G., Izquierdo, M., & Guzman, M. Anti-tumoral action of cannabinoids: Involvement of sustained ceramide accumulation and extracullar signal-regulated kinase activation. *Nature Medicine*, **2000**, *6*, 313.

287. Gan, L., Zhang, S. H., Luo, Q., & Xu, H. B. A polysaccharide–protein complex *Lycium barbarum* upregulates cytokine expression in human peripheral blood mononuclear cells. *European Journal of Pharmacology*, **2003**, *471*, 217–222.

288. Ganjewala, D., Sam, S., & Khan, K. H. Biochemical compositions and antibacterial activities of *Lantana camara* plants with yellow, lavender, red and white flowers. *Eurasian Journal of Biosciences*, **2009**, *3*, 69–77.

289. Ganong, W. *Review of Medical Physiology* (17ᵗʰ edn.). Applenton and Lange, Stamford, CT, USA, **2005**.

290. Gao, X. M., Pu, Z. X., Huang, S. X., & Yang, L. M. Lignans from *Kadsura angustifolia*. *Journal of Natural Products*, **2008**, *71*, 558–563.

291. Gao, X. M., Xu, Z. M., & Li, Z. W. *Traditional Chinese Medicines* (p. 185). People's Health Publishing House, Beijing, **2000**.

292. Gautam, M. K., Goel, S., Ghatule, R. R., Singh, A., & Nath, G. Curative effect of *Terminalia chebula* extract on acetic acid-induced experimental colitis, role of antioxidants, free radicals and acute inflammatory marker. *Inflammopharmacology*, **2012**, *21*, 377–383.

293. GBD. Disease and Injury Incidence and Prevalence, Collaborators. Global, regional, and national incidence, prevalence, and years lived with disability for 310 diseases and injuries, 1990–2015: A systematic analysis for the global burden of disease study. *Lancet*, **2015**, *388*, 1545–1602.

294. Geitmann, M., Petsom, A., & Danielson, U. H. Inhibition of viral proteases by zingiberaceae extracts and flavones isolated from *Kaempferia parviflora*. *Pharmazie*, **2006**, *61*, 717–721.

295. Gericke, N., Albrecht, C. F., Van Wyk, B., Mayeng, B., Mutwa, C., & Hutchings, A. *Sutherlandia frutescens*. *Australian Journal of Medical Herbalism*, **2001**, *13*, 9–15.

296. Ghafoor, K., AL-Juhaimi, F. Y., & Choi, Y. H. Supercritical fluid extraction of phenolic compounds and antioxidants from grape (*Vitis labrusca* B.) seeds. *Plant Foods for Human Nutrition*, **2012**, *67*, 407–414.

297. Ghaisas, M. M., Shaikh, A. S., & Deshpande, D. A. Evaluation of immunomodulatory activity of ethanolic extract of stem bark of *Bauhinia variegates*. *International Journal of Green Pharmacy*, **2009**, *3*, 70–74.

298. Ghani, A. *Medicinal Plants of Bangladesh* (p. 303). The Asiatic Society of Bangladesh, Dhaka, **2003**.

299. Gil, M. I., Tomás-Barberán, F. A., Hess-Pierce, B., Holcroft, D. M., & Kader, A. A. Antioxidant activity of pomegranate juice and its relationship with phenolic composition and processing. *Journal of Agricultural and Food Chemistry*, **2000**, *48*, 4581–4589.

300. Gilbert, N. E., Reilly, J. E., Chang, C. J., Lin, Y. C., & Brueggemeier, R. W. Antiproliferative activity of gossypol and gossypolone on human breast cancer cells. *Life Sciences*, **1995**, *57*, 61–67.

301. Goldsby, R. A., Kindt, T. J., Osborne, B. A., & Kuby, J. *Immunology* (5ᵗʰ edn., pp. 1–25). W.H. Freeman and Co., New York, **2003**.

302. Gomes, A., Das, R., Sarkhel, S., Mishra, R., Mukherjee, S., Bhattacharya, S., & Gomes, A. Herbs and herbal constituents active against snakebite. *Indian Journal of Experimental Biology*, **2010**, *48*, 865–878.

303. González, D., Bejarano, I., Barriga, C., Rodríguez, A. B., & Pariente, J. A. Oxidative stress-induced caspases are regulated in human myeloid HL-60 cells by calcium signal. *Current Signal Transduction Therapy*, **2010**, *5*, 181–186.

304. Gopinathan, G., & Dhiman, K. S. Triphala in eye diseases: A critical review. *Journal of Homeopathy & Ayurvedic Medicine*, **2013**, *2*, 123.

305. Gordon, M. C., & David, J. Plants as a source of anti-cancer agents. *Journal of Ethnopharmacology*, **2005**, *100*, 72–79.

306. Gottlieb, S. US relaxes its guidelines on herbal supplements. *The BMJ*, **2000**, *320*, 207.

307. Government of Republic of Zambia. *Guidelines for Research in Traditional Medicines in Zambia* (p. 34). Lusaka, Ministry of Health, Government of Republic of Zambia, Lusaka, **2008**.

308. Govind, P., & Madhuri, S. Pharmacological activities of *Ocimum Sanctum* (Tulsi): A review. *International Journal of Pharmaceutical Sciences and Research*, **2010**, *5*, 1–65.

309. Govindachari, T. R., Sath, S. S., Viswanathan, N., Pai, B. R., & Srinivasan, M. Chemical constituents of *Cleistanthus collinus* (Roxb.). *Tetrahedron*, **1969**, *25*, 2815–2821.

310. Govindan, S., Viswanathan, S., Vijayasekaran, V., & Alagappan, R. Apilot study on the clinical efficacy of *Solanum xanthocarpum* and *Solanum trilobatum* in bronchial asthma. *Journal of Ethnopharmacology*, **1999**, *66*, 205–210.

311. Granowitz, E. V., Poutsiaka, D. D., & Dinarello, C. A. Effect of interleukin-1 (IL-1) blockade on cytokine synthesis. II. IL-1 receptor antagonist inhibits lipopolysaccharide-induced cytokine synthesis by human monocytes. *Blood*, **1992**, *79*, 2364–2369.

312. Green, M. Method of treating viral infections with amino acid analog. *US Patent-5110, 600*, **1988**, p. 23.

313. Gu, Q., Wang, R. R., & Zhang, X. M. A new benzofuranone and anti HIV constituents from the stems of *Rhus chinensis*. *Planta Medica*, **2007**, *73*, 279–282.

314. Guerin, I., & De Chastellier, C. Pathogenic mycobacteria disrupt the macrophage actin filament network. *Infection and Immunity*, **2000**, *68*, 2655–2662.

315. Gulick, R. M., Mellors, J. W., Havlir, D., Eron, G. C., Mcmahon, D., Richman, D. D., Valentine, F. T., Jonas, L., Meibohm, A., Emini, E. A., & Chodakewit, J. A. Treatment with indinavir, zidovudine and lamivudine in adults with human immunodeficiency virus infection and prior antiretroviral therapy. *The New England Journal of Medicine*, **1997**, *337*, 734–739.

316. Gunasena, H. P. M., & Hughes, A. *Tamarind, Tamarindus indica* L. (p. 30). International Center for Underutilized Crops, Southampton, UK, **2000**.

317. Guo, S., Copps, K. D., Dong, X., Park, S., Cheng, Z., Pocai, A., Rossetti, L., Sajan, M., Farese, R. V., & White, M. F. The Irs1 branch of the insulin signaling cascade plays a dominant role in hepatic nutrient homeostasis. *Molecular and Cellular Biology*, **2009**, *29*, 5070–5083.

318. Guo, S. Insulin signaling, resistance, and the metabolic syndrome: Insights from mouse models into disease mechanisms. *Journal of Endocrinology*, **2014**, *220*, T1–T23.

319. Gupta, S., Ahmad, N., Nieminen, A. L., & Mukhtar, H. Growth inhibition, cell-cycle dysregulation, and induction of apoptosis by green tea constituent (-)-epigallocatechin-3-gallate in androgen-sensitive and androgen-insensitive human prostate carcinoma cells. *Toxicology and Applied Pharmacology*, **2000**, *164*, 82–90.

320. Gupta, S., Pramanik, S., Tiwari, O., Thacker, N., Pande, M., & Upmanyu, N. Immunomodulatory activity of *Gymnema sylvestre* leaves. *International Journal of Pharmaceutics*, **2009**, *8*, 12–16.

321. Gur, S., Turgut-Balik, D., & Gur, N. Antimicrobial activities and some fatty acids of turmeric, ginger root and linseed used in the treatment of infectious diseases. *World Journal of Agricultural Sciences*, **2006**, *2*, 439–442.

322. Gustafson, K. R., Cardellina, J. H., McMahon, J. B., Gulakowsi, R. J., Ishitoya, J., Szallasi, Z., et al.A nonpromoting phorbol from the Samoan medicinal plant *Homolanthus nutans* inhibits cell killing by HIV-1. *Journal of Chemistry*, **1992**, *35*, 1978–1986.

323. Habbal, O., Hasson, S. S., El-Hag, A. H., Al-Mahrooqi, Z., Al-Hashmi, N., Al-Bimani, Z., Al-Balushi, M. S., & Al-Jabri, A. A. Antibacterial activity of *Lawsonia inermis* linn (Henna) against *Pseudomonas aeruginosa*. *Asian Pacific Journal of Tropical Biomedicine*, **2011**, *1*, 173–176.

324. Hackett, C. J. Innate immune activation as a broad-spectrum biodefense strategy: Prospects and research challenges. *Journal of Allergy and Clinical Immunology*, **2003**, *112*, 686–694.

325. Hahm, E. R., Park, S., & Yang, C. H. The 7, 8-dihydroxyflavanone as an inhibitor for Jun-Fos-DNA complex formation and its cytotoxic effect on cultured human cancer cells. *Natural Product Research*, **2003**, *17*, 431–436.

326. Haji, M. M. Y., Chin, W., & Holdsworth, D. Traditional medicinal plants of *Brunei Darussalam*, part III: Sengkurong. *International Journal of Pharmacognosy*, **1992**, *30*, 105–108.

327. Halicka, H. D., Bedner, E., & Darzynkiewicz, Z. Segregation of RNA and separate packaging of DNA and RNA in apoptotic bodies during apoptosis. *Experimental Cell Research*, **2000**, *260*, 248–256.

328. Halliwell, B. Oxidative stress and cancer: Have we moved forward? *Biochemical Journal*, **2007**, *401*, 1–11.

329. Halliwell, B., & Aruoma, O. I. *DNA and Free Radicals* (p. 231). CRC Press, Boca Raton _FL, **1993**.

330. Ham, S. S., Kim, S. H., Moon, S. Y., Chung, M. J., Cui, C. B., Han, E. K., Chung, C. K., & Choe, M. Antimutagenic effects of subfractions of Chaga mushroom (*Inonotus obliquus*) extract. *Mutation Research–Genetic Toxicology and Environmental*, **2009**, *672*, 55–59.

331. Ham, S. S., Kim, S. H., Moon, S. Y., Chung, M. J., Cui, C. B., Han, E. K., Chung, C. K., & Choe, M. Antimutagenic effects of subfractions of Chaga mushroom (*Inonotus obliquus*) extract. *Mutation Research*, **2009**, *672*, 55–59.

332. Hanahan, D., & Weinberg, R. A. The hallmarks of cancer. *Cell*, **2000**, *100*, 57–70.

333. Hannan, A., Ullah, M., Usman, M., Hussain, S., Absar, M., & Javed, K. Anti-mycobacterial activity of garlic (*Allium sativum*) against multi-drug resistant and non-multi-drug resistant *mycobacterium tuberculosis*. *Pakistan Journal of Pharmaceutical Sciences*, **2011**, *24*, 81–85.

334. Hanum, F., & Hamzah, N. The use of medicinal plant species by the Temuan tribe of Ayer Hitam Forest, Selangor, Peninsular Malaysia. *Pertanika Journal of Tropical Agricultural Science*, **1999**, *22*, 85–94.

335. Bringmann, G. J., Schlauer, H., Rischer, M., Wohlfarth, J., Mühlbacher, A., Buske, A., Porzel, J. S., & Adam, G. Revised structure of antidesmone, an unusual alkaloid from tropical *Antidesma* plants (Euphorbiaceae). *Tetrahedron*, **2000**, *56*, 3691–3695.

336. Haq, M. R., Ashraf, S., Malik, C. P., Ganie, A. A., & Shandilya, U. *In vitro* cytotoxicity of *Parthenium hysterophorus* extracts against human cancerous cell lines. *Journal of Chemical and Pharmaceutical Research*, **2011**, *3*, 601–608.

337. Harborne, J. B., & Baxter, H. *Phytochemical Dictionary: A Handbook of Bioactive Compounds from Plants* (p. 993). Taylor & Francis, London, UK, **1993**.

338. Harborne, J. B., Baxter, H., & Moss, G. P. Phenolics, Part IV. Chapter 32, In: *Phytochemical Dictionary: Handbook of Bioactive Compounds from Plants* (2nd edn., pp. 361–374). Taylor & Francis, London, **1999**.

339. Harikrishnan, R., Balasundaram, C., Jawahar, S., & Heo, M. S. Immunomodulatory effect of *Withania somnifera* supplementation diet in the giant freshwater prawn *Macrobrachium rosenbergii* (de Man) against *Aeromonas hydrophila*. *Fish & Shellfish Immunology*, **2012**, *32*, 94–100.

340. Harris, J. C., Cottrell, S. L., Plummer, S., & Lloyd, D. Antimicrobial properties of *Allium sativum* (garlic). *Applied Microbiology and Biotechnology*, **2001**, *57*, 282–286.

341. Hartati, M. S., Mubarika, S., Gandjar, I. G., Hamann, M. T., Rao, K. V., & Wahyuono, S. Phalerin, a new benzophenoic glucoside isolated from the methanolic extract of Mahkota Dewa (*Phaleria macrocarpa* (scheff). Boerl.) leaves. *Majalah Farmasi Indonesia*, **2005**, *16*, 51–57.

342. Harvey, A. L. Medicines from nature: are natural product still relevant to drug discovery? *Trends in Pharmacological Sciences*, **1999**, *20*, 196–198.

343. Hatcher, H., Planalp, R., Cho, J., Torti, F. M., & Torti, S. V. Curcumin: From ancient medicine to current clinical trials. *Cellular and Molecular Life Sciences*, **2008**, *65*, 1631–1652.

344. Hattori, T., Ikematsu, S., Koito, A., Matsushita, S., Maeda, Y., Hada, M., Fujimaki, M., & Takatsuki, K. Preliminary evidence for inhibitory effect of glycyrrhizin on HIV replication in patients with AIDS. *Antiviral Research*, **1989**, *11*, 255–26113.

345. He, J., Chen, X. Q., Zhao, Y., Xu, G., & Peng, Y. Lycojapodine A: A novel alkaloid from *Lycopodium japonicum*. *Organic Letters*, **2009**, *11*, 1397–1400.

346. He, D. X., Ru, X. C., Wen, L., Wen, Y. C., Jiang, H. D., Bruce, I. C., Jin, J., Ma, X., & Xia, Q. Total flavonoids of Flos Chrysanthemi protect arterial endothelial cells against oxidative stress. *Journal of Ethnopharmacology*, **2012**, *139*, 68–73.

347. He, X., & Liu, R. H. Phytochemicals of apple peels: Isolation, structure elucidation, and their anti-proliferative and antioxidant activities. *Journal of Agricultural and Food Chemistry*, **2008**, *56*, 9905–9910.

348. He, Z., Chen, H., Li, G., Zhu, H., Gao, Y., Zhang, L., & Sun, J. Diosgenin inhibits the migration of human breast cancer MDA-MB-231 cells by suppressing Vav2 activity. *Phytomedicine*, **2014**, *21*, 871–876.

349. He, X., & Liu, R. H. Triterpenoids isolated from apple peels have potent antiproliferative activity and may be partially responsible for apple's anticancer activity. *Journal of Agricultural and Food Chemistry*, **2007**, *55*, 4366–4370.

350. Heitzman, M. E., Neto, C. C., Winiarz, E., Vaisberg, A. J., & Hammond, G. B. Ethnobotany, phytochemistry and pharmacology of *Uncaria* (Rubiaceae). *Phytochemistry*, **2005**, *66*, 5–29.

351. Hendra R., Ahmad, S., Sukari, A., Shukor, M. Y., & Oskoueian, E. Flavonoid analyzes and antimicrobial activity of various parts of *Phaleria macrocarpa* (Scheff.) Boerl fruit. *International Journal of Molecular Sciences*, **2011**, *12*, 3422–3431.

352. Hendra, R., Ahmad, S., Oskoueian, E., Sukari, A., & Shukor, M. Y. Antioxidant, anti-inflammatory and cytotoxicity of *Phaleria macrocarpa* (Boerl.) Scheff fruit. *BMC Complementary and Alternative Medicine*, **2011**, *11*, 1472–1482.

353. Hennenfent, K. L., & Govindan, R. Novel formulations of taxanes: A review – Old wine in a new bottle? *Annals of Oncology*, **2006**, *17*, 735–749.

354. Heo, M. Y., Sohn, S. J., & Au, W. W. Anti-genotoxicity of galangin as a cancer chemopreventive agent candidate. *Mutation Research*, **2001**, *488*, 135–150.

355. Heredity and Cancer. American Cancer Society, 2 August **2013**, https://web. archive.org/web/20130805065458/http://www.cancer.org:80/cancer/cancercauses/ geneticsandcancer/heredity-and-cancer (Accessed on 30 September 2019).

356. Herman-Antosiewicz, A., & Singh, S. V. Checkpoint kinase-1 regulates diallyl trisulfide-induced mitotic arrest in human prostate cancer cells. *The Journal of Biological Chemistry*, **2005**, *280*, 28519–28528.

357. Herman-Antosiewicz, A., & Singh, S. V. Signal transduction pathways leading to cell cycle arrest and apoptosis induction in cancer cells by *Alliumvege*. *Mutation Research*, **2004**, *555*, 121–131.

358. Hernandez, P., Rodriguez, P. C., Delgado, R., & Walczak, H. Protective effect of *Mangifera indica* L. polyphenols on human T lymphocytes against activation-induced cell death. *Pharmacological Research*, **2007**, *55*, 167–173.

359. Hewlett, M. J., Begley, M. J., Groenewegen, W. A., Heptinstall, S., Knight, D. W., May, J., Salan, U., & Toplis, D. Sesquiterpene lactones from feverfew, *Tanacetum parthenium*: Isolation, structural revision, activity against human blood platelet function and implications for migraine therapy. *Journal of the Chemical Society, Perkin Trans*, **1996**, *16*, 1979–1986.

360. Ho, Y. C., Yang, S. F., Peng, C. Y., Chou, M. Y., & Chang, Y. C. Epigallocatechin-3-gallate inhibits the invasion of human oral cancer cells and decreases the productions of matrix etalloproteinases and urokinase-plasminogen activator. *Journal of Oral Pathology & Medicine*, **2007**, *36*, 588–593.

361. Ho, Y. T., Lu, C. C., Yang, J. S., Chiang, J. H., Li, T. C., Ip, S. W., Hsia, T. C., Liao, C. L., Lin, J. G., Wood, W. G., & Chung, J. G. Berberine induced apoptosis via promoting the expression of caspase-8, -9 and -3, apoptosis-inducing factor and endonuclease G in SCC-4 human tongue squamous carcinoma cancer cells. *Anticancer Research*, **2009**, *29*, 4063-4070.

362. Hölker, U., Höfer, M., & Lenz, J. Biotechnological advances of laboratory-scale solid-state fermentation with fungi. *Applied Microbiology and Biotechnology*, **2004**, *64*, 175–186.

363. Holland-Frei Cancer Medicine (Chapter 40). Hasan, B., Ortel, A. C. E., Moor, B. W., & Pogue. *Photodynamic Therapy of Cancer* (6th edn., pp. 605–622). B.C. Decker Inc., Hamilton, ON, **2003**.

364. Hong, C. E., & Lyu, S. Y. Anti-inflammatory and anti-oxidative effects of Korean red ginseng extract in human keratinocytes. *Immune Network*, **2011**, *11*, 42–49.

365. Hong, J., Smith, T. J., Ho, C. T., August, D. A., & Yang, C. S. Effects of purified green and black tea polyphenols on cyclooxygenase- and lipoxygenase-dependent metabolism of arachidonic acid in human colon mucosa and colon tumor tissues. *Biochemical Pharmacology*, **2001**, *62*, 1175–1183.

366. Horiuch, M., Murakami, C., Yu, D., & Chen, T. H. Trifordines AC Sesquiterpenes alkaloids from *Tripterygium wilfordi* and structure anti HIV relationships of Tripterygium alkaloids. *Journal of Natural Products*, **2006**, *69*, 1271–1274.

367. Hosono, T., Fukao, T., Ogihara, J., Yoshimasa, I., Hajime, S., Taiichiro, S., & Toyohiko, A. Diallyl trisulfide suppresses the proliferation and induces apoptosis of human colon cancer cells through oxidative modification of beta-tubulin. *The Journal of Biological Chemistry*, **2005**, *280*, 41487–41493.

368. Hou, D. X. Potential mechanisms of cancer chemoprevention by anthocyanins. *Current Molecular Medicine*, **2003**, *3*, 149–159.

369. Hsieh, C. R., Yu, P. J., Chen, L. G., Chaw, S. M., Chang, C. C., & Wang, C. C. Cytotoxic constituents of *Hydrangea angustipetala* on human gastric carcinoma cells. *Botanical Studies*, **2010**, *51*, 45–51.

370. Hsu, C. L., & Yen, G. C. Effect of gallic acid on high fat dietinduced dyslipidaemia, hepatosteatosis and oxidative stress in rats. *British Journal of Nutrition*, **2007**, *98*, 727–735.

371. Hsu, S. C., Kuo, C. L., Lin, J. P., Lee, J. H., Lin, C. C., Su, C. C., Yang, M. D., & Chung, J. G. Crude extracts of *Euchresta formosana radix* inhibit invasion and migration of human hepatocellular carcinoma cells. *Anticancer Research*, **2007**, *27*, 2377–2384.

372. Hsu, Y. L., Kuo, P. L., Chiang, L. C., & Lin, C. C. Involvement of p53, nuclear factor kappaB and Fas/Fas ligand in induction of apoptosis and cell cycle arrest by saikosaponin d in human hepatoma cell lines. *Cancer Letters*, **2004**, *213*, 213–221.

373. Hsu, Y. L., Kuo, P. L., Weng, T. C., Yen, M. H., Chiang, L. C., & Lin, C. C. The antiproliferative activity of saponin-enriched fraction from Bupleurum Kaoi is through Fas-dependent apoptotic pathway in human non-small cell lung cancer A549 Cells. *Biological and Pharmaceutical Bulletin*, **2004**, *27*, 1112–1115.

374. Hu, H., Ahn, N. S., Yang, X., Lee, Y. S., & Kang, K. S. *Ganoderma lucidum* extract induces cell cycle arrest and apoptosis in MCF-7 human breast cancer cell. *International Journal of Cancer*, **2002**, *102*, 250–253.

375. Hu, Z., Yang, Y., Ho, P. C., Chan, S. Y., Heng, P. W., Chan, E., Duan, W., Koh, H. L., & Zhou, S. Herb-drug interactions: A literature review. *Drugs*, **2005**, *65*, 1239–1282.

376. Huang, K. C. *The Pharmacology of Chinese Herbs* (p. 388). CRC Press. Boca Raton, FL, **1992**.

377. Huang, K. C. *The Pharmacology of Chinese Herbs* (p. 185). CRC Press, Boca Raton, FL, **1993**.

378. Huang, K. C. *The Pharmacology of Chinese Herbs* (p. 190). CRC Press, Boca Raton, FL, **1998**.

379. Huang, W., & Zou, K. Cytotoxicity of a plant steroidal saponin on human lung cancer cells. *Asian Pacific Journal of Cancer Prevention*, **2011**, *12*, 513–517.

380. Huang, W. W., Chiu, Y. J., Fan, M. J., Lu, H. F., Yeh, H. F., Li, K. H., Chen, P. Y., Chung, J. G., & Yang, J. S. Kaempferol induced apoptosis via endoplasmic reticulum stress and mitochondria-dependent pathway in human osteosarcoma U-2 OS cells. *Molecular Nutrition & Food Research*, **2010**, *54*, 1585–1595.

381. Huang, Y. T., Huang, D. M., Chueh, S. C., Teng, C. M., & Guh, J. H. Alisol B acetate, a triterpene from Alismatis rhizoma, induces Bax nuclear translocation and apoptosis in human hormoneresistant prostate cancer PC-3 cells. *Cancer Letters*, **2006**, *231*, 270–278.

382. Hudson, J. B. *Antiviral Compounds from Plants* (p. 200). CRC Press, Boca Raton, Florida, **1990**.

383. Hudson, J. B., Lopez-Bazzocchi, I., & Towers, G. H. N. Antiviral activities of hypericin. *Antiviral Research*, **1991**, *15*, 101–112.

384. Huerta-Reyes, M. H., Basualdo, M. C., Abe, F., Estrada, M. J., Soler, C., & Chilpa, R. R. HIV-1 inhibitory compounds from *Calophyllum brasiliense* leaves. *Biological and Pharmaceutical Bulletin*, **2000**, *27*, 1471–1475.

385. Hossain, M. A., Shah, M. D., Sang, S. V., & Sakari, M. Chemical composition and antibacterial properties of the essential oils and crude extracts of *Merremia borneensis*. *Journal of King Saud University Science*, **2012**, *24*, 243–249.

386. Hussain, H., Badawy, A., Elshazly, A., Elsayed, A., Krohn, K., Riaz, M., & Schulz, B. Chemical constituents and antimicrobial activity of *Salix subserrata*. *Records of Natural Products*, **2011**, *5*, 133–137.

387. Hwang, J. T., Ha, J., Park, I. J., Lee, S. K., Baik, H. W., Kim, Y. M., & Park, O. J. Apoptotic effect of EGCG in HT-29 colon cancer cells via AMPK signal pathway. *Cancer Letters*, **2007**, *247*, 115–121.

388. Hwu, J. R., Hsu, M. H., & Huang, R. C. C. New nordihydroguaiaretic acid derivatives as anti-HIV agents. *Bioorganic & Medicinal Chemistry Letters*, **2008**, *18*, 1884–1888.

389. Hyder, P. W., Fredrickson, E. L., Estell, R. E., Tellez, M., & Gibbens, R. P. Distribution and concentration of total phenolics, condensed tannins, and nordihydroguaiaretic acid (NDGA) in creosotebush (*Larrea tridentata*). *Biochemical Systematics and Ecology*, **2002**, *30*, 905–912.

390. Ichimura, T., Watanabe, O., & Maruyama, S. Inhibition of HIV-1 protease by water-soluble lignin-like substance from an edible mushroom. *Bioscience, Biotechnology, and Biochemistry*, **1998**, *62*, 575–577.

391. Iciek, M., Kwiecien, I., Chwatko, G., Sokolowska-Jezewicz, M., Kowalczyk-Pachel, D., & Rokita, H. The effects of garlic-derived sulfur compounds on cell proliferation, caspase 3 activity, thiol levels and anaerobic sulfur metabolism in human hepatoblastoma HepG2 cells. *Cell Biochemistry and Function*, **2011**, *30*, 198–204.

392. Ignacimuthu, S., Ayyanar, M., & Sivaraman, S. K. Ethnobotanical investigations among tribes in Madurai district of Tamil Nadu (India). *Journal of Ethnobiology and Ethnomedicine*, **2006**, *2*, 25–30.

393. Igoli, L. O., Ogaji, O. G., Tor-Aryiin, A., & Igoli, N. P. Traditional medicinal practices amongst the igede people of Nigeria, Part II. *African Journal of Traditional, Complementary and Alternative*, **2005**, *2*, 134–152.

394. Imbert, T. F. Discovery of podophyllotoxins. *Biochimie*, **1998**, *80*, 207–222.

395. Imir, A. G., Lin, Z., Yin, P., Deb, S., Yilmaz, B., Cetin, M., Cetin, A., & Bulun, S. E. Aromatase expression in uterine leiomyomata is regulated primarily by proximal promoters I.3/II. *The Journal of Clinical Endocrinology & Metabolism*, **2007**, *92*, 1979–1982.

396. Imir, N., Aydemir, E., Simsek, R. S., Gokturk, E., Yesilada, & Fiskin, K. Cytotoxic and immunomodulatory effects of *Ebenus boissieri* Barbey on breast cancer cells. *Genetics and Molecular Research*, **2016**, *15*, Article number 15017766.

397. Inamdar, N., Edalat, S., Kotwal, V. B., & Pawar, S. Herbal drugs in milieu of modern drugs. *International Journal of Green Pharmacy*, **2008**, *2*, 2–8.

398. Inui, T., Wang, Y., Deng, S., Smith, D. C., Franzblau, S. G., & Pauli, G. F. Counter-current chromatography based analysis of synergy in an anti-tuberculosis ethnobotanical. *Journal of Chromatography A*, **2007**, *1151*, 211–215.

399. Ip, S. W., Weng, Y. S., Lin, S. Y., Mei, D., Tang, N. Y., Su, C. C., & Chung, J. G. The role of Ca2+ on rhein-induced apoptosis in human cervical cancer Ca Ski cells. *Anticancer Research*, **2007**, *27*, 379–390.

400. Ishikawa, C., Tafuku, S., Kadekaru, T., Sawada, S., Tomita, M., Okudaira, T., Nakazato, T., Toda, T., Uchihara, J. N., Taira, N., Ohshiro, K., Yasumoto, T., Ohta, T., & Mori, N. Anti-adult T-cell leukemia effects of brown algae fucoxanthin and its deacetylated product, fucoxanthinol. *International Journal of Cancer*, **2008**, *123*, 2702–2712.

401. Ismail, N. F., Ismail, A., Chuah, T. G., Abdullah, Z., Li, A. R., Pin, K. Y., & Adawiah, I. Extraction, separation and identification of phalerin from *Phaleria macrocarpa* (Sheff.) Boerl. *Proceedings International Conference on Chemistry Innovation* (pp. 23, 24). Terengganu, Malaysia, **2011**.

402. Ito, M., Nakashima, H., Baba, H., Pauwels, R., De Clercq, E., Shigeta, S., & Yamamoto, N. Inhibitory effect of glycyrrizin on the *in-vitro* infectivity and cytopathic activity of the human immunodeficiency virus (HIV) (HTLV-III/LA-V). *Antiviral Research*, **1987**, *7*, 127–137.

403. Ito, M., Baba, M., Sato, A., Pauwels, R., De Clercq, E., & Shigeta, S. Inhibitory effect of dextran sulfate and heparin on the replication of human immunodeficiency virus (HIV) *in vitro*. *Antiviral Research*, **1987**, *7*, 361–367.

404. Ito, M., Sato, A., Hirabayashi, K., Tanabe, F., Shigeta, S., Baba, M., De Clercq, E., Nakashima, H., & Yamamoto, N. Mechanism of inhibitory effect of glycyrrhizin on replication human immune deficiency virus (HIV). *Antiviral Research*, **1988**, *10*, 289–298.

405. Jagetia, G. C., Baliga, M. S., Malagi, K. L., & Kamath, M. S. The evaluation of the radioprotective effect of Tripala (an ayurvedic rejuvenating drug) in the mice exposed to radiation. *Phytomedicines*, **2002**, *9*, 99–108.

406. Jagtap, U. B., & Bapat, V. A. Artocarpus: A review of its traditional uses, phytochemistry and pharmacology. *Journal of Ethnopharmacology*, **2010**, *129*, 142–166.

407. Jahan, N., Ahmed, W., & Malik, A. New steroidal glycosides from *Mimusops elengi*. *Journal of Natural Products*, **1995**, *8*, 1244–1247.

408. Jain, A., & Jain, A. *Tridax procumbens* (L): A weed with immense medicinal importance: A review. *International Journal of Pharma and Bio Sciences*, **2012**, *3*, 544 552.

409. Jang, A., Srinivasan, P., Lee, N. Y., Song, H. P., Lee, J. W., Lee, M., & Jo, C. Comparison of hypolipidemic activity of synthetic gallic acid-linolenic acid ester with mixture of gallic acid and linolenic acid, gallic acid, and linolenic acid on highfat diet induced obesity in C57BL/6 Cr Slc mice. *Chemico-Biological Interactions*, **2008**, *174*, 109–117.

410. Javaid, A., & Shafique, S. Herbicidal activity of *Withania somnifera* against *Phalaris minor*. *Natural Product Research*, **2010**, *24*, 1457–1468.

411. Jayaprakasam, B., & Nair, M. G. Cyclooxygenase-2 enzyme inhibitory withanolides from *Withania somnifera* leaves. *Tetrahedron*, **2003**, *59*, 841–849.

412. Jayaprakasam, B., Zhang, Y., Seeram, N. P., & Nair, M. G. Growth inhibition of human tumor cell lines by withanolides from *Withania somnifera* leaves. *Life Sciences*, **2003**, *74*, 125–132.

413. Jayasekara, H., MacInnis, R. J., Room, R., & English, D. R. Long-term alcohol consumption and breast, upper aero-digestive tract and colorectal cancer risk: A systematic review and meta-analysis. *Alcohol and Alcoholism*, **2016**, *51*, 315–330.

414. Jedinak, A., Thyagarajan-Sahu, A., Jiang, J., & Sliva, D. Ganodermanontriol, a lanostanoid triterpene from *Ganoderma lucidum*, suppresses growth of colon cancer cells through ss-catenin signaling. *International Journal of Oncology*, **2011**, *38*, 761–767.

415. Jemal, A., Bray, F., Center, M. M., Ferlay, J., Ward, E., & Forman, D. Global cancer statistics. *CA: A Cancer Journal for Clinicians*, **2011**, *61*, 69–90.

416. Ji, C., Ren, F., & Xu, M. Caspase-8 and p38MAPK in DATS-induced apoptosis of human CNE2 cells. *Brazilian Journal of Medical and Biological Research*, **2010**, *43*, 821–827.

417. Jiang, Y., Sorokin, D. Y., Kleerebezem, R., Muyzer, G., & Van Loosdrecht, M. C. M. *Plasticicumulans acidivorans* gen. nov., sp. nov., a polyhydroxyalkanoate-accumulating gammaproteobacterium from a sequencing-batch bioreactor. *International Journal of Systematic and Evolutionary Microbiology*, **2011**, *61*, 2314–2319.

418. Jiang, X. H., Wong, B. C., Lin, M. C., Zhu, G. H., Kung, H. F., Jiang, S. H., Yang, D., & Lam, S. K. Functional p53 is required for triptolide-induced apoptosis and AP-1 and nuclear factor-kappa B activation in gastric cancer cells. *Oncogene*, **2001**, *20*, 8009–8018.

419. Jiménez, J. P., & Saura-Calixto, F. Grape products and cardiovascular disease risk factors. *Nutrition Research Reviews*, **2008**, *21*, 158–173.

420. Jiménez, J. P., Serrano, J., Tabernero, M., Arranz, S., Díaz-Rubio, M. E., García-Diz, L., Goñi, I., & Saura-Calixto, F. Effects of grape antioxidant dietary fiber in cardiovascular disease risk factors. *Nutrition*, **2008**, *24*, 646–653.

421. Johnson, G. *Unearthing Prehistoric Tumors, and Debate*. http://www.nytimes.com/2010/12/28/health/28cancer.html (Accessed on 30 September 2019). The New York Times.

422. Johnson, G. L., & Lapadat, R. Mitogen-activated protein kinase pathways mediated by ERK, JNK, and p38 protein kinases. *Science*, **2002**, *298*, 1911–1912.

423. Jones, S. B., De Primo, S. E., Whitfield, M. L., & Brooks, J. D. Resveratrol-induced gene expression profiles in human prostate cancer cells. *Cancer Epidemiology, Biomarkers & Prevention*, **2005**, *14*, 596–604.

424. Jose, A. K. *Apoptosis in Carcinogenesis and Chemotherapy* (p. 384). Springer, Netherlands, **2009**.

425. Jose, J. K., Kuttan, G., & Kutan, R. Anti-tumor activity of *Emblica officinalis*. *Journal of Ethnopharmacology*, **2001**, *75*, 65–69.

426. Joshi, B. N., Kamat, V. N., & Govindachari, T. R. Structure of tuberosin, a new pterocarpan from *Pueraria tuberose*. *Indian Journal of Chemistry*, **1972**, *10*, 1112–1113.

427. Jung, S. H., Woo, M. S., Kim, S. Y., Kim, W. K., Hyun, J. W., Kim, E. J., Kim, D. H., & Kim, H. S. Ginseng saponin metabolite suppresses phorbol ester-induced matrix metalloproteinase-9 expression through inhibition of activator protein-1 and mitogen-activated protein kinase signaling pathways in human astroglioma cells. *International Journal of Cancer*, **2006**, *15*, 118, 490–497.

428. Kaatsch, P. Epidemiology of childhood cancer. *Cancer Treatment Reviews*, **2010**, *36*, 277–285.

429. Kadota, Y., Taniguchi, C., Masuhara, S., Yamamoto, S., Furusaki, S., Iwahara, M., Goto, K., Matsumoto, Y., & Ueoka, R. Inhibitory effects of extracts from peels of *Citrus natsudaidai* encapsulated in hybrid liposomes on the growth of tumor cells *in vitro*. *Biological and Pharmaceutical Bulletin*, **2004**, *27*, 1465–1467.

430. Kaido, T. L., Veale, D. J. H., Havlik, I., & Rama, D. B. K. Preliminary screening of plants used in South Africa as traditional herbal remedies during pregnancy and labor. *Journal of Ethnopharmacology*, **1987**, *55*, 185–191.

431. Kakiuchi, N., Kusumoto, I. T., Hattori, M., Tsuneo, N., Tsutomu, H., & Takuo, O. Effect of condensed tannins and related compounds on reverse transcriptase. *Phytotherapy Research*, **1991**, *5*, 270–272.

432. Kalvatchev, Z., Walder, R., & Garzard, D. Anti-HIV activity of extracts from *Calendula officinalis* flowers. *Biomedicine & Pharmacotherapy*, **1997**, *51*, 176–180.

433. Kamali, H. H., & Amir, E. L. Antibacterial activity and phytochemical screening of ethanolic extracts obtained from selected sudanese medicinal plants. *Current Research Journal of Biological Sciences*, **2010**, *2*, 143–146.

434. Kamali, M., Khosroyar, S., & Mohammadi, A. Antibacterial activity of various extracts from *Dracocephalum kotschyi* against food pathogenic microorganisms. *International Journal of Pharm. Tech Research*, **2015**, *8*, 158–163.

435. Kamboj, V. P. Herbal medicine. *Current Science*, **2000**, *78*, 35–39.

436. Kane, D. J., Ord, T., Anton, R., & Bredesen, D. E. Expression of bcl-2 inhibits necrotic neural cell death. *Journal of Neuroscience Research*, **1995**, *40*, 269–275.

437. Kang, S. C., Lee, C. M., Choung, E. S., Bak, J. P., Bae, J. J., Yoo, H. S., Kwak, J. H., & Zee, O. P. Anti-proliferative effects of estrogen receptor-modulating compounds isolated from Rheum palmatum. *Archives of Pharmacal Research*, **2008**, *31*, 722–726.

438. Kapil, A., & Sharma, S. Immunopotentiating compounds from *Tinospora cordifolia*. *Journal of Ethnopharmacology*, **1997**, *58*, 89–95.

439. Karmazyn, M., Moey, M., & Gan, X. T. Therapeutic potential of ginseng in the management of cardiovascular disorders. *Drugs*, **2011**, *71*, 1989–2008.

440. Karna, P., Chagani, S., Gundala, S. R., Rida, P. C. G., Asif, G., Sharma, V., & Gupta, M. V. Aneja, R. Benefits of whole ginger extract in prostate cancer. *British Journal of Nutrition*, **2012**, *107*, 473–484.

441. Kashiwada, Y., Hashimoto, F., Cosentino, L. M., Chen, C. H., Garrett, P. E., & Lee, K. H. Betulinic acid and dihydrobetulinic acid derivatives as potent anti-HIV agents. *Journal of Medicinal Chemistry*, **1996**, *39*, 1016–1017.

442. Kashiwada, Y., Nagao, T., Hashimoto, A., Ikeshiro, Y., Okabe, H., Cosentino, M. L., & Lee, K. H. Anti-AIDS agents 38. Anti-HIV activity of 3-Oacyl ursolic acid derivatives. *Journal of Natural Products*, **2000**, *63*, 1619–1622.

443. Kashiwada, Y., Wang, H. K., & Nagao, T. Anti HIV activity of oleanolic acid, pomolic acid and structurally related triterpenoids. *Journal of Natural Products*, **1998**, *61*, 1090–1095.

444. Kasolo, J. N., Bimenya, G. S., Ojok, L., Ochieng, J., & Ogwal-Okeng, J. W. Phytochemicals and uses of *Moringa oleifera* in Ugandan rural communities. *Journal of Medicinal Plants Research*, **2010**, *4*, 753–757.

445. Kasper, D. L., Braunwald, E., Anthony, S. F., et al. Human immunodeficiency virus disease: AIDS and related disorders. In: *Harrison's Principles of Internal Medicine?* (16th edn. Vol. I, pp. 1076–1139). McGraw Hill, New York, **2005**.

446. Kasturi, T. R., Thomas, M., & Abraham, E. M. Essential oil of *Ageratum conyzoides*. Isolation and structure of 2 new constituents. *Indian Journal of Chemistry*, **1973**, *11*, 91–95.

447. Kato, A., Nasu, N., Takebayashi, K., Adachi, I., Minami, Y., Sanae, F., Asano, N., Watson, A. A., & Nash, R. J. Structureactivity relationships of flavonoids as potential inhibitors of glycogen phosphorylase. *Journal of Agricultural and Food Chemistry*, **2008**, *56*, 4469–4473.

448. Katsube, N., Iwashita, K., Tsushida, T., Yamaki, K., & Kobori, M. Induction of apoptosis in cancer cells by Bilberry (*Vaccinium myrtillus*) and the anthocyanins. *Journal of Agricultural and Food Chemistry*, **2003**, *51*, 68–75.

449. Katz, S. R., Newman, R. A., & Lansky, E. P. *Punica granatum* L.: Heuristic treatment for diabetes mellitus. *Journal of Medicinal Food*, **2007**, *10*, 213–217.

450. Kaur, R., & Kharb, R. Anti-HIV potential of medicinally important plants. *International Journal of Pharma and Bio Sciences*, **2011**, *2*, 387–398.

451. Kaur, S., Grover, I. S., Singh, M., & Kaur, S. Antimutagenicity of hydrolyzable tannins from *Terminalia chebula* in *Salmonella typhimurium*. *Mutation Research*, **1998**, *419*, 169–179.

452. Kawase, M., Motohashi, N., Satoh, K., Sakagami, H., Nakashima, H., Tani, S., Shirataki, Y., Kurihara, T., Spengler, G., Wolfard, K., & Molnár, J. Biological activity of persimmon (*Diospyros kaki*) peel extracts. *Phytotherapy Research*, **2003**, *17*, 495–500.

453. Kayombo, E. J., Uiso, F. C., Mbwambo, Z. H., Mahunnah, R. L., Moshi, M. J., & Mgonda, Y. H. Experience of initiating collaboration of traditional healers in managing HIV and AIDS in Tanzania. *Journal of Ethnobiology and Ethnomedicine*, **2007**, *3*, 1–9.

454. Keawpradub, N., Eno-Amooquaye, E., Burke, P. J., & Houghton, P. J. Cytotoxic activity of indole alkaloids from *Alstonia macrophylla*. *Planta Medica*, **1999**, *65*, 311–315.

455. Keawpradub, N., Houghton, P. J., Eno-Amooquaye, E., & Burke, P. J. Activity of extracts and alkaloids of Thai *Alstonia* species against human lung cancer cell lines. *Planta Medica*, **1997**, *63*, 97–101.

456. Kellogg, J., Wang, J., Flint, C., Ribnicky, D., Kuhn, P., De Mejia, E. G., Raskin, I., & Lila, M. A. Alaskan wild berry resources and human health under the cloud of climate change. *Journal of Agricultural and Food Chemistry*, **2010**, *58*, 3884–3900.

457. Keplinger, K., Laus, G., Wurm, M., Dierich, M. P., & Teppner, H. *Uncaria tomentosa* (Willd.) DC.-ethnomedicinal use and new pharmacological, toxicological and botanical results. *Journal of Ethnopharmacology*, **1999**, *64*, 23–24.

458. Keum, Y. S., Kim, J., Le, K. H., Park, K. K., Surh, Y. J., Lee, J. M., Lee, S. S., Yoon, J. H., Joo, S. Y., Cha, I. H., & Yook, J. I. Induction of apoptosis and caspase-3 activation by chemopreventive [6]-paradol and structurally related compounds in KB cells. *Cancer Letters*, **2002**, *177*, 41–47.

459. Khade, K. A. Report on antimicrobial activity and phytochemical screening of *Argemone Mexicana* Linn. *International Journal of Advanced Multidisciplinary Research*, **2015**, *3*, 1–7.

460. Khanh, T. C. Contribution to the studies on botany, chemical composition and biological activities of *Pseuderanthemum palatiferum (Nees) Radlk. Journal of Materials Science: Materials in Medicine*, **1998**, *3*, 37–41.

461. Khanna, D., Sethi, G., Ahn, K. S., Pandey, M. K., Kunnumakkara, A. B., Sung, B., Aggarwal, A., & Aggarwal, B. B. Natural products as a gold mine for arthritis treatment. *Current Opinion in Pharmacology*, **2007**, *7*, 344–351.

462. Khare, C. *Indian Medicinal Plants: An Illustrated Dictionary* (p. 432). Springer, New York, **2007**, E-book.

463. Khonkarn, R., Okonogi, S., Ampasavate, C., & Anuchapreeda, S. Investigation of fruit peel extracts as sources for compounds with antioxidant and antiproliferative activities against human cell lines. *Food and Chemical Toxicology*, **2010**, *48*, 2122–2129.

464. Kikuchi, T., Nihei, M., Nagai, H., Fukushi, H., Tabata, K., Suzuki, T., & Akihisa, T. Albanol A from the root bark of *Morus alba* L. induces apoptotic cell death in HL60 human leukemia cell line. *Chemical and Pharmaceutical Bulletin*, **2010**, *58*, 568–571.

465. Kim, Y., Kim, J. E., Lee, S. D., Lee, T. G., Kim, J. H., Park, J. B., Han, J. M., Jang, S. K., Suh, P. G., & Ryu, S. H. Phospholipase D1 is located and activated by protein kinase C alpha in the plasma membrane in 3Y1 fibroblast cell. *Biochimica et Biophysica Acta*, **1999**, *1436*, 319–330.

466. Kim, G. N., Shin, J. G., & Jang, H. D. Antioxidant and antidiabetic activity of *Dangyuja* (*Citrus grandis Osbeck*) extract treated with *Aspergillus saitoi*. *Food Chemistry*, **2009**, *117*, 35–41.

467. Kim, H. G., Cho, H. G., Jeong, E. Y., Lim, J. H., & Lee, S. H. Growth inhibitory activity of active component of *Terminalia chebula* fruits against intestinal bacteria. *Journal of Food Protection*, **2006**, *69*, 2205–2209.

468. Kim, H. J., Woo, E. R., & Shin, C. G. A new flavonol glycoside gallate ester from Acer okamotoanum and its inhibitory activity against human immunodeficiency virus-1 (HIV-1) integrase. *Journal of Natural Products*, **1998**, *61*, 145–148.

469. Kim, H. K., Lee, H. K., Shin, C. G., & Huh, H. HIV integrase inhibitory activity of *Agastache rugosa. Archives of Pharmacal Research*, **1999**, *22*, 520–523.

470. Kim, K. N., Heo, S. J., Kang, S. M., Ahn, G., & Jeon, Y. J. Fucoxanthin induces apoptosis in human leukemia HL-60 cells through a ROS-mediated Bcl-xL pathway. *Toxicology in Vitro*, **2010**, *24*, 1648–1654.

471. Kim, M. J., Park, H. J., Hong, M. S., Park, H. J., Kim, M. S., Leem, K. H., Kim, J. B., Kim, Y. J., & Kim, H. K. *Citrus reticulate* blanco induces apoptosis in human gastric cancer cells SNU-668. *Nutrition and Cancer*, **2005**, *51*, 78–82.

472. Kim, R. S., Kim, H. W., & Kim, B. K. Suppressive effects of *Ganoderma lucidum* on proliferation of peripheral blood mononuclear cells. *Molecules and Cells*, **1997**, *7*, 52–57.

473. Kim, W. J., Veriansyah, B., Lee, Y. O., Kim, J., & Kim, J. D. Extraction of mangiferin from Mahkota Dewa (*Phaleria macrocarpa*) using subcritical water. *Journal of Industrial and Engineering Chemistry*, **2010**, *16*, 425–430.

474. Kim, Y. A., Dong, X., Hui, X., Anna, A. P., Karen, L. L., Megan, L. R., Yan, Z., Zhou, W., & Shivendra, V. S. Mitochondria-mediated apoptosis by diallyl trisulfide in human prostate cancer cells is associated with generation of reactive oxygen species and regulated by Bax/Bak. *Molecular Cancer Therapeutics*, **2007**, *6*, 1599–1609.

475. Kimura, Y. N., Watari, K., Fotovati, A., Hosoi, F., Yasumoto, K., Izumi, H., Kohno, K., Umezawa, K., Iguchi, H., Shirouzu, K., Takamori, S., Kuwano, M., & Ono, M. Inflammatory stimuli from macrophages and cancer cells synergistically promote tumor growth and angiogenesis. *Cancer Science*, **2007**, *98*, 2009–2018.

476. Kirbag, S., Zengin, F., & Kursat, M. Antimicrobial activities of extracts of some plants. *Pakistan Journal of Botany*, **2009**, *41*, 2067–2070.

477. Kirtikar, K. R., & Basu, B. D. *Indian Medicinal Plants* (2nd edn., pp. 1760–1764). International Book Distributors, Dehradun, India, **1999**.

478. Kirtikar, K. R., & Basu, B. D. *Indian Medicinal Plants* (2nd edn., Vol. 4, pp. 2609–2610). Publisher Ltd., DehraDun, India, **1994**.

479. Kirtikar, K. R., & Basu, B. D. *Indian Medicinal Plants* (pp. 1494–1496). M/s. Bishensingh Mahendra Palsingh, Dehradun, **1995**.

480. Kitamura, K., Honda, M., Yoshizaki, H., Yamamoto, S., Nakane, H., Fukushimam M., Onom, K., & Tokunagam, T. Baicalin, an inhibitor of HIV-1 production *in vitro*. *Antiviral Research*, **1998**, *37*, 131–140.

481. Kiviharju, T. M., Lecane, P. S., Sellers, R. G., & Peehl, D. M. Antiproliferative and proapoptotic activities of triptolide (PG490), a natural product entering clinical trials, on primary cultures of human prostatic epithelial cells. *Clinical Cancer Research*, **2002**, *8*, 2666–2674.

482. Knödler, M., Conrad, J., Wenzig, E. M., Bauer, R., Lacorn, M., Beifuss, U., Carle, R., & Schieber, A. Anti-inflammatory 5-(11'Z-heptadecenyl)-and 5-(8'Z, 11'Z-heptadecadienyl)-resorcinols from mango (*Mangifera indica* L.) peels. *Phytochemistry*, **2008**, *69*, 988–993.

483. Knowles, L. M., & Milner, J. A. Diallyl disulfide increases cyclin B1 protein expression and depresses phosphatase activity in HCT-15 cells. *FASEB Journal*, **1998**, *12*, 656.

484. Ko, J. K., Lam, F. Y., & Cheung, A. P. Amelioration of experimental colitis by *Astragalus membranaceus* through anti-oxidation and inhibition of adhesion molecule synthesis. *World Journal of Gastroenterology*, **2005**, *11*, 5787–5794.

485. Ko, Y. C., Lien, J. C., Liu, H. C., Hsu, S. C., Ji, B. C., Yang, M. D., Hsu, W. H., & Chung, J. G. Demethoxycurcumin induces the apoptosis of human lung cancer NCI-H460 cells through the mitochondrial-dependent pathway. *Oncology Reports*, **2015**, *33*, 2429–2437.

486. Kokate, C. K. *Pharmacognosy* (12th edn., pp. 210–211). Nirali Prakashan, Mumbai, **1999**.

487. Kokate, C. K., Purohit, A. P., & Gokhale, S. B. *Pharmacognosy* (18th edn., pp. 130–138). Nirali Prakashan, Mumbai, **2005**.

488. Konar, A., Shah, N., Singh, R., Saxena, N., Kaul, S. C., Wadhwa, R., & Thakur, M. K. Protective role of *Ashwagandha* leaf extract and its component withanone on scopolamine-induced changes in the brain and brain-derived cells. *PLoS One*, **2011**, *6*, e27265.

489. Kong, C. S., Kim, J. A., Yoon, N. Y., & Kim, S. K. Induction of apoptosis by phloroglucinol derivative from *Ecklonia Cava* in MCF-7 human breast cancer cells. *Food and Chemical Toxicology*, **2009**, *47*, 1653–1658.

490. Konishi, I., Hosokawa, M., Sashima, T., Kobayashi, H., & Miyashita, K. Halocynthiaxanthin and fucoxanthinol isolated from *Halocynthia roretzi* induce apoptosis in human leukemia, breast and colon cancer cells. *Comparative Biochemistry and Physiology – Part C*, **2006**, *142*, 53–59.

491. Konyalioglu, S., & Karamenderes, C. The protective effects of *Achillea* L. species native in Turkey against H_2O_2-induced oxidative damage in human erythrocytes and leucocytes. *Journal of Ethnopharmacology*, **2005**, *102*, 221–227.

492. Koo, B. S., Lee, W. C., Chang, Y. C., & Kim, C. H. Protective effects of alpinae oxyphyllae fructus (*Alpinia oxyphylla* MIQ) water-extracts on neurons from ischemic damage and neuronal cell toxicity. *Phytotherapy Research*, **2004**, *18*, 142–148.

493. Kostova, I. Coumarins as inhibitors of HIV reverse transcriptase. *Current HIV Research*, **2006**, *4*, 347–363.

494. Kosuge, T., Yokota, M., Sugiyama, K., Yamamoto, T., Mure, T., & Yamazawa, H. Studies on bioactive substances in crude drugs used for arthritic diseases in traditional Chinese medicine. II. Isolation and identification of an anti-inflammatory and analgesic principle from the root of *Angelica pubescens* Maxim. *Chemical and Pharmaceutical Bulletin (Tokyo)*, **1985**, *33*, 5351–5354.

495. Kotake-Nara, E., Asai, A., & Nagao, A. Neoxanthin and fucoxanthin induce apoptosis in PC-3 human prostate cancer cells. *Cancer Letters*, **2005**, *220*, 75–84.

496. Kris-Etherton, P. M., Hecker, K. D., Bonanome, A., Coval, S. M., Binkoski, A. E., Hilpert, K. F., Griel, A. E., & Etherton, T. D. Bioactive compounds in foods: Their role in the prevention of cardiovascular disease and cancer. *American Journal of Medicine*, **2002**, *113*, 71S–88S.

497. Kronenberg, F., & Fugh-Berman, A. Complementary and alternative medicine for menopausal symptoms: A review of randomized, controlled trials. *Annals of Internal Medicine*, **2002**, *137*, 805–813.

498. Krysko, D. V., & Vandenabeele, P. *Phagocytosis of Dying Cells: From Molecular Mechanisms to Human Diseases* (p. 447). Springer publication, Netherlands, **2009**.

499. Kumar, A. A., Vikram, S. M., Subhasis, C., & Kishore, C. P. Novel immunomodulatory effect of *Gracilaria verrucosa* and *Potamogeton pectinatus* extracts on *in vitro* activation of T cells. *Life Sciences*, **2012**, *2*, 233–239.

500. Kumar, A. G., Nandagopal, S., Illanjiam, S., Suganthi, K., Dhanalakshmi, D. P., & Joshua, D. E. Anti-proliferative activity and immunomodulator effects of *Andrographis paniculata* (Burm.f.) Wall. ex Nees against cancerous cell line. *Der Pharmacia Lettre*, **2016**, *8*, 17–24.

501. Kumar, S., Hassan, S. A., Dwivedi, S., Kukreja, A. K., Sharma, A., Singh, A. K., Sharma, S., & Tewari, R. Proceedings of the National Seminar on the Frontiers of Research and Development in medicinal plants. *Journal Medicinal and Aromatic Plant Sciences Central Institute of Medicinal and Aromatic Plants (CIMAP)* (Vol. 22/4A and Vol. 23/1A). Lucknow, India, **2000**.

502. Kun, S., Supaporn, S., Parinya, P., Supa, H., & Kiattawee, C. Anti-HIV-1 reverse transcriptase activities of hexane extracts from some Asian medicinal plants. *Journal of Medicinal Plants Research*, **2011**, *5*, 4899–4906.

503. Kuo, C. L., Wu, S. Y., Ip, S. W., Wu, P. P., Yu, C. S., Yang, J. S., Chen, P. Y., Wu, S. H., & Chung, J. G. Apoptotic death in curcumin-reated NPC-TW 076 human nasopharyngeal carcinoma cells is mediated through the ROS, mitochondrial depolarization and caspase-3-dependent signaling responses. *International Journal of Oncology*, **2011**, *39*, 319–328.

504. Kuo, C. P., Lu, C. H., Wen, L. L., Cherng, C. H., Wong, C. S., Borel, C. O., Ju, D. T., Chen, C. M., & Wu, C. T. Neuroprotective effect of curcumin in an experimental rat model of subarachnoid hemorrhage. *Anesthesiology*, **2011**, *115*, 1229–1238.

505. Kuo, P. L., Hsu, Y. L., Ng, L. T., & Lin, C. C. Rhein inhibits the growth and induces the apoptosis of Hep G2 cells. *Planta Medica*, **2004**, *70*, 12–16.

506. Kuo, P. L., Lin, T. C., & Lin, C. C. The ant proliferative activity of aloe-emodin is through p53-dependent andp21-dependent apoptotic pathway in human hepatoma cell lines. *Life Sciences*, **2002**, *71*, 1879–1892.

507. Kuo, Y. H., & Kuo, L. M. Y. Antitumor and anti-AIDS triterpenoids from *Celastrus hindsii*. *Phytochemistry*, **1997**, *44*, 1275–1281.

508. Kurien, J. C. *Plants That Heal* (4th edn., p. 82). Orient Watchman Publishing House, Pune, **2001**.

509. Kushi, L. H., Doyle, C., McCullough, M., Rock, C. L., Demark-Wahnefried, W., Bandera, E. V., Gapstur, S., Patel, A. V., Andrews, K., & Gansler, T. American cancer society guidelines on nutrition and physical activity for cancer prevention: Reducing the risk of cancer with healthy food choices and physical activity. *Cancer*, **2012**, *62*, 30–67.

510. Kusomoto, I. T., Nakabayashi, T., Kida, H., Miyashiro, H., Hattori, M., Namba, T., & Shimotohno, K. Screening of various plant extracts used in ayurvedic medicine for inhibitory effects on human immunodeficiency virus type 1 (HIV-1) protease. *Phytotherapy Research*, **1995**, *9*, 180–184.

511. Lai, H., & Singh, N. P. Oral artemisinin prevents and delays the development of 7,12-dimethylbenz[a]anthracene (DMBA)-induced breast cancer in the rat. *Cancer Letters*, **2006**, *231*, 43–48.

512. Lai, K. C., Hsu, S. C., Kuo, C. L., Yang, J. S., Ma, C. Y., Lu, H. F., Tang, N. Y., Hsia, T. C., Ho, H. C., & Chung, J. G. Diallyl sulfide, diallyl disulfide, and diallyl trisulfide inhibit migration and invasion in human colon cancer colo 205 cells through the inhibition of matrix metalloproteinase-2, -7, and -9 expressions. *Environmental Toxicology*, **2013**, *28*, 479–488.

513. Lalitha, P., Arathi, K. A., Shubashini, K., Sripathi, H. S., & Jayanthi, P. Antimicrobial activity and phytochemical screening of an ornamental foliage plant, *Pothos aurea* (Linden ex Andre). *International Journal of Chemistry*, **2010**, *1*, 63–71.

514. Lall, N., Das, S. M., Hazra, B., & Meyer, J. J. Antimycobacterial activity of diospyrin derivatives and a structural analog of diospyrin against mycobacterium tuberculosis *in vitro*. *Journal of Antimicrobial Chemotherapy*, **2003**, *51*, 435–438.

515. Lall, N., & Meyer, J. J. Inhibition of drug sensitive and drug resistant strains of Mycobacterium tuberculosis by diospyrin, isolated from *Euclea natalensis*. *Journal of Ethnopharmacology*, **2001**, *78*, 213–216.

516. Lambertini, E., Piva, R., Khan, M. T., Lampronti, I., Bianchi, N., Borgatti, M., & Gambari, R. Effects of extracts from Bangladeshi medicinal plants on *in vitro* proliferation of human breast cancer cell lines and expression of estrogen receptor alpha gene. *International Journal of Oncology*, **2004**, *24*, 419–423.

517. Lampronti, I., Khan, M. T., Bianchi, N., Ather, A., Borgatti, M., Vizziello, L., Fabbri, E., & Gambari, R. Bangladeshi medicinal plant extracts inhibiting molecular interactions between nuclear factors and target DNA sequences mimicking NF-kappaB binding sites. *Medicinal Chemistry*, **2005**, *1*, 327–333.

518. Lampronti, I., Bianchi, N., Borgatti, M., Fabbri, E., Vizziello, L., Khan, M. T., Ather, A., Brezena, D., Tahir, M. M., & Gambari, R. Effects of vanadium complexes on cell growth of human leukemia cells and protein-DNA interactions. *Oncology Reports*, **2005**, *14*, 9–15.

519. Lampronti, I., Martello, D., Bianchi, N., Borgatti, M., Lambertini, E., Piva, R., Jabbar, S., Choudhuri, M. S., Khan, M. T., & Gambari, R. *In vitro* antiproliferative effects on human tumor cell lines of extracts from the Bangladeshi medicinal plant *Aegle marmelos* Correa. *Phytomedicine*, **2003**, *10*, 300–308.

520. Lan, H., & Lu, Y. Y. Allitridi induces apoptosis by affecting Bcl-2 expression and caspase-3 activity in human gastric cancer cells. *Acta Pharmaceutica Sinica B*, **2004**, *25*, 219–225.

521. Lan, H., & Lu, Y. Y. Effect of allitridi on cyclin D1 and p27(Kip1) protein expression in gastric carcinoma BGC823 cells. *Ai Zheng*, **2003**, *22*, 1268–1271.

522. Langner, E., Greifenberg, S., & Gruenwald, J. Ginger: History and use. *Advances in Therapy*, **1998**, *15*, 25–44.

523. Lansky, S. B., Lowman, J. T., Gyulay, J., & Briscoe, K. Team approach to coping with cancer. In: Cullen, J., Fox, B., & Isom, R., (eds.), *Cancer: The Behavioral Dimensions* (pp. 295–319). Raven Press, New York, **1976**.

524. Lansky, E. P., & Newman, R. A. *Punica granatum* (pomegranate) and its potential for prevention and treatment of inflammation and cancer. *Journal of Ethnopharmacology*, **2007**, *109*, 177–206.

525. Latha, R. C. R., & Daisy, P. Influence of *Terminalia bellerica* Roxb. fruit extracts on biochemical parameters in streptozotocin diabetic rats. *International Journal of Pharmacology*, **2010**, *6*, 89–96.

526. Lau, F. Y., Chui, C. H., Gambari, R., Kok, S. H., Kan, K. L., Cheng, G. Y., Wong, R. S., Teo, I. T., Cheng, C. H., Wan, T. S., Chan, A. S., & Tang, J. C. Antiproliferative and apoptosis-inducing activity of *Brucea javanica* extract on human carcinoma cells. *International Journal of Molecular Medicine*, **2005**, *16*, 1157–1162.

527. Laure, F., Butard, J. F., & Gaydou, E. M. Screening of anti HIV-1 inophyllums by HPLC-DAD of *Calophyllum inophyllum* leaf extracts from French Polynesia islands. *Analytica Chimica Acta*, **2008**, *624*, 147–153.

528. Lavie, G., Mazur, Y., Lavie, D., & Meruelo, D. The chemical and biological properties of hypericina compound with a broad spectrum of biological activities. *Medicinal Research Reviews*, **1995**, *15*, 111–119.

529. Lavie, G., Valentine, F., Levin, B., Mazur, Y., Gallo, G., Lavie, D., Weiner, D., & Mruelo, D. Studies of the mechanism of action of the antiretroviral agents hypericin and pseudohypericin. *Proceedings of the National Academy of Sciences of USA*, **1989**, *86*, 5963–5967.

530. Lednicer, D., & Sander, K. M. Plants and other organisms as a source of anti-human immunodeficiency virus (HIV) drugs. In: Wagner, H., & Farnsworth, N. R., (eds.), *Economic and Medicinal Plant Research* (pp. 1–20). Academic Press, London, **1991**.

531. Lee, B. C., Park, B. H., Kim, S. Y., & Lee, Y. J. Role of Bim in diallyl trisulfide-induced cytotoxicity in human cancer cells. *Journal of Cellular Biochemistry*, **2011**, *112*, 118–127.

532. Lee, C. C., & Houghton, P. Cytotoxicity of plants from Malaysia and Thailand used traditionally to treat cancer. *Journal of Ethnopharmacology*, **2005**, *100*, 237–243.

533. Lee, E., & Surh, Y. J. Induction of apoptosis in HL-60 cells by pungent vanilloids, [6]-gingerol and [6]-paradol. *Cancer Letters*, **1998**, *134*, 163–168.

534. Lee, H., Itokawa, H., & Kozuka, M. Asian herbal products: The basis for development of high-quality dietary supplements and new medicines. In: Shi, J., Ho, C. T., & Shahidi, F., (eds.), *Asian Functional Foods* (p. 647). CRC Press, Boca Raton – FL, **2005**.

535. Lee, H., & Lin, J. Y. Antimutagenic activity of extracts from anticancer drugs in Chinese medicine. *Mutation Research*, **1988**, *204*, 229–234.

536. Lee, H. K., Oh, S. R., Kim, J. I., Kim, J. W., & Lee, C. O. Agastaquinone, a new cytotoxic diterpenoid quinine from *Agastache rugosa*. *Journal of Natural Products*, **1995**, *58*, 1718–1721.

537. Lee, H. S., & Ku, S. K. Effect of *Picrorrhiza rhizoma* extracts on early diabetic nephropathy in streptozotocin-induced diabetic rats. *Journal of Medicinal Food*, **2008**, *11*, 294–301.

538. Lee, H. Z., Hsu, S. L., Liu, M. C., & Wu, C. H. Effects and mechanisms of aloe-emodin on cell death in human lung squamous cell carcinoma. *European Journal of Pharmacology*, **2001**, *431*, 287–295.

539. Lee, I. K., Jung, J. Y., Yeom, J. H., Ki, D. W., Lee, M. S., Yeo, W. H., & Yun, B. S. Fomitoside K, a new lanostane triterpene glycoside from the fruiting body of *Fomitopsis nigra*. *Mycobiology*, **2012**, *40*, 76–78.

540. Lee, J., Huh, M. S., Kim, Y. C., & Hattori, M. Lignan, sesquilignans and dilignans, novel HIV protease and cytopathic effect of inhibitors purified from the rhizomes of *Saururus chinensis*. *Antiviral Research*, **2010**, *85*, 425–428.

541. Lee, J., Oh, W. K., Ahn, J. S., & Kim, Y. H. Prenyl isoflavonoids from *Erythrina senegalensis* as novel HIV-1 protesae inhibitors. *Planta Medica*, **2009**, *75*, 268–270.

542. Lee, J. S., Kim, H. J., & Lee, Y. S. A new anti-HIV flavonoid glucuronide from *Chrysanthemum morifolium*. *Planta Medica*, **2003**, *69*, 859–861.

543. Lee, J. S., Miyashiro, H., & Nakamura, N. Two new triterpenoids from the rhizome of dryopteris crassirhizoma and inhibitory activities of its constituents on HIV 1 protease. *Chemical and Pharmaceutical Bulletin*, **2008**, *56*, 711–714.

544. Lee, J. W., Hong, H. M., Kwon, D. D., Pae, H. O., & Jeong, H. J. Dimethoxycurcumin, a structural analog of curcumin, induces apoptosis in human renal carcinoma caki cells through the production of reactive oxygen species, the release of cytochrome *c*, and the activation of caspase-3. *Korean Journal of Urology*, **2010**, *51*, 870–878.

545. Lee, K. Y., Chang, W., Qiu, D., Kao, P. N., & Rosen, G. D. PG490 (triptolide) cooperates with tumor necrosis factor-alpha to induce apoptosis in tumor cells. *The Journal of Biological Chemistry*, **1999**, *274*, 13451–13455.

546. Lee, S. H., Jaganath, I. B., Wang, S. M., & Sekaran, S. D. Antimetastatic effects of *Phyllanthus* on human lung (A549) and breast (MCF-7) cancer cell lines. *PloS One*, **2011**, *6*, E-article ID: 20994.

547. Lee, Y. J., Jin, Y. R., Lim, W. C., Park, W. K., Cho, J. Y., Jang, S., & Lee, S. K. Ginsenoside-Rb1 acts as a weak phytestrogen in MCF-7 human breast cancer cells. *Archives of Pharmacal Research*, **2003**, *26*, 58–63.

548. Leger, D. Y., Liagre, B., & Beneytout, J. L. Role of MAPKs and NF-kappaB in diosgenin induced megakaryocytic differentiation and subsequent apoptosis in HEL cells. *International Journal of Oncology*, **2006**, *28*, 201–207.

549. Leger, D. Y., Liagre, B., Cardot, P. J., Beneytout, J. L., & Battu, S. Diosgenin dose-dependent apoptosis and differentiation induction in human erythroleukemia cell line and sedimentation field-flow fractionation monitoring. *Analytical Biochemistry*, **2004**, *335*, 267–278.

550. Leger, D. Y., Liagre, B., Corbiere, C., Cook-Moreau, J., & Beneytout, J. L. Diosgenin induces cell cycle arrest and apoptosis in HEL cells with increase in intracellular calcium level, activation of cPLA2 and COX-2 overexpression. *International Journal of Oncology*, **2004**, *25*, 555–562.

551. Lele, R. D. Ayurveda (Ancient Indian System of Medicine) and modern molecular medicine. *Journal of the Association of Physicians of India*, **1999**, *47*, 625–628.

552. Lepage, C., Liagre, B., Cook-Moreau, J., Pinon, A., & Beneytout, J. L. Cyclooxygenase-2 and 5-lipoxygenase pathways in diosgenin-induced apoptosis in HT-29 and HCT-116 colon cancer cells. *International Journal of Oncology*, **2010**, *36*, 1183–1191.

553. Leu, Y. L., Hwang, T. L., Chung, Y. M., & Hong, P. Y. The inhibition of superoxide anion generation in human neutrophils by *Viscum coloratum*. *Chemical and Pharmaceutical Bulletin (Tokyo)*, **2006**, *54*, 1063–1066.

554. Li, B. Q., Fu, T., Dongyan, Y., Mikovits, J. A., Ruscetti, F. W., & Wang, J. M. Flavonoid baicalin inhibits HIV-1 infection at the level of viral entry. *Biochemical and Biophysical Research Communications*, **2000**, *276*, 534–538.

555. Li, F., Goila-Gaur, R., Salzwedel, K., Kilgore, N. R., Reddick, M., Matallana, C., Castillo, A., Zoumplis, D., Martin, D. E., Orenstein, J. M., Allaway, G. P., Freed, E. O., & Wild, C. T. PA-457: A potent HIV inhibitor that disrupts core condensation by targeting a late step in Gag processing. *Proceedings of the National Academy of Sciences – USA*, **2003**, *100*, 13555–13560.

556. Li, H. Y., Sun, N. J., Kashiwada, Y., Sun, L., Snider, J. V., Cosentino, L. M., & Lee, K. H. Anti-AIDS agents, 9. Suberosol, a new C 31 lanostane-type triterpene and anti-HIV principle from *Ployalthia suberosa*. *Journal of Natural Products*, **1993**, *56*, 1130–1133.

557. Li, L. Opportunity and challenge of traditional Chinese medicine in face of the entrance to WTO (World Trade Organization). *Journal of Traditional Chinese Medicine*, **2000**, *7*, 7–8 (in Chinese).

558. Li, M. O., Sarkisian, M. R., Mehal, W. Z., Rakic, P., & Flavell, R. A. Phosphatidylserine receptor is required for clearance of apoptotic cells. *Science*, **2003**, *302*, 1560–1563.

559. Li, N., Guo, R., Li, W., Shao, J., Li, S., Zhao, K., Chen, X., Xu, N., Liu, S., & Lu, Y. A proteomic investigation into a human gastric cancer cell line BGC823 treated with diallyl trisulfide. *Carcinogenesis*, **2006**, *27*, 1222–1231.

560. Li, Q. Y. *Healthy Functions and Medicinal Prescriptions of Lycium Barbarum (Gou Ji Zi)* (pp. 1–205). Jindun Press, Beijing, **2001**.

561. Li, X. N., Pu, J. X., Du, X., Yang, L. M., & Lei, C. Lignans with anti HIV activity from *Schisandra propinqua*. *Journal of Natural Products*, **2009**, *72*, 1133–1141.

562. Li, Y. L., Gan, G. P., Zhang, H. Z., Wu, H. Z., Li, C. L., Huang, Y. P., Liu, Y. W., & Liu, J. W. A flavonoid glycoside isolated from *Smilax china*L. rhizome *in vitro* anticancer effects on human cancer cell lines. *Journal of Ethnopharmacology*, **2007**, *113*, 115–124.

563. Liagre, B., Bertrand, J., Leger, D. Y., & Beneytout, J. L. Diosgenin, a plant steroid, induces apoptosis in COX-2 deficient K562 cells with activation of the p38 MAP kinase signaling and inhibition of NF-kappaB binding. *International Journal of Molecular Medicine*, **2005**, *16*, 1095–1101.

564. Liang, X. S., Rogers, A. J., Webber, C. L., Ormsby, T. J., Tiritan, M. E., Maltin, S. A., & Benz, C. C. Developing the gossypol derivatives with enhanced antitumor activity. *Investigational New Drugs*, **1995**, *13*, 181–186.

565. Lieu, D. K., Pappone, P. A., & Barakat, A. I. Differential membrane potential and ion current responses to different types of shear stress in vascular endothelial cells. *American Journal of Physiology-Cell Physiology*, **2004**, *286*, C1367–C1375.

566. Lieu, C. W., Lee, S. S., & Wang, S. Y. The effect of *Ganoderma-Lucidum* on induction of differentiation in leukemic U937-cells. *Anticancer Research*, **1992**, *12*, 1211–1216.

567. Lin, C. C., Yang, J. S., Chen, J. T., Fan, S., Yu, F. S., Yang, J. L., Lu, C. C., Kao, M. C., Huang, A. C., Lu, H. F., & Chung, J. G. Berberine induces apoptosis in human HSC-3 oral cancer cells via simultaneous activation of the death receptor-mediated and mitochondrial pathway. *Anticancer Research*, **2007**, *27*, 3371–3378.

568. Lin, C. Y., Lin, C. J., Chen, K. H., Wu, J. C., Huang, S. H., & Wang, S. M. Macrophage activation increases the invasive properties of hepatoma cells by destabilization of the adherens junction. *FEBS Letters*, **2006**, *580*, 3042–3050.

569. Lin, M. L., Chen, S. S., Lu, Y. C., Liang, R. Y., Ho, Y. T., Yang, C. Y., & Chung, J. G. Rhein induces apoptosis through induction of endoplasmic reticulum stress and Ca21-dependent mitochondrial death pathway in human nasopharyngeal carcinoma cells. *Anticancer Research*, **2007**, *27*, 3313–3322.

570. Lin, N., Sato, T., & Ito, A. Triptolide, a novel diterpenoid triepoxide from *Tripterygium wilfordii* Hook. F., suppresses the production and gene expression of pro-matrix metalloproteinases 1 and 3 and augments those of tissue inhibitors of metalloproteinases 1 and 2 in human synovial fibroblasts. *Arthritis & Rheumatology*, **2002**, *44*, 2193–2200.

571. Lin, T. S., Schinazi, R., Griffith, B. B., August, E. M., Eriksson, B. F., Zheng, D. K., Huang, L. A., & Prusoff, W. H. Selective inhibition of human immunodeficiency virus type-1 replication by the (–) but not (+) enatiomer of gossypol. *Antimicrobial Agents and Chemotherapy*, **1989**, *33*, 2149–2151.

572. Lin, X. F., Min, W., & Luo, D. Anticarcinogenic effect of ferulic acid on ultraviolet-B irradiated human keratinocyte HaCaT cells. *Journal of Medicinal Plants Research*, **2010**, *4*, 1686–1694.

573. Lin, Y. M., Anderson, H., Flavin, M. T., & Pai, Y. S. H. *In vitro* anti-HIV activity of biflavonoids isolated from *Rhus succedanea* and *Garcinia multiflora*. *Journal of Natural Products*, **1997**, *60*, 884–888.

574. Lin, Y. T., Yang, J. S., Lin, S. Y., Tan, T. W., Ho, C. C., Hsia, T. C., Chiu, T. H., Yu, C. S., Lu, H. F., Weng, Y. S., & Chung, J. G. Diallyl disulfide (DADS) induces apoptosis in human cervical cancer Ca Ski cells *via* reactive oxygen species and Ca2+-dependent mitochondria-dependent pathway. *Anticancer Research*, **2008**, *28*, 2791–2800.

575. Liu, D. D., Ye, Y. L., Zhang, J., Xu, J. N., Qian, X. D., & Zhang, Q. Distinct pro-apoptotic properties of Zhejiang saffron against human lung cancer via a caspase-8–9–3 cascade. *Asian Pacific Journal of Cancer Prevention*, **2014**, *15*, 6075–6080.

576. Liu, H. K., Perrier, S., Lipina, C., Finlay, D., McLauchlan, H., Hastie, C. J., Hundal, H. S., & Sutherland, C. Functional characterization of the regulation of CAAT enhancer binding protein alpha by GSK-3 phosphorylation of Threonines 222/226. *BMC Molecular Biology*, **2006**, *7*, 14–18.

577. Liu, K. C., Huang, A. C., Wu, P. P., Lin, H. Y., Chueh, F. S., Yang, J. S., Lu, C. C., Chiang, J. H., Meng, M., & Chung, J. G. Gallic acid suppresses the migration and invasion of PC-3 human prostate cancer cells via inhibition of matrix metalloproteinase-2 and -9 signaling pathways. *Oncology Reports*, **2011**, *26*, 177–184.

578. Liu, L. F., Liang, C. H., Shiu, L. Y., Lin, W. L., Lin, C. C., & Kuo, K. W. Action of solamargine on human lung cancer cells-enhancement of the susceptibility of cancer cells to TNFs. *FEBS Letters*, **2004**, *577*, 67–74.

579. Liu, L. Z., Fang, J., Zhou, Q., Hu, X., Shi, X., & Jiang, B. H. Apigenin inhibits expression of vascular endothelial growth factor and angiogenesis in human lung cancer cells implication of chemoprevention of lung cancer. *Molecular Pharmacology*, **2005**, *68*, 635–643.

580. Liu, M. J., Wang, Z., Ju, Y., Wong, R. N., & Wu, Q. Y. Diosgenin induces cell cycle arrest and apoptosis in human leukemia K562 cells with the disruption of Ca^{2+} homeostasis. *Cancer Chemotherapy and Pharmacology*, **2005**, *55*, 79–90.

581. Liu, R. M., & Zhong, J. J. Ganoderic acid Mf and S induce mitochondria mediated apoptosis in human cervical carcinoma HeLa cells. *Phytomedicine*, **2011**, *18*, 349–355.

582. Liu, R. M., Li, Y. B., & Zhong, J. J. Cytotoxic and pro-apoptotic effects of novel ganoderic acid derivatives on human cervical cancer cells *in vitro*. *European Journal of Pharmacology*, **2012**, *681*, 23–33.

583. Liu, W. K., Xu, S. X., & Che, C. T. Anti-proliferative effect of ginseng saponins on human prostate cancer cell line. *Life Sciences*, **2000**, *67*, 1297–1306.

584. Liu, Y., Ye, F., Qiu, G. Q., Zhang, M., Wang, R., He, Q. Y., & Cai, Y. Effects of lactone I from *Atractylodes macrocephala* Koidz on cytokines and proteolysis-inducing factors in cachectic cancer patients. *Di Yi Jun Yi Da Xue Xue Bao*, **2005**, *25*, 1308–1311.

585. Lo, W. L., Wu, C. C., & Chang, F. R. Antiplatelet and anti HIV constituents from *Euchresta formosana*. *Natural Product Research*, **2003**, *17*, 91–97.

586. Lo, Y. T., Tsai, Y. H., Wu, S. J., Chen, J. R., & Chao, J. C. Ginsenoside Rb1 inhibits cell activation and liver fibrosis in rat hepatic stellate cells. *Journal of Medicinal Food*, **2011**, *14*, 1135–1143.

587. Lobo, V., Anita, P., & Naresh, C. Comparative evaluation of antioxidant activity of aqueous and alcoholic extract of *Cassia Tora* Linn leaves. *Asian Journal of Experimental Biological Sciences*, **2011**, *2*, 826–832.

588. Locher, C. P., Witvrouw, M., De Béthune, M. P., Burch, M. T., Mower, H. F., Davis, H., Lasure, A., Pauwels, R., De Clercq, E., & Vlietinck, A. J. Antiviral activity of Hawalian medicinal plants against human immunodeficiency virus type-1 (HIV-1). *Phytomedicine*, **1996**, *2*, 259–264.

589. Lodish, H., Harvey, L., Arnold, B. S. L. Z., Paul, M., David, B., & James, D. *Molecular Cell Biology* (p. 1280). W.H. Freedman and Company, New York, **2004**.

590. Lozano, R., Naghavi, M., Foreman, K., Lim, S., Shibuya, K., Aboyans, V., et al. Global and regional mortality from 235 causes of death for 20 age groups in 1990 and 2010: A systematic analysis for the global burden of disease study 2010. *Lancet*, **2012**, *380*, 2095–2128.

591. Lu, C. C., Yang, J. S., Huang, A. C., Hsia, T. C., Chou, S. T., Kuo, C. L., Lu, H. F., Lee, T. H., Wood, W. G., & Chung, J. G. Chrysophanol induces necrosis through the production of ROS and alteration of ATP levels in J5 human liver cancer cells. *Molecular Nutrition & Food Research*, **2010**, *54*, 967–976.

592. Lu, H. F., Chen, Y. L., Yang, J. S., Yang, Y. Y., Liu, J. Y., Hsu, S. C., Lai, K. C., & Chung, J. G. Antitumor activity of capsaicin on human colon cancer cells *in vitro* and colo 205 tumor xenografts *in vivo*. *Journal of Agricultural and Food Chemistry*, **2010**, *58*, 12999–13005.

593. Lu, K. H., Lue, K. H., Chou, M. C., & Chung, J. G. Paclitaxel induces apoptosis via caspase-3 activation in human osteogenic sarcoma cells (U-2 OS). *Journal of Orthopaedic Research*, **2005**, *23*, 988–994.

594. Lu, K. W., Tsai, M. L., Chen, J. C., Hsu, S. C., Hsia, T. C., Lin, M. W., Huang, A. C., Chang, Y. H., Ip, S. W., Lu, H. F., & Chung, J. G. Gypenosides inhibited invasion and migration of human tongue cancer SCC4 cells through down-regulation of NFkappaB and matrix metalloproteinase-9. *Anticancer Research*, **2008**, *28*, 1093–1099.

595. Lu, L., Shu-Wen, L., Shi-Bo, J., & Shu-Guang, W. Tannin inhibits HIV-1 entry by targeting gp41. *Acta Pharmacologica Sinica*, **2004**, *25*, 213–218.

596. Lu, Y., Jiang, F., Jiang, H., Wu, K., Zheng, X., & Cai, Y. Gallic acid suppresses cell viability, proliferation, invasion and angiogenesis in human glioma cells. *European Journal of Pharmacology*, **2010**, *641*, 102–107.

597. Luo, W., Zhao, M. M., Yang, B., Shen, G. L., & Rao, G. H. Identification of bioactive compounds in *Phyllanthus emblica* L. fruit and their free radical scavenging activities. *Food Chemistry*, **2009**, *114*, 499–504.

598. Lüscher-Mattli, M. Polyanions – a lost chance in the fight against HIV and other virus diseases? *Antiviral Chemistry and Chemotherapy*, **2000**, *11*, 249–259.

599. Luthra, P. M., Kumar, R., & Prakash, A. Demethoxycurcumin induces Bcl-2 mediated G2/M arrest and apoptosis in human glioma U87 cells. *Biochemical and Biophysical Research Communications*, **2009**, *384*, 420–425.

600. Ma, C. M., Nakamura, N., Miyashiro, H., Hattori, M., Komatsu, K., Kawahata, T., & Otake, T. Screening of Chinese and Mongolian herbal drugs for anti human immuno-deficiency virus type-1 (HIV-1) activity. *Phytotherapy Research*, **2002**, *16*, 186–189.

601. Ma, T., Gao, Q., Chen, Z., Wang, L., & Liu, G. Chemical resolution of calanolide A, cordatolide A and their 11-demethyl analogs. *Bioorganic & Medicinal Chemistry Letters*, **2008**, *18*, 1079–1083.

602. Machana, S., Weerapreeyakul, N., Barusrux, S., Nonpunya, A., Sripanidkulchai, B., & Thitimetharoch, T. Cytotoxic and apoptotic effects of six herbal plants against the human hepatocarcinoma (HepG2) cell line. *Chinese Medicine*, **2011**, *6*, 39.

603. Madeleine, K., Imreh, G., & Hallberg, E. Sequential degradation of proteins from the nuclear envelope during apoptosis. *Journal of Cell Science*, **2001**, *114*, 3643–3653.

604. Madhyastha, R., Madhyastha, H., Nakajima, Y., Omura, S., & Maruyama, M. Curcumin facilitates fibrinolysis and cellular migration during wound healing by modulating urokinase plasminogen activator expression. *Pathophysiology of Haemostasis and Thrombosis*, **2010**, *37*, 59–66.

605. Magadula, J. J., & Tewtrakul, S. Anti-HIV-1 protease activities of crude extracts of some *Garcinia* species growing in Tanzania. *African Journal of Biotechnology*, **2010**, *9*, 1848–1852.

606. Malckzadeh, F., Ehsanifar, H., Shahamat, N., Levin, M., & Colwell, R. R. Antibacterial activity of black myrobalan (*Terminalia chebula* Retz.) against *Helicobacter pyroli*. *International Journal of Antimicrobial Agents*, **2001**, *18*, 85–88.

607. Malik, J. K., Manvi, F. V., Nanjwade, B. K., Alagawadi, K. R., & Singh, S. Immunomodulatory activity of *Gymnema sylvestre* leaves on *in-vitro* human neutrophils. *Journal of Pharma Research*, **2009**, *2*, 1284–1286.

608. Manandhar, N. P. Traditional medicinal plants used by tribals of Lamjung District, Nepal. *International Journal of Crude Drug Research*, **1987**, *25*, 236–240.

609. Mandal, S., Deb Mandal, M., & Pal, N. K. Synergistic anti-*Staphylococcus aureus* activity of amoxicillin in combination with *Emblica officinalis* and *Nymphae odorata* extracts. *Asian Pacific Journal of Tropical Medicine*, **2010**, *3*, 711–714.

610. Mandel, S. A., Weinreb, O., Amit, T., & Youdim, M. B. The importance of the multiple target action of green tea polyphenols for neuroprotection. *Frontiers in Bioscience*, **2012**, *4*, 581–598.

611. Manosroi, A., Jantrawut, P., Akihisa, T., Manosroi, W., & Manosroi, J. *In vitro* anti-aging activities of *Terminalia chebula* gall extract. *Pharmaceutical Biology*, **2010**, *48*, 469–481.

612. Mantle, D., & Wilkins, R. Medicinal plants in the prevention and therapy of cancer. In: Yaniv, Z., Bachrach, U., (eds.), *Handbook of Medicinal Plants* (p. 501). Pub.: Food Products Press-Haworth Press, New York, **2005**.

613. Mantovani, A., & Sica, A. Macrophages, innate immunity and cancer: Balance, tolerance, and diversity. *Current Opinion in Immunology*, **2010**, *22*, 231–237.

614. Marculescu, A., Vlase, L., Hanganu, D., Dragulescu, C., Antonie, I., & Neli-Kinga, O. Polyphenols analyzes from *Thymus* species. *Proceedings of the Romanian Academy-Series B*, **2007**, *3*, 117–121.

615. Gharagozloo, M., & Ghaderi, A. Immunomodulatory effect of concentrated lime juice extract on activated human mononuclear cells. *Journal of Ethnopharmacology*, **2001**, *77*, 85–90.

616. Masood, E. Medicinal plants threatened by over-use. *Nature*, **1997**, *385*, 570–580.

617. Matthée, G., Wright, A. D., & König, G. M. HIV reverse transcriptase inhibitors of natural origin. *Planta Medica*, **1999**, *65*, 493–506.

618. Matuse, I. T., Lim, Y. A., & Hattori, M. A search for antiviral properties in Panamian medicinal plants. The effects on HIV and its essential enzymes. *Journal of Ethnopharmacology*, **1999**, *64*, 15–22.

619. Mazid, M., Khan, T. A., & Mohammada, F. Medicinal plants of rural India – a review of use by Indian folks. *Indo Global Journal of Pharmaceutical Sciences*, **2012**, *2*, 286–304.

620. Mazur, W. M., Duke, J. A., Wahala, K., Rasku, S., & Adlercreutz, H. Isoflavonoids and lignans in legumes: Nutritional and health aspects in humans. *The Journal of Nutritional Biochemistry*, **1998**, *6*, 193–200.

621. Mbah, A. U., Udeinya, I. J., Shu, E. N., Chijioke, C. P., Nubila, T., Udeinya, F., Muobuike, A., Mmuobieri, A., & Obioma, M. S. Fractionated neem leaf extract is safe and increases CD4+ cell levels in HIV/AIDS patients. *American Journal of Therapeutics*, **2007**, *14*, 369–374.

622. McDaniel, H. R., Pulse, T., Watson, T., & Mc Analley, B. H. Prediction and results obtained using oral acemannan (ace-m) in 41 symptomatic HIV patients. In: *Proceedings of the IV International Conference on AIDS*, Stockholm, Abstract 3566, **1988**.

623. Meeran, S. M., & Katiyar, S. K. Proanthocyanidins inhibit mitogenic and survival-signaling *in vitro* and tumor growth *in vivo*. *Frontiers in Bioscience*, **2008**, *13*, 887–897.

624. Mehmood, N., Pizza, C., & Aquino, R. Inhibition of HIV infection by flavonoids. *Antiviral Research*, **1993**, *22*, 189–199.

625. Mekkawy, S., Meselhy, M. R., Kusumoto, I. T., Kadota, S., Hattori, M., & Namba, T. Inhibitory effects of Egyption folk medicines on human immunodeficiency virus (HIV) reverse transcriptase. *Chemical and Pharmaceutical Bulletin*, **1995**, *43*, 641–648.

626. Mendelson, R., & Balick, M. J. The value of undiscovered pharmaceuticals in tropical forests. *Economic Botany*, **1995**, *49*, 223–228.

627. Meragelman, K. M., McKee, T. C., & Boyd, M. R. Anti HIV prenylated flavonoids from *Monotes africanus*. *Journal of Natural Products*, **2001**, *64*, 546–548.

628. Meragelman, K. M., Mckee, T. C., & Boyd, M. R. Siamenol a new carbazole alkaloid from *Murraya siamensis*. *Journal of Natural Products*, **2000**, *63*, 427–428.

629. Meruelo, D., Lavie, G., & Lavie, D. Therapeutic agents with dramatic antiretroviral activity and little toxicity at effective doses, aromatic polycyclic diones hypericin and pseudhypericin. *Proceedings of the National Academy of Science – USA*, **1988**, *85*, 5230–5234.

630. Meyer, A. S., Heinonen, M., & Frankel, E. N. Antioxidant interactions of Catechin cyaniding caffeic acid quercetin and ellagic acid on human LDL oxidation. *Food Chemistry*, **1998**, *61*, 71–75.

631. Miccadei, S., Di Venere, D., Cardinali, A., Romano, F., Durazzo, A., Foddai, M. S., Fraioli, R., Mobarhan, S., & Maiani, G. Antioxidative and apoptotic properties of polyphenolic extracts from edible part of artichoke (*Cynara scolymus* L.) on cultured rat hepatocytes and on human hepatoma cells. *Nutrition and Cancer*, **2008**, *60*, 276–283.

632. Michèle, F., Karim, D., & Angelita, R. CD^{4+} T cell depletion in human immunodeficiency virus (HIV) infection: Role of apoptosis. *Viruses*, **2011**, *3*, 586–612.

633. Miller, L. E. In: Ludke, H. R., Peacock, J. E., & Tomar, R. H., (ed.), *Manual of Laboratory Immunology* (pp. 1–18). Lea and Febiger, London, **1991**.

634. Mills, E., Cooper, C., Seely, D., & Kanfer, I. African herbal medicines in the treatment of HIV: Hypoxis and Sutherlandia. An overview of evidence and pharmacology. *Nutrition Journal*, **2005**, *4*, 4–19.

635. Min, D. S., Ahn, B. H., & Jo, Y. H. Differential tyrosine phosphorylation of phospholipase D isozymes by hydrogen peroxide and the epidermal growth factor in A431 epidermoid carcinoma cells. *Molecules and Cells*, **2001**, *11*, 369–378.

636. Min, B. S., Jung, H. J., & Lee, J. S. Inhibitory effect of triterpenes from *Crataegus pinatifida* on HIV-1 protease. *Planta Medica*, **1999**, *5*, 374–375.

637. Min, B. S., Kim, Y. H., Tomiyama, M., Nakamura, N., Miyashiro, H., Otake, T., & Hattori, M. Inhibitory effects of Korean plants on HIV-1 activities. *Phytotherapy Research*, **2001**, *15*, 481–486.

638. Min, B. S., Lee, H. K., Lee, S. M., Kim, Y. H., Bae, K. H., Otake, T., Nakamura, N., & Hattori, M. Anti-human immunodeficiency virus-type 1 activity of constituents from *Juglans mandshurica*. *Archives of Pharmacal Research*, **2002**, *25*, 441–445.

639. Min, M., Bae, K. H., Kim, Y. H., Miyashiro, H., Hattori, M., & Shimotohno, K. Screening of Korean plants against human immunodeficiency virus type 1 protease. *Phytotherapy Research*, **1999**, *13*, 680–682.

640. Mishra, A., Rajesh, N., Saumya, S., Kiran, T., Poonam, C., & Satish, K. V. *In vitro* cytotoxicity of *Moringa oleifera* against different human cancer cell lines. *Asian Journal of Pharmaceutical and Clinical Research*, **2012**, *5*, 271–272.

641. Mishra, V., Agrawal, M., Onasanwo, S. A., Madhur, G., Rastogi, P., Pandey, H. P., Palit, G., & Narender, T. Antisecretory and cyto-protective effects of chebulinic acid isolated from the fruits of *Terminalia chebula* on gastric ulcers. *Phytomedicine*, **2013**, *20*, 506–511.

642. Mitchell, D. E., & Madore, M. A. Patterns of assimilate production and translocation in Muskmelon (*Cucumis melo* L.): II. Low temperature effects. *Plant Physiology*, **1992**, *99*, 966–971.

643. Miyagoshi, M., Amagaya, S., & Ogihara, Y. Choleretic actions of iridoid compounds. *Journal of pharmacobio-Dynamics*, **1988**, *11*, 186–190.

644. Mlinaric, A., Kreft, S., Umek, A., & Strukelf, B. Screening of selected plants extracts for *in-vitro* inhibitory activity on HIV reverse transcriptase. *Pharmazie*, **2000**, *55*, 75–77.

645. Mo, S., Wang, S., Zhou, G., Yang, Y., Li, Y., Chen, X., & Shi, J. Phelligridins C-F: Cytotoxic pyrano[4, 3-c][2]benzopyran-1, 6-dione and furo[3, 2-c]pyran-4-one

derivatives from the fungus *Phellinus igniarius*. *Journal of Natural Products*, **2004**, *67*, 823–828.

646. Moalic, S., Liagre, B., Corbiere, C., Bianchi, A., Dauca, M., Bordji, K., & Beneytout, J. L. A plant steroid, diosgenin, induces apoptosis, cell cycle arrest and COX activity in osteosarcoma cells. *FEBS Letters*, **2001**, *506*, 225–230.

647. Mohan, S., Bustamam, A., Abdelwahab, S. I., Al-Zubairi, A. S., Aspollah, M., Abdullah, R., Elhassan, M. M., Ibrahim, M. Y., & Syam, S. *Typhonium flagelliforme* induces apoptosis in CEMss cells via activation of caspase-9, PARP cleavage and cytochrome c release: Its activation coupled with G0/G1 phase cell cycle arrest. *Journal of Ethnopharmacology*, **2010**, *131*, 592–600.

648. Möller, M., Herzer, K., Wenger, T., Herr, I., & Wink, M. The alkaloid emetine as a promising agent for the induction and enhancement of drug-induced apoptosis in leukemia cells. *Oncology Reports*, **2007**, *14*, 737–744.

649. Moongkarndi, P., Kosem, N., Luanratana, O., Jongsomboonkusol, S., & Pongpan, N. Antiproliferative activity of Thai medicinal plant extracts on human breast adenocarcinoma cell line. *Fitoterapia*, **2004**, *75*, 375–377.

650. Moonjal, H. K., Young, S., & Somi, K. The chloroform fraction of guava (*Psidium cattlianum* sabine) leaf extract inhibits human gastric cell proliferation via induction of apoptosis. *Food Chemistry*, **2011**, *125*, 369–375.

651. Moore, J. P., Lindsey, G. G., Farrant, J. M., & Brandt, W. F. An overview of the biology of the desiccation-tolerant resurrection plant *Myrothamnus flabellifolia*. *Annals of Botany*, **2007**, *99*, 211–217.

652. Moore, P. S., & Pizza, C. Observations on the inhibition of HIV-1 reverse transcriptase by catechins. *Biochemical Journal*, **1992**, *288*, 717–719.

653. Mora, L. B., Buettner, R., Seigne, J., Diaz, J., Ahmad, N., Garcia, R., Bowman, T., Falcone, R., Fairclough, R., Cantor, A., Muro-Cacho, C., Livingston, S., Karras, J., Pow-Sang, J., & Jove, R. Constitutive activation of Stat3 in human prostate tumors and cell lines: Direct inhibition of Stat3 signaling induces apoptosis of prostate cancer cells. *Cancer Research*, **2002**, *62*, 6659–6666.

654. Morebise, F., Makinde, M. A., Olajide, J. M., & Awe, O. A. Antiinflammatory property of the leaves of *Gongronema latifolium*. *Phytotherapy Research*, **2002**, *1*, 75–77.

655. Mosihuzzaman, M., & Choudhary, M. I. Protocols on safety, efficacy, standardization, and documentation of herbal medicine. *Pure and Applied Chemistry*, **2008**, *80*, 2195–2230.

656. Mousa, O., Vuorela, P., Kiviranta, J., Abdel, W. S., Hiltunen, R., & Vuorela, H. Bioactivity of certain Egyptian *Ficus* species. *Journal of Ethnopharmacology*, **1994**, *41*, 71–76.

657. Mujovo, S. F. Antimicrobial activity of compounds isolated from *Lippia javanica* (Burm.f.) Spreng and *Hoslundia opposita* against *Mycobacterim tuberculosis* and HIV-1 reverse transcriptase. *PhD Thesis*, University of Pretoria, **2009**, p. 218.

658. Mukherjee, P. K., Rai, S., Kumar, V., Mukherjee, K., Hylands, P. J., & Hider, R. C. Plants of Indian origin in drug discovery. *Expert Opinion on Drug Discovery*, **2007**, *2*, 633–657.

659. Murphy, K. M., Ranganathan, V., Farnsworth, M. L., Kavallaris, M., & Lock, R. B. Bcl-2 inhibits Bax translocation from cytosol to mitochondria during drug-induced apoptosis of human tumor cells. *Cell Death & Differentiation*, **2000**, *7*, 102–211.

660. Murray, M. T., & Pizzorno, J. E. Jr. Botanical medicine-a modern perspective. In: Pizzorno, J. E. Jr., & Murray, M. T., (eds.), *Text book of Natural Medicine* (Vol. 1, pp. 267–279). Churchill, Livingstone, **2000**.

661. Murtaza, I., Mara, G., Schlapbach, R., Patrignani, A., Kunzli, M., Wagner, U., Sabates, J., & Dutt, A. A preliminary investigation demonstrating the effect of quercetin on the expression of genes related to cell-cycle arrest, apoptosis and xenobiotic metabolism in human CO115 colon adenocarcinoma cells using DNA microarray. *Biotechnology and Applied Biochemistry*, **2006**, *45*, 29–36.

662. Murthy, K. N., Reddy, V. K., Veigas, J. M., & Murthy, U. D. Study on wound healing activity of *Punica granatum* peel. *Journal of Medicinal Food*, **2004**, *7*, 256–259.

663. Nadkarni, A. K., & Nadkarni, K. M. *Indian Materia Medica* (pp. 263–269). Popular Prakashan: Bombay, India, **1976**.

664. Nadkarni, K. M. *Indian Materia Medica* (p. 541). Popular Prakashan: Bombay, India, **1993**.

665. Naeini, A., Khosravi, A. R., Chitsaz, M., Shokri, H., & Kamlnejad, M. Anti-Candida albicans activity of some Iranian plants used in traditional medicine. *Japanese Journal of Medical Mycology*, **2009**, *9*, 168–172.

666. Nagata, S. Apoptotic DNA fragmentation. *Experimental Cell Research*, **2000**, *256*, 12–18.

667. Naghibi, F., Mosaddegh, M., Mohammadi, M. M., & Ghorbani, A. Labiatae family in folk medicine in Iran: From ethnobotany to pharmacology. *Iranian Journal of Pharmaceutical Research*, **2010**, *4*, 63–79.

668. Nagina-Makker, P., Hogan, V., Honjo, Y., Baccarini, S., Tait, L., Bresalier, R., & Raz, A. Inhibition of human cancer cell growth and metastasis in Nude mice by oral intake of modified citrus pectin. *Journal of the National Cancer Institute*, **2002**, *94*, 1854–1862.

669. Nair, H. K., Rao, K. V. K., Aalinkeel, R., Mahajan, S., Chawda, R., & Schwartz1, S. A. Inhibition of prostate cancer cell colony formation by the flavonoid quercetin correlates with modulation of specific regulatory genes. *Clinical and Diagnostic Laboratory Immunology*, **2004**, *11*, 63–69.

670. Nair, R., & Chanda, S. *In-vitro* antimicrobial activity of *Psidium guajava* L. leaf extracts against clinically important pathogenic microbial strains. *Brazilian Journal of Microbiology*, **2007**, *38*, 452–458.

671. Nakatani, K., Nakahata, N., Arakawa, T., Yasuda, H., & Ohizumi, Y. Inhibition of cyclooxygenase and prostaglandin E2 synthesis by gamma-mangostin, a xanthone derivative in mangosteen, in C6 rat glioma cells. *Biochemical Pharmacology*, **2002**, *63*, 73–79.

672. Nam, S. Y., Yi, H. K., Lee, J. C., Kim, J. C., Song, C. H., Park, J. W., Lee, D. Y., Kim, J. S., & Hwang, P. H. *Cortex Mori* extract induces cancer cell apoptosis through inhibition of microtubule assembly. *Archives of Pharmacal Research*, **2002**, *25*, 191–196.

673. Namsa, N. D., Tag, H., Mandal, M., Kalita, P., & Das, A. K. An ethnobotanical study of traditional anti-inflammatory plants used by the Lohit community of Arunachal Pradesh, India. *Journal of Ethnopharmacology*, **2009**, *125*, 234–245.

674. Nathan, C. F. Secretory products of macrophages. *Journal of Clinical Investigation*, **1987**, *79*, 319–326.

675. Nathan, C. F., & Hibbs, Jr. J. B. Role of nitric oxide synthesis in macrophage antimicrobial activity. *Current Opinion in Immunology,* **1991,** *3,* 65–70.

676. National AIDS Control Organization (NACO), Ministry of Health & Family Welfare, Govt. of India. Female sex workers & their clients, National behavioral surveillance survey, New Delhi, **2006.** http/www.Naco.Online.Org.upload/naco/20PDF/female. Sexworkers(FSWS)_and_theirclients.pdf(Accessed on 30 September 2019).

677. National Cancer Institute. *Targeted Cancer Therapies.* www.cancer.gov (Accessed on 30 September 2019).

678. Naveena, B. M., Sen, A. R., Vaithiyanathan, S., Babji, Y., & Kondaiah, N. Comparative efficacy of pomegranate juice, pomegranate rind powder extract and BHT as antioxidants in cooked chicken patties. *Meat Science,* **2008,** *80,* 1304–1308.

679. NCI: Targeted Therapy tutorials, **2014.** https://www.xsieve.com/search/Cancer (Accessed on 30 September 2019).

680. Negi, P. S., Jayaprakasha, G. K., & Jena, B. S. Antioxidant and antimutagenic activities of pomegranate peel extracts. *Food Chemistry,* **2003,** *80,* 393–397.

681. Netala, V. R., Ghosh, S. B., Bobbu, P., Anitha, D., & Tartte, V. *Triterpenoid saponins*: A review on biosynthesis, applications and mechanism of their action. *International Journal of Pharmacy and Pharmaceutical* Sciences, **2015,** *7,* 24–28.

682. Ng, T. B., Huang, B., Fong, W. P., & Yeung, H. W. Anti-human immunodeficiency virus (anti-HIV) natural products with special emphasis on HIV reverse transcriptase inhibitors. *Life Sciences,* **1997,** *61,* 933–949.

683. Ngamkitidechakul, C., Jaijoy, K., Hansakul, P., Soonthornchareonnon, N., & Sireeratawong, S. Antitumor effects of *Phyllanthus emblica* L.: Induction of cancer cell apoptosis and Inhibition of *in vivo* tumor promotion and *in vitro* invasion of human cancer cells. *Phytotherapy Research,* **2010,** *24,* 1405–1413.

684. Ngan, B. T. H., Kiat, T. K., Kheng, L. S., Lin, C. S., Suan, T. A., Nagammal, S., Khim, T. S., & Kei, S. L. New nursing student's perceptions towards HIV transmission and aids. *The World's Knowledge,* **1999,** *27,* 26–33.

685. Ngan, R. S., Chang, R. S., Tabba, H. D., & Smith, K. M. Isolation, purification and partial characterization of an active Anti HIV compound from Chinese medicinal herb *Viola yedoensis. Antiviral Research,* **1989,** *10,* 107–116.

686. Ni, C. H., Chen, P. Y., Lu, H. F., Yang, J. S., Huang, H. Y., Wu, S. H., Ip, S. W., Wu, C. T., Chiang, S. Y., Lin, J. G., Wood, W. G., & Chung, J. G. Chrysophanol-induced necrotic-like cell death through an impaired mitochondrial ATP synthesis in Hep3B human liver cancer cells. *Archives of Pharmacal Research,* **2012,** *35,* 887–895.

687. Ni, X. Z., Wang, B. Q., Zhi, N. N., Wei, N. N., Zhang, X., Li, S. S., Tai, G. H., Zhou, Y. F., & Zhao, J. M. Total fractionation and analysis of polysaccharides from leaves of *Panax ginseng* C. A. Meyer. *Chemical Research in Chinese Universities,* **2010,** *26,* 230–234.

688. Ni, W., Zhang, X., Wang, B., Chen, Y., Han, H., Fan, Y., Zhou, Y., & Tai, G. Antitumor activities and immunomodulatory effects of ginseng neutral polysaccharides in combination with 5-fluorouracil. *Journal of Medicinal Food,* **2010,** *13,* 270–277.

689. Nigam, P. S. Production of bioactive secondary metabolites. In: Nigam, P. S., & Pandey, A., (eds.), *Biotechnology for Agro-Industrial Residues Utilization* (1st edn., pp. 129–145). Springer, Netherlands, **2009.**

690. Nigam, P. S., Gupta, N., & Anthwal, A. Pre-treatment of agro-industrial residues. In: Nigam, P. S., & Pandey, A.,(eds.), *Biotechnology for Agro-Industrial Residues Utilization* (1st edn., pp. 13–33). Springer, Netherlands, **2009**.

691. Nigam, P. S., & Pandey, A. Solid-state fermentation technology for bioconversion of biomass and agricultural residues. In: Nigam, P. S., & Pandey, A., (eds.), *Biotechnology for Agro-Industrial Residues Utilization* (1st edn., pp. 197–221). Springer, Netherlands, **2009**.

692. Nino, J., Navaez, D. M., Mosquera, O. M., & Correa, Y. M. Antibacterial, antifungal and cytotoxic activities of eight Asteraceae and two Rubiaceae plants from Colombian biodiversity. *Brazilian Journal of Microbiology*, **2006**, *37*, 566–570.

693. Nishikawa, T., Nakajima, T., Moriguchi, M., Jo, M., Sekoguchi, S., Ishii, M., Takashima, H., Katagishi, T., Kimura, H., Minami, M., Itoh, Y., Kagawa, K., & Okanoue, T. A green tea polyphenol, epigalocatechin-3-gallate, induces apoptosis of human hepatocellular carcinoma, possibly through inhibition of Bcl-2 family proteins. *Journal of Hepatology*, **2006**, *44*, 1074–1082.

694. Nishino, H., Tsushima, M., Matsuno, T., Tanaka, Y., Okuzumi, J., Murakoshi, M., Satomi, Y., Takayasu, J., Tokuda, H., Nishino, A., & Iwashima, A. Anti-neoplastic effect of halocynthiaxanthin, a metabolite of fucoxanthin. *Anticancer Drug*, **1992**, *3*, 493–497.

695. Nita, M., Nagawa, H., Tominago, O., Tsunol, N., Fujii, S., Sasak, S., Fu, C. G., Takenouel, T., Tsuruo, T., & Mutol, T. 5-Fluorouracil induces apoptosis in human colon cancer cell lines with modulation of Bcl-2 family proteins. *British Journal of Cancer*, **1998**, *78*, 986–992.

696. Nizamutdinova, I. T., Lee, G. W., Lee, J. S., Cho, M. K., Son, K. H., Jeon, S. J., Kang, S. S., Kim, Y. S., Lee, J. H., Seo, H. G., Chang, K. C., & Kim, H. J. Tanshinone I suppresses growth and invasion of human breast cancer cells, MDA-MB-231, through regulation of adhesion molecules. *Carcinogenesis*, **2008**, *29*, 1885–1892.

697. Nores, M. M., Courreges, M. C., Benencia, F., & Coulombie, F. C. Immunomodulatory activity of *Cedrela lilloi* and *Trichilia elegans* aqueous leaf extracts. *Journal of Ethnopharmacology*, **1997**, *55*, 99–106.

698. Oanh, L. T. Investigation of some biochemical characters of proteolytic activity of *Pseuderanthemum palatiferum*. *Journal of Materials Science*, **1999**, *4*, 13–17.

699. Obesity and Cancer Risk. National Cancer Institute, **2012**. https://www.cancer.gov/about-cancer/causes-prevention/risk/obesity/obesity-fact-sheet (Accessed on 30 September 2019).

700. Obi, C. L., Potgieter, N., Masebe, T., Mathebula, H., & Molobela, P. *In vitro* antibacterial activity of Venda medicinal plants. *South African Journal of Botany*, **2003**, *69*, 199–203.

701. Obi, C. L., Potgieter, N., Randima, L. P., Mavhungu, N. J., Musie, E., Bessong, P. O., Mabogo, D. E. N., & Mashimbye, J. Antibacterial activity of five medicinal plants against some medically significant human bacteria. *South African Journal of Plant and Soil*, **2002**, *98*, 25–28.

702. Oboh, G. Hepatoprotective property of ethanolic extract of fluted pumpkin leaves against garlic-Induced oxidative stress. *Journal of Medicinal Food*, **2005**, *8*, 560–563.

703. O'Keefe, B. R. Biologically active proteins from natural product extracts. *Journal of Natural Products*, **2001**, *64*, 1373–1381.

704. Olivera, A., Moore, T. W., Hu, F., Brown, A. P., Sun, A., & Liotta, D. C. Inhibition of the NF-kappaB signaling pathway by the curcumin analog, 3, 5-Bis(2-pyridinylmethylidene)-4-piperidone (EF31): Anti-inflammatory and anticancer properties. *International Immunopharmacology*, **2011**, *12*, 368–377.

705. Oliver-Bever, B. *Medicinal Plants in Tropical West Africa* (p. 356). Cambridge University Press, Great Britain, **1986**.

706. Olusesan, A., Ebele, L., Onwuegbuchulam, O., & Olorunmola, E. Preliminary *in-vitro* Antibacterial activities of ethanolic extracts of *Ficus sycomorus* Linn. and *Ficus platyphylla* Del. (Moraceae). *African Journal of Microbiology Research*, **2010**, *4*, 598–601.

707. Oslund, R., & Sherman, W. Pinotol and derivatives thereof for the treatment of metabolic disorders. *US Patent-5*, 882, 896, **1998**, p. 16.

708. Otake, T., Mori, M., Ueba, N., Sutardjo, S., Kusumoto, I. T., Hattori, M., & Namba, T. Screening of Indonesian plant extracts for antihuman immunodeficiency virus-type1 (HIV-1) activity. *Phytotherapy Research*, **1995**, *9*, 6–10.

709. Oulmouden, F., Ghalim, N., El Morhit, M., Benomar, H., Daoudi, E. M., & Amrani, S. Hypolipidemic and anti-atherogenic effect of methanol extract of fennel (*Foeniculum vulgare*) in hypercholesterolemic mice. *International Journal of Knowledge and Systems Science*, **2014**, *3*, 42–52.

710. Oulmouden, F., Saïle, R., Gnaoui, N., Benomar, H., Lkhider, M., Amrani, S., et al. Hypolipidemic and anti-atherogenic effect of aqueous extract of fennel (*Foeniculum vulgare*) extract in an experimental model of atherosclerosis induced by triton WR-1339. *European Journal of Scientific Research*, **2011**, *52*, 91–99.

711. Ozsoy, M., & Ernst, E. How effective are complementary therapies for HIV and AIDS ? a systematic review. *Int. J. STD & AIDS*, **1999**, *10*, 629–635.

712. Padam, S. K., Grover, I. S., & Singh, M. Antimutagenic effects of polyphenols isolated from *Terminalia bellerica* myroblan in *Salmonella typhimurium*. *Indian Journal of Experimental Biology*, **1996**, *34*, 98–102.

713. Pakalapati, G., Wink, M., Li, L., Gretz, N., & Koch, E. Influence of red clover (*Trifolium pratense*) isoflavones on gene and protein expression profiles in liver of ovariectomized rats. *Phytomedicine*, **2009**, *16*, 845–855.

714. Paliwal, S., Sundaram, J., & Mitragotri, S. Induction of cancer-specific cytotoxicity towards human prostate and skin cells using quercetin and ultrasound. *British Journal of Cancer*, **2005**, *92*, 499–502.

715. Palomino, S. S., Abad, M. J., Bedoya, L. M., García, J., Gonzales, E., Chiriboga, X., Bermejo, P., & Alcami, J. Screening of South American plants against human immunodeficiency virus: Preliminary fractionation of aqueous extract from *Baccharis trinervis*. *Biological and Pharmaceutical Bulletin*, **2002**, *25*, 1147–1150.

716. Palozza, P., Torelli, C., Boninsegna, A., Simone, R., Catalano, A., Mele, M. C., & Picci, N. Growth-inhibitory effects of the astaxanthin-rich alga *Haematococcus pluvialis* in human colon cancer cells. *Cancer Letters*, **2009**, *283*, 108–117.

717. Pandey, A. K., & Chowdhry, P. K. Propagation techniques and harvesting time on productivity and root quality of *Withania somnifera*. *Journal of Tropical Medicinal Plants*, **2006**, *7*, 79–81.

718. Pandey, K., Sharma, P. K., & Dudhe, R. Antioxidant and anti-inflammatory activity of ethanolic extract of *Parthenium hysterophorus* L. *Asian Journal of Pharmaceutical and Clinical Research*, **2012**, *5*, 28–31.

719. Pandurangan, A., Khosa, R. L., & Hemalatha, S. Evaluation of anti inflammatory activity of leaf extract of *Solanum trilobatum*. *Journal of Pharmaceutical Sciences and Research*, **2009**, *1*, 16–21.

720. Pandurangan, A., Khosa, R. L., & Hemalatha, S. Evaluation of Anti-inflammatory activity of *Solanum trilobatum* root. *Oriental Pharmacy and Experimental Medicine*, **2008**, *8*, 416–422.

721. Pandurangan, A., Kosha, R. L., & Hemalatha, S. Anti-inflammatory activity of alkaloid from *Solanum trilobatum* on acute and chronic inflammation models. *Natural Product Research*, **2011**, *25*, 1132–1141.

722. Parada-Turska, J., Paduch, R., Majdan, M., Kandefer-Szerszeń, M., & Rzeski, W. Antiproliferative activity of parthenolide against three human cancer cell lines and human umbilical vein endothelial cells. *Pharmacological Reports*, **2007**, *59*, 233–237.

723. Parekh, J., & Chanda, S. Antibacterial and phytochemical studies on twelve species of Indian medicinal plants. *African Journal of Biomedical Research*, **2007**, *10*, 175–181.

724. Parida, A. K., Das, A. B., & Das, P. NaCl stress causes changes in photosynthetic pigments, proteins and other metabolic components in the leaves of a true mangrove, *Bruguiera parviflora*, in hydroponic cultures. *Journal of Plant Biology*, **2002**, *45*, 28–36.

725. Park, E. J., Oh, H., Kang, T. H., Sohn, D. H., & Kim, Y. C. An isocoumarin with hepatoprotective activity in Hep G2 and primary hepatocytes from *Agrimonia pilosa*. *Archives of Pharmacal Research*, **2004**, *27*, 944–946.

726. Park, H. J., Lee, J. Y., Chung, M. Y., Park, Y. K., Bower, A. M., Koo, S. I., Giardina, C., & Bruno, R. S. Green tea extract suppresses NF-kappaB activation and inflammatory responses in diet-induced obese rats with nonalcoholic steatohepatitis. *The Journal of Nutrition*, **2012**, *142*, 57–63.

727. Park, I. W., Han, C., Song, X., Green, L. A., Wang, T., Liu, Y., Cen, C., Song, X., Yang, B., Chen, G., & He, J. J. Inhibition of HIV-1 entry by extracts derived from traditional Chinese medicinal herbal plants. *BMC Complementary and Alternative Medicine*, **2009**, *9*, 29.

728. Park, J. W., Choi, Y. J., Jang, M. A., Lee, Y. S., Jun, D. Y., Suh, S. I., Baek, W. K., Suh, M. H., Jin, I. N., & Kwon, T. K. Chemopreventive agent resveratrol, a natural product derived from grapes, reversibly inhibits progression through S and G2 phases of the cell cycle in U937 cells. *Cancer Letters*, **2001**, *163*, 43–49.

729. Park, S., Yeo, M., Jin, J. H., Lee, K. M., Jung, J. Y., Choue, R., Cho, S. W., & Hahm, K. B. Rescue of *Helicobacter pylori*-induced cytotoxicity by red ginseng. *Digestive Diseases and Sciences*, **2005**, *50*, 1218–1227.

730. Park, S. H., Jang, J. H., Chen, C. Y., Na, H. K., & Surh, Y. J. A formulated red ginseng extract rescues PC12 cells from PCB-induced oxidative cell death through Nrf2-mediated upregulation of heme oxygenase-1 and glutamate cysteine ligase. *Toxicology*, **2010**, *278*, 131–139.

731. Parkin, D. M., Boyd, L., & Walker, L. C. The fraction of cancer attributable to lifestyle and environmental factors in the UK in 2010. *British Journal of Cancer*, **2011**, *105*, S77–S81.

732. Parvathy, K. S., Negi, P. S., & Srinivas, P. Antioxidant, antimutagenic and antibacterial activities of curcumin-β-diglusoside. *Food Chemistry*, **2009**, *115*, 265–271.

733. Patil, A., Khyati, V., Darshana, P., Anita, P., Aarti, J., & Naresh, C. *In vitro* anticancer activity of *Argemone mexicana* l. seeds and *Alstonia scholaris* (l.) r. br. bark on different human cancer cell lines. *World Journal of Pharmaceutical Sciences*, **2014**, *3*, 706–722.

734. Patil, J. K., Jalalpure, S. S., Hamid, S., & Ahirrao, R. A. *In-vitro* Immunomodulatory activity of extracts of *Bauhinia vareigata* Linn stem Bbark on human neutrophils, *Iranian Journal of Pharmacology and Therapeutics*, **2010**, *9*, 42–46.

735. Patil, J. R., Murthy, K. N. C., Jayaprakasha, G. K., Chetti, M. B., & Patil, B. S. Bioactive compounds from Mexican lime (*Citrus aurantifolia*) juice induce apoptosis in human pancreatic cells. *Journal of Agricultural and Food Chemistry*, **2009**, *57*, 10933–10942.

736. Patil, K. S., Jalalpure, S. S., & Wadekar, R. R. Effect of *Baliospermum montanum* root extract on phagocytosis by human neutrophils. *Indian Journal of Pharmaceutical Sciences*, **2009**, *71*, 68–71.

737. Pattnaik, S., & Sharma, G. D. Antibacterial nature of some common and indigenous plant extracts. *J. Sci. Technol.*, **2004**, *XVI*, 7–10.

738. Patwardhan, B., Kalbag, D., Patki, P. S., & Nagasampagi, B. A. Search of Immunomodulatory agents: A review. *Indian Drugs*, **1990**, *28*, 348–358.

739. Pawan, K., Khan, R. A., & Agrawal, K. R. S. Puetuberosanol an epoxychalcanol from *Pueraria tuberose*. *Phytochemistry*, **1996**, *42*, 243–244.

740. Pawelec, G., Derhovanessian, E., & Larbi, A. Immunosenescence and cancer. *Critical Reviews in Oncology/Hematology*, **2010**, *75*, 165–172.

741. Peng, G., Dixon, D. A., Muga, S. J., Smith, T. J., & Wargovich, M. J. Green tea polyphenol (-)-epigallocatechin-3-gallate inhibits cyclooxygenase-2 expression in colon carcinogenesis. *Molecular Carcinogenesis Journal*, **2006**, *45*, 309–319.

742. Peng, X. M., Qi, C. H., Tian, G. Y., & Zhang, Y. X. Physico-chemical properties and bioactivities of a glycoconjugate LbGp5B from *Lycium barbarum* L. *Chinese Journal of Chemistry*, **2001**, *19*, 842–846.

743. Percival, S. S., Talcott, S. T., Chin, S. T., Mallak, A. C., Lounds-Singleton, A., & Pettit-Moore, J. Neoplastic transformation of BALB/3T3 cells and cell cycle of HL-60 cells are inhibited by mango (*Mangifera indica* L.) juice and mango juice extracts. *The Journal of Nutrition*, **2006**, *136*, 1300–1304.

744. Perelson, A. S., Essunger, P., Cao, Y., Vesamen, M., Hurley, A., Sakela, K., Markowitz, M., & Ho, D. D. Decay characteristics of HIV-1 infected compartments during combination therapy. *Naturition*, **1997**, *387*, 88–191.

745. Perumal, S. R., & Ignacimuthu, S. Screening of 34 Indian medicinal plants for antibacterial properties. *Journal of Ethnopharmacology*, **1998**, *62*, 173–182.

746. Pham-Huy, L. A., He, H., & Pham-Huy, C. Free radicals, antioxidants in disease and health. *International Journal of Biomedical Science*, **2008**, *4*, 89–96.

747. Phetkate, P., Kummalue, T., U-pratya, Y., & Kietinun, S. Significant increase in cytotoxic T lymphocytes and natural killer cells by triphala: A clinical phase I study. *Evidence-Based Complementary and Alternative Medicine*, **2012**, *2012*, 1–6.

748. Piacente, S., Pizza, C., De Tommasi, N., & Mahmmod, N. Constituents of *Ardista japonica* and there *in vitro* anti HIV activity. *Journal of Natural Products*, **1996**, *59*, 565–569.

749. Piccinelli, A. C., Morato, P. N., Dos Santos, B. M., Croda, J., Sampson, J., Kong, X., Konkiewitz, E. C., Ziff, E. B., Amaya-Farfan, J., & Kassuya, C. A. Limonene reduces hyperalgesia induced by gp120 and cytokines by modulation of IL-1 β and protein expression in spinal cord of mice. *Life Sciences*, **2017**, *174*, 28–34.

750. Piccinelli, A. L., Cuesta-Rubio, O., Chica, M. B., Mahmood, N., Pagano, B., Pavone, M., Barone, V., & Rastrelli, L. Structural revision of clusianone and 7-epi-clusianone and anti-HIV-1 activity of polyisoprenylated benzophenones. *Tetrahedron*, **2005**, *61*, 8206–8211.

751. Piddock, K. J. V., & Wise, R. Mechanisms of resistance to quinolones and clinical perspective. *Journal of Antimicrobial Chemotherapy*, **1989**, *23*, 475–483.

752. Pietta, P. Flavonoids as antioxidants. *Journal of Natural Products*, **2000**, *63*, 1035–1042.

753. Pinmai, K., Chunlaratthanabhorn, S., Ngamkitidechakul, C., Soonthornchareon, N., & Hahnvajanawong, C. Synergistic growth inhibitory effects of *Phyllanthus emblica* and *Terminalia bellerica* extracts with conventional cytotoxic agents: Doxorubicin and cisplatin against human hepatocellular carcinoma and lung cancer cells. *World Journal of Gastroenterology*, **2008**, *14*, 1491–1497.

754. Pinto, J. T., & Rivlin, R. S. Antiproliferative effects of allium derivatives from garlic. *The Journal of Nutrition*, **2001**, *131*, S1058–S1060.

755. Pisha, E., Chai, H., Lee, I. S., Chagwedera, T. E., Farnsworth, N. R., Cordell, G. A., et al. Discovery of betulinic acid as a selective inhibitor of human melanoma that functions by induction of apoptosis. *Nature Medicine*, **1995**, 1, 1046–1051.

756. Poltanov, E. A., Shikov, A. N., Dorman, H. J. D., Pozharitskaya, O. N., Makarov, V. G., Tikhonov, V. P., & Hiltunen, R. Chemical and antioxidant evaluation of Indian gooseberry (*Emblica officinalis* Gaertn., syn. *Phyllanthus emblica* L.) supplements. *Phytotherapy Research*, **2009**, *23*, 1309–1315.

757. Pommier, Y., Johnson, A. A., & Marchand, C. Integrase inhibitors to treat HIV/AIDS. *Nature Reviews Drug Discovery*, **2005**, *4*, 236–248.

758. Ponnusankara, S., Pandita, S., Babub, R., Bandyopadhya, A., & Mukherjee, P. K. Cytochrome P450 inhibitory potential of Triphala-A rasayana from ayurveda. *Journal of Ethnopharmacology*, **2011**, *133*, 120–125.

759. Porchezhian, E., & Dobriyal, R. M. An overview on the advances of *Gymnema sylvestre*: Chemistry, pharmacology and patents. *Pharmazie*, **2003**, *58*, 5–12.

760. Potenza, M. A., Marasciulo, F. L., Tarquinio, M., Tiravanti, E., Colantuono, G., Federici, A., Kim, J. A., Quon, M. J., & Montagnani, M. EGCG, a green tea polyphenol, improves endothelial function and insulin sensitivity, reduces blood pressure, and protects against myocardial I/R injury in SHR. *American Journal of Physiology-Endocrinology and Metabolism*, **2007**, *292*, E1378-E1387.

761. Prabu, P. N., Ashokkumar, P., & Sudhandiran, G. Antioxidative and antiproliferative effects of astaxanthin during the initiation stages of 1, 2-dimethyl hydrazine-induced experimental colon carcinogenesis. *Fundamental & Clinical Pharmacology*, **2009**, *23*, 225–234.

762. Pradhan, D., Panda, K. P., & Thripathi, G. Evaluation of immunomodulatory activity of the methanolic extract of *Couroupita guinensis* flowers in rats. *Natural Product Radiance*, **2009**, *8*, 37–42.

763. Prakash, P., & Gupta, N. Therapeutic uses of *Ocimum sanctum* Linn (Tulsi) with a note on Eugenol and its pharmacological actions: A review. *Indian Journal of Physiology and Pharmacology*, **2005**, *49*, 125–131.

764. Prasad, A. V. K., Kapil, R. S., & Polpi, S. P. Structure of Pterocarponoids anhydrotuberosin 3-O methylanhydrotuberosin and tuberostan from *Pueraria tuberose*. *Indian Journal of Chemistry*, **1985**, *24*, 236–239.

765. Prasad, L., Husain, K. T., Jahangir, T., & Sultana, S. Chemomodulatory effects of *Terminalia chebula* against nickel chloride induced oxidative stress and tumor promotion response in male Wistar rats. *Journal of Trace Elements in Medicine and Biology*, **2006**, *20*, 233–239.

766. Prassas, I., & Diamandis, E. P. Novel therapeutic applications of cardiac glycosides. *Nature Reviews Drug Discovery*, **2008**, *7*, 926–935.

767. Pratheeba, M., Rajalakshmi, G., Ramesh. B. Hepatoprotective and antibacterial activity of leaf extract of *Solanum trilobatum*. *International Journal of Pharma and Bio* Sciences, **2012**, *2*, 17–28.

768. Prathibha, S., Nambisan, B., & Leelamma, S. Enzyme inhibitors in tuber crops and their thermal stability. *Plant Foods for Human Nutrition*, **1995**, *48*, 247–257.

769. Premanathan, M., Nakashima, H., Kathiresan, K., Rajendran, N., & Yamamoto, N. *In-vitro* anti-human immunodeficiency virus activity of Mangrove plants. *The Indian Journal of Medical Research*, **1996**, *103*, 278–281.

770. Premanathan, M., Arakaki, R., Izumi, H., Kathiresan, K., Nakano, M., Yamamoto, N., & Nakashima, H. Antiviral properties of a mangrove plant *Rhizophora apiculata* Blume against human immunodeficiency virus. *Antiviral Research*, **1999**, *44*, 113–122.

771. Premanathan, M., Rajendran, S., Ramanathan, T., Kathiresan, K., Nakashima, H., & Yamamoto, N. A survey of some Indian medicinal plants for anti-human immunodeficiency virus (HIV) activity. *The Indian Journal of Medical Research*, **2000**, *112*, 73–77.

772. Price, K. R., Johnson, I. T., & Fenwick, G. R. The chemistry and biological significance of saponins in foods and feed stuffs. *CRC Critical Reviews in Food Science and Nutrition*, **1987**, *26*, 27–135.

773. Punturee, K., Wild, C. P., Kasinrerk, W., & Vinitketkumnuen, U. Immunomodulatory activities of *Centella asiatica* and *Rhinacanthus nasutus* extracts. *Asian Pacific Journal of Cancer Prevention*, **2005**, *6*, 396–400.

774. Puspangadan, P., & Atal, C. K. Ethnomedico-botanical investigation in Kerala, I: Some primitive tribals of Western Ghats and their herbal medicine. *Journal of Ethnopharmacology*, **1984**, *11*, 59–77.

775. Puspita, N. A. Isolation and characterization of medicinal compounds from *Phyllanthus niruri* L. *PhD Thesis*, University of Salford, Salford, UK, **2014**, p. 217.

776. Qi, C. H., Zhang, X. Y., Zhao, X. N., Huang, L. J., Wei, C. H., & Ru, X. B. Immuno-activity of the crude polysaccharides from the fruit of *Lycium barbarum* L. *Chinese Journal of Pharmacology and Toxicology*, **2001**, *15*, 180–184.

777. Qiu, G. Q., Liu, Y., Wang, R., & Nan, K. J. Study on the effects of a traditional Chinese preparation Fu-Zheng-Yi-Liu-Yin combine with chemotherapy on gastric cancer. *Xi'an Yike Daxue Xuebao*, **2002**, *23*, 322–323.

778. Rabi, T., Wang, L., & Banerjee, S. Novel triterpenoid 25-hydroxy-3-oxoolean-12-en-28-oic acid induces growth arrest and apoptosis in breast cancer cells. *Breast Cancer Research and Treatment*, **2007**, *101*, 27–36.

779. Rafi-Uz-Zaman, M. D. *Phytochemical and Pharmacological Investigations of Genoderma Lucidum* (p. 88). East West University, Aftabnagar, Dhaka, **2012**.

780. Rahman, M. A., Kim, N. H., Yang, H., & Huh, S. O. Angelicin induces apoptosis through intrinsic caspase-dependent pathway in human SH-SY5Y neuroblastoma cells. *Molecular and Cellular Biochemistry*, **2012**, *369*, 95–104.

781. Raimondi, F., Santoro, P., Maiuri, L., Londei, M., Annunziata, S., Ciccimarra, F., & Rubino, A. Reactive nitrogen species modulate the effects of rhein, an active component of senna laxatives, on human epithelium *in vitro*. *Journal of Pediatric Gastroenterology and Nutrition*, **2002**, *34*, 529–534.

782. Rajkumar, V., Guha, G., & Kumar, R. A. Antioxidant and anti-neoplastic activities of *Picrorhiza kurroa* extracts. *Food and Chemical Toxicology*, **2011**, *49*, 363–369.

783. Raju, J., & Bird, R. P. Diosgenin, a naturally occurring steroid [corrected] saponin suppresses 3-hydroxy-3-methylglutaryl CoA reductase expression and induces apoptosis in HCT-116 human colon carcinoma cells. *Cancer Letters*, **2007**, *255*, 194–204.

784. Raju, J., Patlolla, J. M., Swamy, M. V., & Rao, C. V. Diosgenin, a steroid saponin of *Trigonella foenum graecum* (Fenugreek), inhibits azoxymethane-induced aberrant crypt foci formation in F344 rats and induces apoptosis in HT-29 human colon cancer cells. *Cancer Epidemiology, Biomarkers & Prevention*, **2004**, *13*, 1392–1398.

785. Ram, A. J., Bhakshu, L. M., & Raju, R. R. V. *In vitro* antimicrobial activity of certain medicinal plants from Eastern Ghats, India, used for skin diseases. *Journal of Ethnopharmacology*, **2003**, *90*, 333–357.

786. Rama-Bhat, S. P., Shwetha, R. B., & Sadananda, A. Studies on immunomodulatory effects of *Salacia chinensis* L. on albino rat. *Journal of Applied Pharmaceutical Science*, **2012**, *2*, 98–107.

787. Ramakrishna, K. V., Khan, R. A., & Kapil, R. S. A new isoflavone and Coumestan from *Pueraria tuberose*. *Indian Journal of Chemistry*, **1998**, *27*, 285.

788. Ramakrishna, V., Sridhar, R., Hewage, E. K. K., Lavanya, R., & Jairam, K. P. V. Indian gooseberry (*Emblica officinalis* Gaertn.) suppresses cell proliferation and induces apoptosis in human colon cancer stem cells independent of p53 status via suppression of c-Myc and cyclin D1. *Journal of Functional Foods*, **2016**, *25*, 267–278.

789. Ramakrishnappa, K. Impact of cultivation and gathering of medicinal plants on biodiversity: Case studies from India. In: *Biodiversity and the Ecosystem Approach in Agriculture, Forestry and Fisheries* [online], FAO, **2002**. Available at: http://www.fao.org/DOCREP/005/AA021E/AA021e00.htm (Accessed on 30 September 2019).

790. Ramawat, K. G., & Goyal, S. The Indian herbal drugs scenario in global perspectives. In: Ramawat, K. G., & Merillon, J. M., (eds.), *Bioactive Molecules and Medicinal Plants* (p. 323). Springer, Berlin – Heidelberg, **2008**.

791. Ramsey, S. D., Spencer, A. C., Topolski, T. D., Belza, B., & Patrick, D. L. Use of alternative therapies by older adults with osteoarthritis. *Arthritis & Rheumatology*, **2001**, *45*, 222–227.

792. Rani, P., & Khullar, N. Antimicrobial evaluation of some medicinal plants for their anti-enteric potential against multi-drug resistant *Salmonella typhi*. *Phytotherapy Research*, **2004**, *18*, 670–673.

793. Rastogi, R. P., & Malhotra, B. N. *Compendium a Medicinal Plants* (pp. 522–523). Central drug Research Institute Lucknow and National Institute of Science Communication, New Delhi, India, **2001**.

794. Ratsimbazafy, A. M., Michael, P., Rakotoarivelo, S. D., Ratiarson, A., Lariviere, C. B., & Nkongolo, K. K. Cytotoxicity activities against cervical and skin cancer cell lines of *Smilax kraussiana* and *Anthocleista rhizophoroides,* two endemic species widely used in traditional medicine in Madagascar. *International Journal of Herbal Medicine*, **2017**, *5*, 85–91.

795. Rawls, R. Europe's strong herbal brew. *Chem. Eng. News*, **1996**, 53–60.

796. Reboul, E., Borel, P., Mikail, C., Abou, L., Charbonnier, M., Caris-Veyrat, C., Goupy, P., Portugal, H., Lairon, D., & Amiot, M. J. Enrichment of tomato paste with 6% tomato peel increases lycopene and beta-carotene bioavailability in men. *The Journal of Nutrition*, **2005**, *135*, 790–794.

797. Reddy, S., Gopal, G., & Sit, G. *In vitro* multiplication of *Gymnema sylvestre R* Br: An important medicinal plant. *Current Science*, **2004**, *10*, 1–4.

798. Ren, W., Qiao, Z., Wang, H., Zhu, L., & Zhang, L. Flavonoids: Promising anticancer agents. *Medicinal Research Reviews*, **2003**, *23*, 519–534.

799. Reutrakul, V., Chanakul, W., & Jaipetch, T. Anti-HIV-1 constituents from the leaves and twigs of *Cratoxylum arborescens*. *Planta Medica*, **2006**, *72*, 1433–1435.

800. Reutrakul, V., Jaipetch, T., Yoosook, C., & Sophasan, S. Cytotoxic and anti HIV-1caged xanthones from the resins and fruits of *Garcinia hanburyi*. *Planta Medica*, **2007**, *73*, 33–40.

801. Rieux-Laucat, F., Fischer, A., & Le Deist, F. Cell-death signaling and human disease. *Current Opinion in Immunology*, **2003**, *15*, 325–331.

802. Rios, J. L., & Waterman, P. G. A review of the pharmacology and toxicology of Astragalus. *Phytotherapy Research*, **1997**, *11*, 411–418.

803. Rodríguez, J., Di Peirro, D., Gioia, M., Monaco, S., Delgado, R., Coletta, M., & Marini, S. Effects of a natural extract from *Mangifera indica* L., and its active compound, mangiferin, on energy state and lipid peroxidation of red blood cells. *Biochimica et Biophysica Acta*, **2006**, *1760*, 1333–1342.

804. Roselló-Soto, E., Koubaa, M., Moubarik, A., Lopes, R. P., Saraiva, J. A., Boussetta, N., Grimi, N., & Barba, F. J. Emerging opportunities for the effective valorization of wastes and by-products generated during olive oil production process: Non-conventional methods for the recovery of high-added value compounds. *Trends in Food Science & Technology*, **2015**, *2*, 296–310.

805. Rosenkranz, V., & Wink, M. Induction of apoptosis by alkaloids, non-protein amino acids, and cardiac glycosides in human promyelotic HL-60 cells. *Zeitschrift für Naturforschung – Section C: Journal of Biosciences*, **2007**, *62*, 458–466.

806. Ross, H. A., McDougall, G. J., & Stewart, D. Antiproliferative activity is predominantly associated with ellagitannins in raspberry extracts. *Phytochemistry*, **2007**, *68*, 218–228.

807. Ross, J. J., Arnason, J. T., & Birnboim, H. C. Low concentrations of the feverfew component parthenolide inhibit *in vitro* growth of tumor lines in a cytostatic fashion. *Planta Medica*, **1999**, *65*, 126–129.

808. Rothwell, P. M., Fowkes, F. G., Belch, J. F., Ogawa, H., Warlow, C. P., & Meade, T. W. Effect of daily aspirin on long-term risk of death due to cancer: analysis of individual patient data from randomized trials. *Lancet*, **2011**, *377*, 31–41.

809. Rowinsky, E. K., & Calvo, E. Novel agents that target tubulin and related elements. *Seminars in Oncology*, **2006**, *33*, 421–435.

810. Rowley, D. C., Hansen, M. S., & Rhodes, D. Thalassiolins A-C, new marine derived inhibitors of HIV Integrase. *Bioorganic & Medicinal Chemistry Letters*, **2002**, *10*, 3619–3625.

811. Roy, M. K., Nakahara, K., Na, T. V., Trakoontivakorn, G., Takenaka, M., Isobe, S., & Tsushida, T. Baicalein, a flavonoid extracted from a methanolic extract of *Oroxylum indicum* inhibits proliferation of a cancer cell line *in vitro* via induction of apoptosis. *Pharmazie*, **2007**, *62*, 149–153.

812. Rukachaisirikul, V., Pailee, P., Hiranrat, A., Tuchinda, P., Yoosook, C., Kasisit, J., Taylor, W. C., & Reutrakul, V. Anti-HIV-1 protostane triterpenes and digeranylbenzophenone from trunk bark and stems of *Garcinia speciosa*. *Planta Medica*, **2003**, *69*, 1141–1146.

813. Rukunga, G. M., Kofi-Tsekpo, M. W., Kurokawa, M., Kageyama, S., Mungai, G. M., Muli, J. M., Tolo, F. M., Kibaya, R. M., Muthaura, C. N., Kanyara, J. N., Tukei, P. M., & Shiraki, K. Evaluation of the HIV 1 reverse transcriptase inhibitory properties of extracts from some medicinal plants in kenya. *African Journal of Health Sciences*, **2000**, *9*, 81–90.

814. Runowicz, C. D., Wiernik, P. H., Einzig, A. I., Goldberg, G. L., & Horwitz, S. B. *Taxol in Ovarian Cancer* (pp. 67–78). National Conference on Gynecologic Cancers, Orlando, Florida, **1992**.

815. Ryu, N. H., Park, K. R., Kim, S. M., Yun, H. M., Nam, D., Lee, S. G., Jang, H. J., Ahn, K. S., Kim, S. H., Shim, B. S., Choi, S. H., Mosaddik, A., Cho, S. K., & Ahn, K. S. A hexane fraction of guava leaves (*Psidium guajava* L.) induces anticancer activity by suppressing AKT/mammalian target of rapamycin/ribosomal p70 S6 kinase in human prostate cancer cells. *Journal of Medicinal Foods*, **2012**, *15*, 231–241.

816. Sagwan, S., Rao, D. V., & Sharma, R. A. *In-vitro* and *in-vivo* antioxidant activity and total phenolic content of *Pongamia pinnata* (L.) Pierre: An important medicinal plant. *International Journal of Biotechnology*, **2011**, *4*, 568–574.

817. Saidi, H., Nasreddine, N., Jenabian, M. A., Lecerf, M., Schols, D., Krief, C., Balzarini, J., & Belec, L. Differential *in vitro* inhibitory activity against HIV-1 of alpha-(1–3)- and alpha-(1–6)-D mannose specific plant lectins: Implication for microbicide development. *Journal of Translational Medicine*, **2007**, *5*, 28.

818. Saikali, M., Ghantous, A., Halawi, R., Talhouk, S. N., Saliba, N. A., & Darwiche, N. Sesquiterpene lactones isolated from indigenous Middle Eastern plants inhibit tumor promoter-induced transformation of JB6 cells. *BMC Complementary and Alternative Medicine*, **2012**, *12*, 89.

819. Sainis, K. B., Sumariwalla, P. F., Goel, A., Chintalwar, G. J., Sipahimalani, A. T., & Banerji, A. Immunomodulatory properties of stem extracts of Tinospora cordifolia: Cell targets and active principles. In: Upadhyay, S. N., (eds.), *Immunomodulation* (p. 95). Narosa Publishing House, New Delhi, India, **1997**.

820. Saito, H. Regulation of herbal medicines in Japan. *Pharmacological Research*, **2000**, *41*, 515–519.

821. Sakai, S., Kawamata, H., Kogure, T., Mantani, N., Terasawa, K., Umatake, M., & Ochiai, H. Inhibitory effect of ferulic acid and isoferulic acid on the production of macrophage inflammatory protein-2 in response to respiratory syncytial virus infection in RAW264.7 cells. *Mediators of Inflammation*, **1999**, *8*, 173–175.

822. Sakamoto, K., Lawson, L. D., & Milner, J. A. Allyl sulfides from garlic suppress the *in vitro* proliferation of human A549 lung tumor cells. *Nutrition and Cancer*, **1997**, *29*, 152–156.

823. Sakurai, N., & Nagai, M. Chemical constituents of original plants of *Cimicifugae rhizoma* in Chinese medicine. *Yakugaku Zasshi*, **1996**, *116*, 850–865.

824. Saleem, A., Husheem, M., & Härkönen, P. K. Inhibition of cancer cell growth by crude extract and the phenolics of *Terminalia chebula* Retz. fruit. *Journal of Ethnopharmacology*, **2002**, *81*, 327–336.

825. Saleem, A., Jyrki, L., Kalevi, P., & Elina, O. Effects of long-term open-field ozone exposure on leaf phenolics of European silver birch *(Betula pendula* Roth). *Journal of Chemical Ecology*, **2001**, *27*, 1049–1062.

826. Salucci, M., Stivala, L. A., Maiani, G., Bugianesi, R., & Vannini, V. Flavoids uptake and their effect on cell cycle of human colon adenocarcinoma cells (Caco2). *British Journal of Cancer*, **2002**, *86*, 1645–1651.

827. Samarakoon, S. R., Kotigala, S. B., Gammana-Liyanage, I., Thabrew, I., Tennekoon, K. H., & Siriwardana, A. Cytotoxic and apoptotic effect of the decoction of the aerial parts of *Flueggea leucopyrus* on human endometrial carcinoma (AN3CA) cells. *Tropical Journal of Pharmaceutical Research*, **2014**, *13*, 873–880.

828. Sanches, N. R., Cortez, D. A. G., Schiavini, M. S., Nakamura, C. V., & Dias, F. B. P. An evaluation of antibacterial activities of *Psidium guajava* (L.). *Brazilian Archives of Biology and Technology*, **2005**, *48*, 429–436.

829. Sandhya, T., Lathika, K. M., Pandey, B. N., & Mishra, K. P. Potential of traditional ayurvedic formulation, Triphala, as a novel anticancer drug. *Cancer Letters*, **2006**, *231*, 206–214.

830. Sandhya, T., & Mishra, K. P. Cytotoxic of breast cancer cell line, MCF 7 and T 47D to triphala and its modification by antioxidants. *Cancer Letters*, **2006**, *238*, 304–313.

831. Sanodiya, B. S., Thakur, G. S., Baghel, R. K., Prasad, G. B., & Bisen, P. S. *Ganoderma lucidum*: a potent pharmacological macrofungus. *Current Pharmaceutical Biotechnology*, **2009**, *10*, 717–742.

832. Sa-Nunnes, A., Rogerio, A. P., Medeiros, A. I., Fabris, V. E., Andreu, G. P., Rivera, D. G., Delgado, R., & Faccioli, L. H. Modulation of eosinophil generation and migration by *Mangifera indica* L. extract (Vimang). *International Immunopharmacology*, **2006**, *6*, 1515–1523.

833. Sapna, S., & Ravi, T. K. Approaches towards development and promotion of herbal drugs. *Pharmacognosy Reviews*, **2007**, *1*, 180–184.

834. Saraf, S., Pathak, A. K., & Dixit, V. K. Hair growth promoting activity of *Tridax procumbens*. *Fitoterapia*, **1991**, *62*, 495–498.

835. Sasidharan, S., Chen, Y., Saravanan, D., Sundram, K. M., & Latha, L. Y. Extraction, isolation and characterization of bioactive compounds from plants. *The African Journal of Traditional, Complementary and Alternative Medicines*, **2011**, *8*, 1–10.

836. Sathishkumar, N., Sathiyamoorthy, S., Ramya, M., Yang, D. U., Lee, H. N., & Yang, D. C. Molecular docking studies of anti-apoptotic BCL-2, BCL-XL, and MCL-1 proteins with ginsenosides from *Panax ginseng*. *Journal of Enzyme Inhibition and Medicinal Chemistry*, **2011**, *27*, 685–692.

837. Satish, K. V., Santosh, K. S., & Abhishek, M. *In vitro* cytotoxicity of *Calotropis procera* and *Trigonella foenum graecum* against human cancer cell lines. *Journal of Chemical and Pharmaceutical Research*, **2010**, *2*, 861–865.

838. Sato, T., Vries, R. G., Snippert, H. J., Van De Wetering, M., Barker, N., Stange, D. E., Van Es, J. H., Abo, A., Kujala, P., Peters, P. J., & Clevers, H. Single Lgr5 stem cells build crypt-villus structures *in vitro* without a mesenchymal niche. *Nature*, **2009**, *459*, 262–265.

839. Satpute, K. L., Jadhav, M. M., Karodi, R. S., Katare, Y. S., Patil, M. J., Rub, R., & Bafna, A. R. Immunomodulatory activity of fruits of *Randia dumetorum* Lamk. *Journal of Pharmacognosy and Phytotherapy*, **2009**, *1*, 1–5.

840. Savill, J., Gregory, C., & Haslett, C. Eat me or die. *Science*, **2003**, *302*, 1516–1517.

841. Scharfenberg, K., Wagner, R., & Wagner, K. G. The cytotoxic effect of ajoene, a natural product from garlic, investigated with different cell lines. *Cancer Letters*, **1990**, *53*, 103–108.

842. Schieber, A., Stintzing, F. C., & Carle, R. By-products of plant food processing as a source of functional compounds-recent developments. *Trends in Food Science and Technology*, **2001**, *12*, 401–413.

843. Schilter, B., Andersson, C., Anton, R., Cons, A., Kleiner, J., Brien, J. O., Renwick, A. G., Korver, O., Smit, F., & Walker, R. Guidance for the safety assessment of botanicals and botanical preparations for use in food and food supplements. *Food and Chemical Toxicology*, **2003**, *41*, 1625–1649.

844. Schinazi, R. F., Chu, C. K., Babu, J. R., Oswald, B. J., Saalmann, V., Cannon, D. L., Eriksson, B. H. F., & Nasar, M. Anthraquinones as a new class of antiviral agents against human immunodeficiency virus. *Antiviral Research*, **1990**, *13*, 265–272.

845. Schmeller, T., & Wink, M. Utilization of alkaloids in modern medicine. In: Roberts, M. F., & Wink, M., (eds.), *Alkaloids: Biochemistry, Ecology and Medicinal Applications* (pp. 435–459). Plenum: New York, NY, USA, **2008**.

846. Schneider, Y., Vincent, F., Duranton, B., Badolo, L., Gosse, F., Bergmann, C., Seiler, N., & Raul, F. Anti-proliferative effect of resveratrol, a natural component of grapes and wine, on human colonic cancer cells. *Cancer Letters*, **2000**, *158*, 85–91.

847. Schulz, V., Hänsel, R., & Tyler, V. E. *Rational Phytotherapy: A Physician's Guide to Herbal Medicine* (4th edn.). Springer-Verlag, Berlin, **2001**.

848. Seeram, N. P., Adams, L. S., Henning, S. M., Niu, Y., Zhang, Y., Nair, M. G., & Heber, D. *In vitro* antiproliferative, apoptotic and antioxidant activities of punicalagin, ellagic acid and a total pomegranate tannin extract are enhanced in combination with other polyphenols as found in pomegranate juice. *Journal of Nutrition and Biochemistry*, **2005**, *16*, 360–367.

849. Seeram, N. P., Adams, L. S., Zhang, Y., Lee, R., Sand, D., Scheuller, H. S., & Heber, D. Blackberry, black raspberry, blueberry, cranberry, red raspberry, and strawberry extracts inhibit growth and stimulate apoptosis of human cancer cells *in vitro*. *Journal of Agriculture Food Chemistry*, **2006**, *54*, 9329–9339.

850. Seham, M. A. M., Bassem, M. M., Gamila, M. W., Khaled, M., & Marwa, M. M. Screening of some plants in Egypt for their cytotoxicity against four human cancer cell lines. *International Journal of Pharm. Tech. Research*, **2014**, *6*, 1074–1084.

851. Selles, A. J., Rodriguez, M. D., Balseiro, E. R., Gonzalez, L. N., Nicolais, V., & Rastrelli, L. Comparison of major and trace element concentrations in 16 varieties of Cuban mango stem bark (*Mangifera indica* L.). *Journal of Agricultural and Food Chemistry*, **2007**, *55*, 2176–2181.

852. Seshadri, P., Rajaram, A., & Rajaram, R. Plumbagin and juglone induce caspase-3-dependent apoptosis involving the mitochondria through ROS generation in human peripheral blood lymphocytes. *Free Radical Biology & Medicine*, **2011**, *51*, 2090–2107.

853. Setyawan, A. D. Review: Natural products from genus *Selaginella* (Selaginellaceae). *Nusantara Bioscience*, **2011**, *3*, 44–58.

854. Setyawan, A. D. Traditionally utilization of *Selaginella*, field research and literature review. *Nusantara Bioscience*, **2009**, *1*, 146–155.

855. Shah, K. A., Patel, M. B., Patel, R. J., & Parmar, P. K. *Mangifera indica* (Mango). *Pharmacognosy Reviews*, **2010**, *4*, 42–48.

856. Shaikh, M. Recent advance on ethanomedicinal plants as immunomodulator agent. *Ethnomedicine*, **2010**, 227–244.

857. Shamon, L. A., Pezzuto, J. M., Graves, J. M., Mehta, R. R., Wangcharoentrakul, S., Sangsuwan, R., Chaichana, S., Tuchinda, P., Cleason, P., & Reutrakul, V. Evaluation of the mutagenic, cytotoxic, and antitumor potential of TPL, a highly oxygenated diterpene isolated from *Tripterygium wilfordii*. *Cancer Letters*, **1997**, *112*, 113–117.

858. Sharma, A., & Kumar, K. Chemo protective role of Triphala against 1, 2-dimeth-ylhydrazine dihydrochloride induced carcinogenic damage to mouse liver. *Indian Journal of Clinical Biochemistry*, **2011**, *26*, 290–295.

859. Sharma, H., Parihar, L., & Parihar, P. Review on cancer and anticancerous properties of some medicinal plants. *Journal of Medicinal Plants Research*, **2011**, *5*, 1818–1835.

860. Sharma, P. C., Yelne, M. B., & Dennis, T. J. *Database on Medicinal Plants Used in Ayurveda* (Vol. 3, pp. 11–56, 282–313, 387–397). Central Council for Research in Ayurveda & Siddha, Department of ISM &H, Ministry of Health & Family Welfare, Govt. of India, New Delhi, **2005**.

861. Shi, J., Arunasalam, K., Yeung, D., Kakuda, Y., Mittal, G., & Jiang, Y. Saponins from edible legumes: Chemistry, processing, and health Benbefits. *Journal of Medical Food*, **2004**, *7*, 67–78.

862. Shi, L., Ren, A., Mu, D., & Zhao, M. Current progress in the study on biosynthesis and regulation of ganoderic acids. *Applied Microbiology and Biotechnology*, **2010**, *88*, 1243–1251.

863. Shieh, D. E., Chen, Y. Y., Yen, M. H., Chiang, L. C., & Lin, C. C. Emodin-induced apoptosis through p53-dependent pathway in human hepatoma cells. *Life Sciences*, **2004**, *74*, 2279–2290.

864. Shikishima, Y., Takaishi, Y., & Honda, G. Chemical constituents of *Pragnos tschiniganica*, structure elucidation and absolute configuration of coumarin and furanocoumarin derivatives with anti HIV activity. *Chemical and Pharmaceutical Bulletin*, **2001**, *49*, 877–880.

865. Shin, J. E., Han, M. J., Song, M. C., Baek, N. I., & Kim, D. H. 5-Hydroxy-7-(4'-hydroxy-3'-methoxyphenyl)-1-phenyl-3-heptanone: A pancreatic lipase inhibitor isolated from *Alpinia officinarum. Biological and Pharmaceutical Bulletin*, **2004**, *27*, 138–140.

866. Shin, J. E., Joo Han, M., & Kim, D. H. 3-Methylethergalangin isolated from *Alpinia officinarum* inhibits pancreatic lipase. *Biological and Pharmaceutical Bulletin*, **2003**, *26*, 854–857.

867. Shin, T. Y., Jeong, H. J., Kim, D. K., Kim, S. H., Lee, J. K., Kim, D. K., Chae, B. S., Kim, J. H., Kang, H. W., Lee, C. M., Lee, K. C., Park, S. T., Lee, E. J., Lim, J. P., Kim, H. M., & Lee, Y. M. Inhibitory action of water soluble fraction of *Terminalia chebula* on systemic and local anaphylaxis. *Journal of Ethnopharmacology*, **2001**, *74*, 133–140.

868. Shirin, H., Pinto, J. T., Kawabata, Y., Soh, J. W., Delohery, T., Moss, S. F., Murty, V., Rivlin, R. S., Holt, P. R., & Weinstein, I. B. Antiproliferative effects of S-allylmercaptocysteine on colon cancer cells when tested alone or in combination with sulindac sulfide. *Cancer Research*, **2001**, *61*, 725–731.

869. Shishodia, S., & Aggarwal, B. B. Diosgenin inhibits osteoclastogenesis, invasion, and proliferation through the down-regulation of Akt, I kappa B kinase activation and NF-kappa B-regulated gene expression. *Oncogenesis*, **2006**, *25*, 1463–1473.

870. Shivaprasad, H. N., Kharya, M. D., Rana, A. C., & Mohan, S. Preliminary immuno-modulatory activities of the aqueous extract of *Terminalia chebula. Pharmaceutical Biology*, **2006**, *44*, 32–34.

871. Shobana, S., Vidhya, V. G., & Ramya, M. Antibacterial activity of garlic varieties (*Ophioscordon* and *Sativum*) on enteric pathogens. *Current Research Journal of Biological Sciences*, **2009**, *1*, 123–126.

872. Shoeb, M., Celik, S., Jaspars, M., Kumarasamy, Y., MacManus, S., Nahar, L., Thoo-Lin, P. K., & Sarker, S. D. Isolation, structure elucidation and bioactivity of schischkiniin, a unique indole alkaloid from the seeds of *Centaurea schischkinii. Tetrahedron*, **2005**, *61*, 9001–9006.

873. Shoemaker, M., Hamilton, B., Dairkee, S. H., Cohen, I., & Campbell, M. J. *In vitro* anticancer activity of twelve Chinese medicinal herbs. *Phytotherapy Research*, **2005**, *19*, 649–651.

874. Shu, B., Duan, W., Yao, J., Huang, J., Jiang, Z., & Zhang, L. Caspase 3 is involved in the apoptosis induced by triptolide in HK-2 cells. *Toxicology In Vitro*, **2009**, *23*, 598–602.

875. Shukla, B., Visen, P. K., Patnaik, G. K., & Dhawan, B. N. Choleretic effect of picroliv, the hepatoprotective principle of *Picrorhiza kurroa. Planta Medica*, **1991**, *57*, 29–33.

876. Shukla, S., Jadon, A., & Bhadauria, M. Protective effect of *Terminalia bellerica* Roxb., and gallic acid against carbon tetra chloride induced damage in albino rats. *Journal of Ethnopharmacology*, **2006**, *109*, 214–218.

877. Shukla, S., Mehta, A., Kim, M., & Ayyadurai, N. *In vivo* immunomodulatory activities of ethanolic leaf extract of *Stevia rebaudiana* in albino rats. *Research Journal of Biotechnology*, **2011**, *6*, 27–31.

878. Singh, M., Chaudhry, M. A., Yadava, J. N. S., & Sanyal, S. C. The spectrum of antibiotic resistance in human and veterinary isolates of *Escherichia coli* collected

from 1984–1986 in Northern India. *Journal of Antimicrobial and Chemotherapy*, **1992**, *29*, 159–168.

879. Singh, V. K., Padmanabh, D., Chaudhary, B. R., & Ramesh, S. Immunomodulatory effect of *Gymnema sylvestre* (R.Br.) leaf extract: An *in vitro* study in rat model. *Plos One*, 10, e0139631.

880. Singh, M. K., Chetri, K., Pandey, U. B., Mittal, B., Kapoor, V. K., & Choudhuri, G. Mutational spectrum of Kras oncogene among Indian patients with gallbladder cancer. *Journal of Gastroenterology and Hepatology*, **2004**, *19*, 916–912.

881. Soares De Oliveira, J., Pereira, B. D., Teixeira de Freitas, C. D., Delano, B. M. F. J., Odorico de Moraes, M., Pessoa, C., Costa-Lotufo, L. V., & Ramos, M. V. *In vitro* cytotoxicity against different human cancer cell lines of laticifer proteins of *Calotropis procera* (Ait.) R. Br. *Toxicology In Vitro*, **2007**, *21*, 1563–1573.

882. Sofowora, A., Ogunbodede, E., & Onayade, A. The role and place of medicinal plants in the strategies for disease prevention. *The African Journal of Traditional, Complementary and Alternative Medicines*, **2013**, *10*, 210–229.

883. Sohi, K. K., Mittal, N., Hundal, M. K., & Khanduja, K. L. Gallic acid, an antioxidant, exhibits anti apoptotic potential in normal human lymphocytes: a Bcl-2 independent mechanism. *Journal of Nutritional Science and Vitaminology*, **2003**, *49*, 221–227.

884. Son, J. K., Jung, J. H., Lee, C. S., Moon, D. C., Choi, S. W., Min, B. S., & Woo, M. H. DNA Topoisomerases I and II inhibition and cytotoxicity of constituents from the roots of *Rubia cordifolia*. *Chem. Inform.*, **2006**, *38*, 15–18.

885. Son, J. K., Jung, J. H., Lee, C. S., Moon, D. C., Choi, S. W., Min, B. S., & Woo, M. H. DNA topoisomerases I and II inhibition and cytotoxicity of constituents from the roots of *Rubia cordifolia*. *Bulletin of the Korean Chemical Society*, **2006**, *27*, 1231–1234.

886. Song, W. Y., Ma, Y. B., Bai, X., & Zhang, X. M. Two new compounds and anti HIV active constituents from *Illicium verum*. *Planta Medica*, **2007**, *73*, 372–375.

887. Spedding, G., Ratty, A., & Middleton, E. Jr. Inhibition of reverse transcriptases by flavonoids. *Antiviral Research*, **1989**, *12*, 99–110.

888. Spencer, C. M., & Wilde, M. I. *Diacerein. Drugs*, **1997**, *53*, 98–108.

889. Spinella, F., Rosano, L., Decandia, S., Di, C. V., Albini, A., Elia, G., Natali, P. G., & Bagnato, A. Antitumor effect of green tea polyphenol epigallocatechin-3-gallate in ovarian carcinoma cells: Evidence for the endothelin-1 as a potential target. *Experimental Biology and Medicine*, **2006**, *231*, 1123.

890. Spinella, F., Rosano, L., Di, C. V., Decandia, S., Albini, A., Nicotra, M. R., Natali, P. G., & Bagnato, A. Green tea polyphenol epigallocatechin-3-gallate inhibits the endothelin axis and downstream signaling pathways in ovarian carcinoma. *Molecular Cancer Therapeutics*, **2006**, *5*, 1483–1492.

891. Sreeramulu, D., & Raghunath, M. Antioxidant activity and phenolic content of roots, tubers and vegetables commonly consumed in India. *Food Research International*, **2010**, *43*, 1017–1020.

892. Srikumar, R., Parthasarathy, N. J., Shankar, E. M., Manikandan, S., Vijayakumar, R., Thangaraj, R., Vijayananth, K., Sheeladevi, R., & Rao, U. A. Evaluation of the growth inhibitory activities of triphala against common bacterial isolates from HIV infected patients. *Phytotheraphy Research*, **2007**, *21*, 476–480.

893. Srinivas, G., John, A. R., Srinivas, P., Vidhyalakshmi, S., Priya, S. V., & Karunagaran, D. Emodin induces apoptosis of human cervical cancer cells through poly (ADP-ribose) polymerase cleavage and activation of caspase-9. *European Journal of Pharmocology*, **2003**, *473*, 117–125.

894. Srinivasan, S., Koduru, S., Kumar, R., Venguswamy, G., Kyprianou, N., & Damodaran, C. Diosgenin targets Akt-mediated prosurvival signaling in human breast cancer cells. *International Journal of Cancer*, **2009**, *125*, 961–967.

895. Srivastava, R. M., Singh, S., Dubey, S. K., Misra, K., & Khar, A. Immunomodulatory and therapeutic activity of curcumin. *International Immunopharmacology*, **2011**, *11*, 331–341.

896. Srivastava, S. K., Xiao, D., Lew, K. L., Hershberger, P., Kokkinakis, D. M., Johnson, C. S., Trump, D. L., & Singh, S. V. Allyl isothiocyanate, a constituent of cruciferous veges, inhibits growth of PC-3 human prostate cancer xenografts *in vivo*. *Carcinogenesis*, **2003**, *24*, 1665–1670.

897. Stahelin, H. F., & Wartburg, A. V. The chemical and biological route form podophyllotoxin glucoside to etoposide: Ninth Cain Memorail award lecture. *Cancer Research*, **1991**, *51*, 5–15.

898. Su, J. H., & Wen, Z. H. Bioactive cembrane-based diterpenoids from the soft coral *Sinularia triangular*. *Marine Drugs*, **2011**, *9*, 944–951.

899. Su, C. C., Chen, G. W., Tan, T. W., Lin, J. G., & Chung, J. G. Crude extract of garlic induced caspase-3 gene expression leading to apoptosis in human colon cancer cells. *In Vivo*, **2006**, *20*, 85–90.

900. Su, Y. T., Chang, H. L., Shyue, S. K., & Hsu, S. L. Emodin induces apoptosis in human lung adenocarcinoma cells through a reactive oxygen species-dependent mitochondrial signaling pathway. *Biochemical Pharmacology*, **2005**, *70*, 229–241.

901. Subhashini, J., Mahipal, S. V. K., Reddy, M. C., Reddy, M. M., Rachamallu, A., & Reddanna, P Molecular mechanisms in C-Phycocyanin induced apoptosis in human chronic myeloid leukemia cell line-K562. *Biochemical Pharmacology*, **2004**, *68*, 453–462.

902. Sugawara, T., Yamashita, K., Sakai, S., Asai, A., Nagao, A., Shiraishi, T., Imai, I., & Hirata, T. Induction of apoptosis in DLD-1 human colon cancer cells by peridinin isolated from the dinoflagellate, *Heterocapsa triquetra*. *Bioscience, Biotechnology, and Biochemistry*, **2007**, *71*, 1069–1072.

903. Sugnanam, S. D. P., Kunjam, M., & Ranjala, V. Screening of anti-bacterial activity of *Solanum Trilobatum* Linn. seed extract against dental pathogens. *Asian Journal of Plant Science and Research*, **2015**, *5*, 34–37.

904. Sultana, S., Ahmed, S., Sharma, S., & Jahangir, T. *Emblica officinalis* reverses thioacetamide-induced oxidative stress and early promotional events of primary hepatocarcinogenesis. *Journal of Pharmacy and Pharmacology*, **2004**, *56*, 1573–1579.

905. Sumner, H., Salan, U., Knight, D. W., & Hoult, J. R. Inhibition of 5-lipoxygenase and cyclo-oxygenase in leukocytes by feverfew. Involvement of sesquiterpene lactones and other components. *Biochemical Pharmacology*, **1992**, *43*, 2313–2320.

906. Sun, J., Chu, Y. F., Wu, X. Z., & Liu, R. H. Antioxidant and antiproliferative activities of fruits. *Journal of Agricultural and Food Chemistry*, **2002**, *50*, 7449–7454.

907. Sun, Q. Z., Chen, D. F., Ding, P. L., & Ma, C. M. Three new lignans longipedunins A-C from *Kadsura longipedunculata* and their inhibitory activity against HIV-1 protease. *Chemical and Pharmaceutical Bulletin*, **2006**, *54*, 129–132.

908. Sun, Y. F., & Wink, M. Tetrandrine and fangchinoline, bisbenzylisoquinoline alkaloids from *Stephania tetrandra*, are can reverse multidrug resistance by inhibiting P-glycoprotein activity in multidrug resistant human cancer cells. *Phytomedicine*, **2014**, *21*, 1110–1119.

909. Susin, S., Daugas, E., Ravagnan, L., Samejima, K., Zamzami, N., Loeffler, M., et al. Two distinct pathways leading to nuclear apoptosis. *The Journal of Experimental Medicine*, **2000**, *192*, 571–580.

910. Susin, S. A., Lorenzo, H. K., & Zamzami, N. Molecular characterization of mitochondrial apoptosis-inducing factor. *Nature*, **1999**, *397*, 441–446.

911. Suzuki, H., Okubo, L., Yamazaki, S., Suzuki, K., Mitsuya, H., & Toda, S. Inhibition of the infectivity and cytopathic effect of the human immunodeficiency virus by water-soluble lignin in an extract of the culture medium of Lentinus edodes mycelia (LEM). *Biochemical and Biophysical Research Communications*, **1989**, *160*, 367–373.

912. Swanson, M. D., Winter, H. C., Goldstein, I. J., & Markovitz, D. M. A lectin isolated from bananas is a potent inhibitor of HIV replication. *The Journal of Biological Chemistry*, **2010**, *285*, 8646–8655.

913. Syed, B., & Sreedharamurthy, S. Bioprospecting of endophytic Bacterial plethora from medicinal plants. *Journal: Plant Sciences Feed*, **2013**, *3*, 42–45.

914. Tabassum, N., & Ahmad, F. Role of natural herbs in the treatment of hypertension. *Pharmacognosy Reviews*, **2011**, *5*, 30–40.

915. Tabata, K., Motani, K., Takayanagi, N., Nishimura, R., Asami, S., Kimura, Y., Ukiya, M., Hasegawa, D., Akihisa, T., & Suzuki, T. Xanthoangelol, a major chalcone constituent of *Angelica keiskei*, induces apoptosis in neuroblastoma and leukemia cells. *Biological and Pharmaceutical Bulletin*, **2005**, *28*, 1404–1407.

916. Tager, M., Dietzmann, J., Thiel, U., Hinrich, N. K., & Ansorge, S. Restoration of the cellular thiol status of peritoneal macrophages from CAPD patients by the flavonoids silibinin and silymarin. *Free Radical Research*, **2001**, *34*, 137–151.

917. Tajudin, T. J., Mat, N., Siti-Aishah, A. B., Yusran, A. A., Alwi, A., & Ali, A. M. Cytotoxicity, antiproliferative effects, and apoptosis induction of methanol extract of *Cynometra cauliflora* Linn. whole fruit on human promyelocytic leukemia HL-60 cells. *Evidence-Based Complementary and Alternative Medicine*, **2012**, 1–6.

918. Takahashi, I., Nakanishi, S., Kobayashi, E., Nakano, H., Suzuki, T., & Tomaoki, T. Hypericin and pseudo – hypericin specifically inhibit protein kinase C: Possible relation to their anti-retroviral activity. *Biochemical and Biophysical Research Communications*, **1989**, *165*, 1209–1212.

919. Takeda, Y., Togashi, H., Matsuo, T., Shinzawa, H., Takeda, Y., & Takahashi, T. Growth inhibition and apoptosis of gastric cancer cell lines by *Anemarrhena asphodeloides* Bunge. *Journal of Gastroenterology*, **2001**, *36*, 79–90.

920. Taleb-Contini, S. H., Kanashiro, A., Kabeya, L. M., Polizello, A. C., Lucisano-Valim, Y. M., & Oliveira, D. C. Immunomodulatory effects of methoxylated flavonoids from two *Chromolaena* species: Structure-activity relationships. *Phytotherapy Research*, **2006**, *20*, 573–575.

921. Tamura, S., Shiomi, A., & Kinmura, T. Halogenated analog of 1'acetoxychavicolac-etate rev-export inhibitor from Alpiniagalanga, designed from mechanism of action. *Bioorganic & Medicinal Chemistry Letters*, **2010**, *20*, 2082–2085.

922. Tamvakopoulos, C., Dimas, K., Sofianos, Z. D., Hatziantoniou, S., Han, Z., Liu, Z. L., Wyche, J. H., & Pantazis, P. Metabolism and anticancer activity of the curcumin analog, dimethoxycurcumin. *Clinical Cancer Research*, **2007**, *13*, 1269–1277.

923. Tan, C. J., Di, Y. T., Wang, Y. H., Zhang, Y., Si, Y. K., & Gao, S. Three new indole alkaloids from *Trigonos temonlii*. *Organic Letters*, **2010**, *12*, 2370–2373.

924. Tan, Y., Yu, R., & Pezzuto, J. M. Betulinic acid-induced programed cell death in human melanoma cells involves mitogen-activated protein kinase activation. *Clinical Cancer Research*, **2003**, *9*, 2866–2875.

925. Tanaka, H., Sato, M., & Fujiwara, S. Antibacterial activity of isoflavonoids isolated from *Erythrina variegata* against methicillin resistant *Staphylococcus aureus*. *Letters in Applied Microbiology*, **2002**, *35*, 228–489.

926. Tanaka, T., Iinuma, M., Yuki, K., Fujii, Y., & Mizuno, M. Flavonoids in root bark of *Pongamia pinnata*. *Phytochemistry*, **1992**, *31*, 993–998.

927. Tang, J., Calacino, J. M., Larsen, S. H., & Spitzer, W. Virucidal activity of hypericin against enveloped and non-enveloped DNA and RNA viruses. *Antiviral Research*, **1990**, *13*, 313–326.

928. Tang, P. M. K., Chan, J. Y. W., Zhang, D. M., Au, S. W. N., Fong, W. P., Kong, S. K., Tsui, S. K. W., Waye, M. M. Y., Mak, T. C. W., & Fung, K. P. Pheophorbide *a*, an active component in *Scutellaria barbata*, reverses P-glycoprotein-mediated multidrug resistance on a human hepatoma cell line R-HepG2. *Cancer Biology & Therapy*, **2007**, *6*, 504–509.

929. Tang, W., Liu, J. W., Zhao, W. M., Wei, D. Z., & Zhong, J. J. Ganoderic acid T from *Ganoderma lucidum* mycelia induces mitochondria mediated apoptosis in lung cancer cells. *Life Sciences*, **2006**, *80*, 205–211.

930. Tang, Y. Q., Jaganath, I. B., & Sekaran, S. D. *Phyllanthus* spp. induces selective growth inhibition of PC-3 and MeWo human cancer cells through modulation of cell cycle and induction of apoptosis. *Plos One*, **2010**, *5*, E-article: 12644.

931. Tantillo, C., Ding, J., Jacobo-Molina, A., Nanni, R. G., Boyer, P. L., Hughes, S. H., Pauwels, R., Andries, K., Janssen, P. A. J., & Arnold, E. J. Locations of anti-AIDS drug binding sites and resistance mutations in the three-dimensional structure of HIV-1 reverse transcriptase. Implications for mechanisms of drug inhibition and resistance. *Journal of Molecular Biology*, **1994**, *243*, 369–387.

932. Tao, X., & Lipsky, P. E. The Chinese anti-inflammatory and immunosuppressive herbal remedy *Tripterygium wilfordii* Hook F. *Rheumatic Disease Clinics of North America*, **2000**, *26*, 29–50.

933. Tao, X., Younger, J., Fan, F. Z., Wang, B., & Lipsky, P. E. Benefit of an extract of *Tripterygium Wilfordii* Hook F in patients with rheumatoid arthritis: A double-blind, placebo-controlled study. *Arthritis & Rheumatology*, **2002**, *46*, 1735–1743.

934. Tao, X. L. Mechanism of treating rheumatoid arthritis with *Tripterygium wilfordii* hook. II. Effect on PGE2 secretion. *Zhongguo Yi Xue Ke Xue Yuan Xue Bao*, **1989**, *11*, 36–40.

935. Tao, X. L., Cai, J. J., & Lipsky, P. E. The identity of immunosuppressive components of the ethyl acetate and chloroform methanol extract (T2) of *Tripterygium wilfordii*

Hook F. *Journal of Pharmacology and Experimental Therapeutics*, **1995**, *272*, 1305–1312.

936. Targeted Cancer Therapies. NCI, 2014. https://www.cancer.gov/about-cancer/treatment/types/targeted-therapies/targeted-therapies-fact-sheet (Accessed on 30 September 2019).

937. Taylor, L. *Plant Based Drugs and Medicines* (p. 465). Raintree Nutrition Inc., Carson City, NV, **2000**.

938. Teuscher, E., & Lindequist, U. *Biogene Gifte-Biologie, Chemie, Pharmakologie, Toxikologie (Biogenic Poisons: Biology, Chemistry, Pharmacology, Toxicology)* (3rd edn., p. 310). Wissenschaftliche Verlagsgesellschaft: Stuttgart, Germany, **2010**.

939. Teuscher, C., Subramanian, M., Noubade, R., Gao, J. F., Offner, H., Zachary, J. F., & Blankenhorn, E. P. Central histamine H3 receptor signaling negatively regulates susceptibility to autoimmune inflammatory disease of the CNS. *Proceedings of the National Academy of Sciences of the United States*, **2007**, *104*, 10146–10151.

940. Teuscher, E., Melzig, M. F., & Lindequist, U. *Biogene Arzneimittel. Ein Lehrbuch der Pharmazeutischen Biologie (Biogenic Medicines: A Textbook of Pharmaceutical Biology)* (7th edn., p. 233). Wissenschaftliche Verlagsgesellschaft: Stuttgart, Germany, **2012**.

941. Thakong, K. A. Study on the antimalarial constituents and chemical composition of *Eupatorium odoratum* (L.). *Thesis M.Sc. (Pharmaceutical Chemistry and Phytochemistry)* (p. 217). Faculty of Graduate studies, Mahidol University, **1999**.

942. Thakur, C. P., Thakur, B., Singh, S., Sinha, P. K., & Sinha, S. K. The Ayurvedic medicines Haritaki, Amala and Bahira reduce cholesterol-induced atherosclerosis in rabbits. *International Journal of Cardiology*, **1988**, *21*, 167–175.

943. The Wealth of India. *A dictionary of Indian Raw Materials and Industrial Products* (Vol. 2, pp. 56, 57). CSIR, Govt. of India, New Delhi, **1959**.

944. Thoe, A., Masebe, T., Suzuki, Y., Wada, S., & Tohoku, J. *Peltophorum africanum*, a traditional South African medicinal plant contains an anti HIV-1 constituent betulinic acid. *The Journal of Experimental Medicine*, **2009**, *217*, 93–99.

945. Thomas, M. P., Liu, X., Whangbo, J., McCrossan, G., Sanborn, K. B., Basar, E., Walch, M., & Lieberman, J. Apoptosis triggers specific, rapid, and global mRNA decay with 3' uridylated intermediates degraded by DIS3L2. *Cell Reports*, **2015**, *11*, 1079–1089.

946. Thomsen, A., & Kolesar, J. M. Chemoprevention of breast cancer. *American Journal of Health-System Pharmacy*, **2008**, *65*, 2221–2228.

947. Thoppil, R. J., Harlev, E., Mandal, A., Nevo, E., & Bishayee, A. Antitumor activities of extracts from selected desert plants against HepG2 human hepatocellular carcinoma cells. *Pharmaceutical Biology*, **2013**, *51*, 668–674.

948. Thornhill, S. M., & Kelly, A. M. Natural treatment of perennial allergic rhinitis. *Alternative Medicine Review*, **2000**, *5*, 448–454.

949. Tian, Q., & Zang, Y. H. Antiproliferative and apoptotic effects of the ethanolic herbal extract of *Achillea falcata* in human cervical cancer cells are mediated via cell cycle arrest and mitochondrial membrane potential loss. *Journal of the Balkan Union of Oncology*, **2015**, *20*, 1487–1496.

950. Tian, Z., Yang, M., Huang, F., Li, K., Si, J., Shi, L., Chen, S., & Xiao, P. Cytotoxicity of three cycloartane triterpenoids from *Cimicifuga dahurica. Cancer Letters*, **2005**, *226*, 65–75.

951. Tillekeratne, L. M. V., Sherette, A., Fulmer, J. A., Hupe, D., Gabbara, S., Peliska, J. A., & Hudson, R. A. Differential inhibition of polymerase and strand transfer activities of HIV-1 reverse transcriptase. *Bioorganic & Medicinal Chemistry Letters*, **2002**, *12*, 525–528.

952. Timyan, J. *Important Trees in Haiti* (p. 34). Southeast Consortium for International Development, Washington DC, **1996**.

953. Tippo, O., & Stern, W. L. *Humanistic Botany* (p. 216). W.W. Norton, New York, **1977**.

954. Tjandrawinata, R., & Hughes-Fulford, M. Up-regulation of cyclooxyenase-2 by product-prostaglandin E2. *Advances in Experimental Medicine and Biology*, **1997**, *407*, 163–170.

955. Tjandrawinata, R. R., Susanto, L. W., & Nofiarny, D. The use of *Phyllanthus niruri* L. as an immunomodulator for the treatment of infectious diseases in clinical settings. *Asian Pacific Journal of Tropical Medicine*, **2017**, *7*, 132–140.

956. Tjandrawinata, R. R., Arifin, P. F., Tandrasasmita, O. M., Rahmi, D., & Aripin, A. DLBS1425, a *Phaleria macrocarpa* (Scheff.) Boerl. extract confers anti prolifera- tive and proapoptosis effects via eicosanoid pathway. *Journal of Experimental Ther- apeutics and Oncology*, **2010**, *8*, 187–201.

957. Toker, A., Özyurtkan, M. O., Kaba, E., & Nova, G. Da vinci robotic system in the surgery for mediastinal bronchogenic cyst: A report on five patients. *The Journal of Visualized Surgery*, **2015**, *1*, 23.

958. Toker, G. FABAD. *Journal of Pharmacological Sciences*, **1995**, *20*, 75–79.

959. Tomlinson, T. R., & Akerele, O. *Medicinal Plants: Their Role in Health and Biodi- versity* (p. 180). University of Pennsylvania Press, Philadelphia, **1998**.

960. Tresina, P. S., Koilpitchai, P., & Veerabahu, R. M. Immunomodulatory activity of ethanol extract of *Sonerila tinnevelliensis* Fischer (Melastomataceae) whole plant in mice. *Journal of Pharmaceutical Innovation*, **2016**, *5*, 94–97.

961. Tripathi, K. D. *Anti-Rheumatoid and Anti-Gout Drugs, Essentials of Medical Pharmacology* (7th edn., pp. 210–212). Jaypee brother publishers, New Delhi, **2013**.

962. Tsubura, A., Lai, Y. C., Kuwata, M., Uehara, N., & Yoshizawa, K. Anticancer effects of garlic and garlic-derived compounds for breast cancer control. *Anti-Cancer Agents in Medicinal Chemistry*, **2011**, *11*, 249–253.

963. Turano, A., Scura, G., Caruso, A., Bonfanti, C., Luzzati, R., Basetti, D., & Manca, N. Inhibitory effects of papaverine on HIV replication *in vitro. AIDS Research and Human Retroviruses*, **1989**, *5*, 83–191.

964. Turano, A., Scura, G., Caruso, A., Bonfanti, C., Luzzati, R., Bassetti, D., & Manca, N. Inhibitory effects of pepaverine on HIV replication *in vitro. AIDS Research and Human Retroviruses*, **1989**, *5*, 183–192.

965. Tyler, V. E. Phytomedicines: Back to the future. *Journal of Natural Products*, **1999**, *62*, 1589–1592.

966. Udeinya, I. J., Mbah, A. U., Chijioke, C. P., & Shu, E. N. An antimalarial extract from neem leaves is antiretroviral. *Transactions of the Royal Society of Tropical Medicine and Hygiene*, **2004**, *98*, 435–437.

967. Udgirkar, R. F., Kadam, P., & Kale, N. Antibacterial activity of some Indian medicinal plant: A review. *International Journal of Pharma and Bio Sciences*, **2012**, *1*, 1–8.

968. Ueda, J. Y., Tezuka, Y., Banskota, A. H., Tran, Q. L., Tran, Q. K., Saiki, I., & Kadota, S. Constituents of the Vietnamese medicinal plant *Streptocaulon juventas* and their antiproliferative activity against the human HT-1080 fibrosarcoma cell line. *Journal of Natural Products*, **2003**, *66*, 1427–1433.

969. Ueno, Y., Umemori, K., Niimi, E., Tanuma, S., Nagata, S., Sugamata, M., Ihara, T., Sekijima, M., Kawai, K., Ueno, I., & Tashiro, F. Induction of apoptosis by T-2 toxin and other natural toxins in HL-60 human promyelotic leukemia cells. *Natural Toxins*, **1995**, *3*, 129–137.

970. Une, H. D., Pal, S. C., Kasture, V. S., & Kasture, S. B. Phytochemical constituents and pharmacological profile of *Albizzia lebbeck*. *Journal of Natural Remedies*, **2001**, *1*, 1–5.

971. Usha, P. T. A., Jose, S., & Nisha, A. R. Antimicrobial drug resistance–a global concern. *Veterinary World*, **2010**, *3*, 138–139.

972. Vaghasiya, Y., Nair, R., & Chanda, S. Antibacterial and preliminary phytochemical and physico-chemical analysis of *Eucalyptus citriodora* Hk leaf. *Natural Product Research*, **2008**, *22*, 754–762.

973. Vaidya, A. B., Antarkar, D. S., Doshi, J. C., Bhatt, A. D., Ramesh, V., Vora, P. V., Perissond, D., Baxi, A. J., & Kale, P. M. *Picrorhiza kurroa* (Kutaki) Royle ex Benth as a hepatoprotective agent-experimental & clinical studies. *Journal of Postgraduate Medicine*, **1996**, *42*, 105–108.

974. Vaidya, J. S., Vyas, J. J., Chinoy, R. F., Merchant, N., Sharma, O. P., & Mittra, I. Multicentricity of breast cancer: Whole-organ analysis and clinical implications. *British Journal of Cancer*, **1996**, *74*, 820–824.

975. Vaidyaratnam, P. S. *Varier's Indian Medicinal Plants–a Compendium of 500 Species* (pp. 256–260). Orient Longman Ltd., Hyderabad, India, **1994**,

976. Valko, M., Leibfritz, D., Moncol, J., Cronin, M. T. D., Mazur, M., & Telser, J. Free radicals and antioxidants in normal physiological functions and human disease. *The International Journal of Biochemistry & Cell Biology*, **2007**, *39*, 44–84.

977. Valsaraj, R., Pushpangadan, P., Smitt, U. W., Adsersen, A., & Nyman, U. Antimicrobial screening of selected medicinal plants from India. *Journal of Ethnopharmacology*, **1997**, *58*, 75–83.

978. Valsaraj, R., Pushpangadan, P., Smitt, U. W., Adsersen, A., Christensen, S. B., Sittie, A., Nyman, U., Nielsen, C., & Olsen, C. E. New anti-HIV-1, antimalarial, and antifungal compounds from *Terminalia bellerica*. *Journal of Natural Products*, **1997**, *60*, 739–742.

979. Van Der Hem, L. G., Van Der Vliet, J. A., Bocken, C. F., Kino, K., Hoitsma, A. J., & Tax, W. J. Ling Zhi-8: Studies of a new immune modulating agent. *Transplantation*, **1995**, *60*, 438–443.

980. Van Der Sijis, H., & Wiltink, H. Antiviral drugs: Present status and future prospects. *International Journal of Biochemistry*, **1994**, *26*, 621–630.

981. Van Der Woulde, H., Gliszczynska-Swiglo, A., Struijs, K., Smeets, A., Alink, G. M., & Rietjens, I. M. Biphasic modulation of cell proliferation by quercetin at

concentrations physiologically relevant in humans. *Cancer Letters*, **2003**, *200*, 41–47.

982. Van Erk, M. J., Roepman, P., Van Der Lende, T. R., Stierum, R. H., Aarts, J. M., Van Bladeren, P. J., & Van Ommen, B. Integrated assessment by multiple gene expression analysis of quercetin bioactivity on anticancer-related mechanisms in colon cancer cells *in vitro*. *European Journal of Nutrition*, **2005**, *44*, 143–156.

983. Van Wyk, B. V., & Gericke, N. *People's Plants: A Guide to Useful Plants of Southern Africa* (pp. 58–61). Briza Publications, Pretoria, South Africa, **2002**.

984. Van Wyk, B. E., Wink, C., & Wink, M. *Handbuch der Arzneipflanzen (Handbook of Medicinal Plants)* (3rd edn., p. 878). Wissenschaftliche Verlagsgesellschaft: Stuttgart, Germany, **2015**.

985. Van Wyk, B. E., & Wink, M. *Medicinal Plants of the World* (p. 451). Timber Press: Portland, OR, USA, **2004**.

986. Van Wyk, B. E., De Wet, H., & Van Heerden, F. R. An ethnobotanical survey of medicinal plants in the southern Karro, South Africa. *South African Journal of Botany*, **2008**, *74*, 696–704.

987. Van Wyk, B. E., & Wink, M. *Phytomedicines, Herbal Drugs and Poisons* (p. 321). Briza, Kew Publishing, Cambridge University Press: Cambridge, UK, **2015**.

988. Vargas-Arispuro, I., Reyes-Báez, R., Rivera-Castañeda, G., Martínez-Téllez, M. A., & Rivero-Espejel, I. Antifungal lignans from the creosote bush (*Larrea tridentata*). *Industrial Crops and Products*, **2005**, *22*, 101–107.

989. Vattem, D. A., & Shetty, K. Biological function of Ellagic acid: A review. *Journal of Food Biochemisry*, **2005**, *29*, 234–266.

990. Vedavanam, K., Srijayanta, S., O'Reilly, J., Raman, A., & Wiseman, H. Antioxidant action and potential antidiabetic properties of an isoflavonoid-containing soyabean phytochemical extract (SPE). *Phytotherapy Research*, **1999**, *13*, 601–608.

991. Veljko, V., Jean-Francois, M., Nevena, V., Sanja, G., & Zeger, D. Simple criterion for selection of flavonoid compounds with anti-HIV activity. *Bioorganic & Medicinal Chemistry Letters*, **2007**, *17*, 1226–1232.

992. Veluri, R., Singh, R. P., Liu, Z., Thompson, J. A., Agarwal, R., & Agarwal, C. Fractionation of grape seed extract and identification of gallic acid as one of the major active constituents causing growth inhibition and apoptotic death of DU145 human prostate carcinoma cells. *Carcinogenesis*, **2006**, *27*, 1445–1453.

993. Verma, N. K., & Singh, A. P. History of plants and animal products in the treatment of human disease: A review. *Research Journal of Phytomedicine*, **2015**, *1*, 60–62.

994. Verma, S. K., Singh, S. K., Mathur, A., & Singh, S. *In vitro* cytotoxicity of *Argemone mexicana* against different human cancer cell lines. *International Journal of Chemical, Environmental and Pharmaceutical Research*, **2010**, *1*, 37–39.

995. Vermani, K., & Garg, S. Herbal medicines for sexually transmitted diseases and AIDS. *Journal of Ethnopharmacology*, **2002**, *80*, 49–66.

996. Vijayababu, M. R., Kanagaraj, P., Arunkumar, A., Ilangovan, R., Aruldhas, M. M., & Arunakaran, J. Quercetin-induced growth inhibition and cell death in prostatic carcinoma cells (PC-3) are associated with increase in p21 and hypophosphorylated retinoblastoma proteins expression. *Journal of Cancer Research and Clinical Oncology*, **2005**, *131*, 765–771.

997. Vishnukanta, R. A. C. Evaluation of hydroalcoholic extract of *Melia azedarach* Linn roots for analgesic and anti inflammatory activity. *International Journal of Phytomedicine*, **2010**, *2*, 341–344.

998. Vishwakarma, A. P., Vishwe, A., Sahu, P., & Chaurasiya, A. Magical remedies of *Terminalia arjuna* (ROXB.). *International Journal of Pharmaceutical and Biological Archive*, **2013**, *2*, 189–201.

999. Vivekanandan, P., Gobianand, K., Priya, S., Vijayalakshmi, P., & Karthikeyan, S. Protective effect of picroliv against hydrazine-induced hyperlipidemia and hepatic steatosis in rats. *Drug and Chemical Toxicology*, **2007**, *30*, 241–252.

1000. Wagner, H., Vollmar, A., & Bechthold, A. *Pharmazeutische Biologie 2. Biogene Arzneistoffe und Grundlagen von Gentechnik und Immunologie (Pharmaceutical Biology II. Biogenic Drugs and Fundamentals of Genetic Engineering and Immunology)* (7th edn., p. 398). Wissenschaftliche Verlagsgesellschaft: Stuttgart, Germany, **2007**.

1001. Wajant, H. Connection map for Fas signaling pathway. *Science STKE*, **2007**, tr1. doi: 10.1126/stke.3802007tr1.

1002. Wajant, H. The Fas signaling pathway: More than a paradigm. *Science*, **2002**, *296*, 1635–1636.

1003. Waldmann, T. A. Immunotherapy: Past, present and future. *Nature Medicine*, **2003**, *9*, 269–277.

1004. Wall, N. R., Mohammad, R. M., Reddy, K. B., & Al-Katib, A. M. Bryostatin 1 induces ubiquitination and proteasome degradation of Bcl-2 in the human acute lymphoblastic leukemia cell line. Reh. *International Journal of Molecular Medicine*, **2000**, *5*, 165–171.

1005. Wang, X. J., Hayes, J. D., Henderson, C. J., & Wolf, C. R. Identification of retinoic acid as an inhibitor of transcription factor Nrf2 through activation of retinoic acid receptor a. *Proceedings of the National Academy of Sciences*, **2007**, *104*, 19589–19594.

1006. Wang, B. H., & Ou-Yang, J. P. Pharmacological actions of sodium ferulate in cardiovascular system. *Cardiovascular Drug Reviews*, **2005**, *23*, 161–172.

1007. Wang, C. C., Chen, L. G., Lee, L. T., & Yang, L. L. Effects of 6-gingerol, an antioxidant from ginger, on inducing apoptosis in human leukemic HL-60 cells. *In Vivo*, **2003**, *17*, 641–645.

1008. Wang, H., Bian, S., & Yang, C. S. Green tea polyphenol EGCG suppresses lung cancer cell growth through upregulating miR-210 expression caused by stabilizing HIF-1alpha. *Carcinogenesis*, **2011**, *32*, 1881–1889.

1009. Wang, H. X., & Ng, T. B. Isolation and characterization of velutin, a novel lowmolecular – weight ribosome inactivating protein from winter mushroom (*Flammulina velutipes*) fruiting bodies. *Life Sciences*, **2003**, *68*, 2151–2158.

1010. Wang, J. H., Wang, H. Z., Zhang, M., & Zhang, S. H. Effect of anti-aging *Lycium barbarum* polysaccharide. *Acta Nutrimenta Sinica*, **2002**, *24*, 189–191.

1011. Wang, J. N., Hou, C. Y., Liu, Y. L., Lin, L. Z., Gil, R. R., & Cordell, G. A. Swertifrancheside, an HIV-reverse transcriptase inhibitor and the first flavone-xanthone dimer from *Swertia franchetiana*. *Journal of Natural Products*, **1994**, *57*, 211–217.

1012. Wang, Q., Ding, Z. H., Liu, J. K., & Zheng, Y. T. Xanthohumol, a novel anti-HIV-1 agent purified from hops *Humulus lupulus*. *Antiviral Research*, **2004**, *64*, 189–194.

1013. Wang, Q. F., Chen, J. C., Hsieh, S. J., Cheng, C. C., & Hsu, S. L. Regulation of Bcl-2 family molecules and activation of caspase cascade involved in gypenosides-induced apoptosis in human hepatoma cells. *Cancer Letters*, **2002**, *183*, 169–178.

1014. Wang, S. Y., Feng, R., Bowmank, L., Lu, Y., Ballington, J. R., & Ding, M. Antioxidant activity of *Vaccinium stamineum*, exhibition of anticancer capability in human lung and leukemia cells. *Planta Medica*, **2007**, *73*, 451.

1015. Wang, S. Y., Hsu, M. L., Hsu, H. C., Tzeng, C. H., Lee, S. S., Shiao, M. S., & Ho, C. K. The anti-tumor effect of *Ganoderma lucidum* is mediated by cytokines released from activated macrophages and T lymphocytes. *International Journal of Cancer*, **1997**, *70*, 699–705.

1016. Wang, W., Heideman, L., Chung, C. S., Pelling, J. C., Koehler, K. J., & Birt, D. F. Cell-cycle arrest at G2/M and growth inhibition by apigenin in human colon carcinoma cell lines. *Molecular Carcinogenesis*, **2000**, *28*, 02–110.

1017. Wang, X., Wu, Y. C., Fadok, V. A., Lee, M. C., Gengyo-Ando, K., Cheng, L. C., Ledwich, D., Hsu, P. K., Chen, J. Y., Chou, B. K., Henson, P., Mitani, S., & Xue, D. Cell corpse engulfment mediated by C. elegans phosphatidylserine receptor through CED-5 and CED-12. *Science*, **2003**, *302*, 1563–1566.

1018. Wang, Y., Bao, L., Yang, X., Li, L., Li, S., Gao, H., Yao, X. S., Wen, H., & Liu, H. W. Bioactive sesquiterpenoids from the solid culture of the edible mushroom *Flammulina velutipes* growing on cooked rice. *Food Chemistry*, **2012**, *132*, 1346–1353.

1019. Wang, Y. X., Neamat, N., Jacob, J., Palmer, I., Stahl, S. J., Kaufman, J. D., Huang, P. L., Huang, P. L., Winslow, H. E., Pommier, Y., Wingfield, P. T., Lee-Huang, S., Bax, A., & Torchia, D. A. Solution structure of anti-HIV and anti tumor protein MAP-30: Structural insights in to its multiple functions. *Cell*, **1999**, *99*, 433–442.

1020. Wang, Z. G., & Ren, J. Current status and future direction of Chinese herbal medicine. *Trends Pharmacol. Sci.,* **2002**, *23*, 347–348.

1021. Wani, M. C., Taylor, H. L., & Wall, M. E. Plant antitumor agents VI. The isolation and structure of taxol, a novel antileukemic and antitumor agent from *Taxus breuifolia. Journal of the American Chemical Society*, **1971**, *93*, 2325–2327.

1022. Ward, E. M., Thun, M. J., Hannan, L. M., & Jemal, A. Interpreting cancer trend. *Annals of the New York Academy of Sciences*, **2006**, *1076*, 29–53.

1023. Warrier, P. K., Nambiar, V. P. K., & Ramakutty, C. *Indian Medicinal Plants: A Compendium of 500 sps.* (Vol. 3, pp. 339–344). Orient Longman, Hyderabad, India, **1993**.

1024. Warrier, P. K., Nambbiar, V. P. K., & Kutty, R. C. *Indian Medicinal Plants* (Vol. 1, p. 160). Orient Longman, Aryavaidyasala Publication, Hyderbad, India, **1996**.

1025. Warrier, P. K., Nambiar, V. P. K., & Ramankutty, C. *Indian Medicinal Plants, a Compendium of 500 Species* (pp. 10–12). Orient Longman Ltd., Hyderabad, India, **1995**.

1026. Watabe M., Hishikawa, K., Takayanagi, A., Shimizu, N., & Nakaki, T. Caffeic acid phenethyl ester induces apoptosis by inhibition of NFκB and activation of Fas in human breast cancer MCF-7 cells. *The Journal of Biological Chemistry*, **2004**, *279*, 6017–6026.

1027. Watson, J. L., Hill, R., Lee, P. W., Giacomantonio, C. A., & Hoskin, D. W. Curcumin induces apoptosis in HCT-116 human colon cancer cells in a p21-independent manner. *Experimental and Molecular Pathology*, **2008**, *84*, 230–233.

1028. Wei, Y., Ma, C. M., & Hattori, M. Anti-HIV protease triterpenoids from the acid hydrolysate of *Panax ginseng*. *Phytochemistry Letters*, **2009**, *2*, 63–66.

1029. Wei, Y., Ma, C. M., Chen, D. Y., & Hattori, M. Anti HIV 1 protease triterpenoid from *Stauntonia obovatifolia* Hayata intermedia. *Phytochemistry*, **2008**, *69*, 1875–1879.

1030. Westfall, S. D., Nilsson, E. E., & Skinner, M. K. Role of triptolide as an adjunct chemotherapy for ovarian cancer. *Chemotherapy*, **2008**, *54*, 67–76.

1031. Westh, R., Zinn, C. S., & Rosdahl, V. T. An international multicenler study of antimicrobial consumption and resistance in *staphylococcus aures* isolates from 15 hospitals in 14 countries. *Microb. Drug Resist.*, **2004**, *10*, 169–176.

1032. Wetzel, I., & Brachei, F. Revised structure of alkaloid drymaritin. *Journal of Natural Products*, **2009**, *72*, 1908–1910.

1033. WHA (World Health Assembly). *Resolution*, **1977**, *30*, 49.

1034. What is CAM? National Center for Complementary and Alternative Medicine, https://www.cancer.gov/about-cancer/treatment/cam (Accessed on 30 September 2019).

1035. WHO. Progress Report by the Director General, Document No. A44/20, 22 March 1991, World Health Organization, Geneva, **1991**, p. 30.

1036. WHO (World Health Organization). Cancer, **2010**. www.who.int/news-room/fact-sheets/detail/cancerCached (Accessed on 30 September 2019).

1037. WHO. *Monographs on Selected Medicinal Plants* (Vol. 1, p. 112). World Health Organization, Geneva, **1999**.

1038. WHO (World Health Organization). *Research Guidelines for Evaluating the Safety and Efficacy of Herbal Medicines* (p. 110). Manila, **1993**.

1039. WHO (World Health Organization). *The World Health Report Reducing Risks and Promoting Healthy Life* (p. 32). World Health Organization, Geneva, Switzerland, **2002**.

1040. WHO (World Health Organization). *The World Health Report, Health Systems: Improving Performance* (p. 42). WHO, Geneva, **2000**.

1041. WHO (World Health Organization). *Trends in Smoking and Lung Cancer Mortality in Japan, by birth cohort, 1949–2010* (p. 41). **2013**.

1042. WHO (World Health Organization). *World Cancer Report 2014*. World Health Organization, **2014**, Chapter 1.3. ISBN 9283204298.

1043. Wild, S., Roglic, G., Green, A., Sicree, R., & King, H. Global prevalence of diabetes: Estimates for 2000 and projections for 2030. *Diabetes Care*, **2004**, *27*, 1047–1053.

1044. Wilken, R., Veena, M. S., Wang, M. B., & Srivatsan, E. S. Curcumin: A review of anti-cancer properties and therapeutic activity in head and neck squamous cell carcinoma. *Molecular Cancer*, **2011**, *10*, 12–20.

1045. Willaman, J. J., & Hui-Lin, I. Alkaloid bearing plants and their contained alkaloids. *Lloydia*, **1970**, *33*, 139–182.

1046. William, C. S. C., & Kwok, L. N. *In vitro* and *in vivo* anti-tumor effects of *Astragalus membranaceus*. *Cancer Letters*, **2007**, *252*, 43–54.

1047. Wink, M. Molecular modes of action of cytotoxic alkaloids-From DNA intercalation, spindle poisoning, topoisomerase inhibition to apoptosis and multiple drug resistance. In: Cordell, G., (ed.), *The Alkaloids* (Vol. 64, pp. 1–48). Elsevier: Amsterdam, The Netherlands, **2007**.

1048. Wink, M. Introduction: Biochemistry, role and biotechnology of secondary products. In: Wink, M., (ed.), *Biochemistry of Secondary Product Metabolism* (pp. 1–16). CRC Press, Boca Raton, FL, **1999**.

1049. Wink, M., & Van Wyk, B. E. *Mind-Altering and Poisonous Plants of the World* (p. 212), Timber Press: Portland, OR, USA, **2010**.

1050. Witvrouw, M., Este, J. A., Mateu, M. Q., Reymen, D., Andrei, G., Snoeck, R., Ikeda, S., Pauwels, R., Bianchini, N. V., Desmyter, J., & De Clercq, E. Activity of a sulfated polysaccharide extracted from the red seaweed *Aghardhiella tenera* against human immunodeficiency virus and other enveloped viruses. *Antiviral Chemistry and Chemotherapy*, **1994**, *5*, 297–303.

1051. Wong, W. W., & Puthalakath, H. Bcl-2 family proteins: The sentinels of the mitochondrial apoptosis pathway. *IUBMB Life*, **2008**, *60*, 390–397.

1052. World Health Organization. *In vitro* screening of traditional medicines for anti-HIV activity: Memorandum from a WHO meeting. *Bulletin of the World Health Organization*, **1989**, *67*, 613–618.

1053. Wu, G., Qian, Z., Guo, J., Hu, D., Bao, J., Xie, J., Xu, W., Lu, J., Chen, X., & Wang, Y. *Ganoderma lucidum* extract induces G1 cell cycle arrest, and apoptosis in human breast cancer cells. *The American Journal of Chinese Medicine*, **2012**, *40*, 631–642.

1054. Wu, J., Yi, Y. H., Tang, H. F., Zou, Z. R., & Wu, H. M. Structure and cytotoxicity of a new lanostane-type triterpene glycoside from the sea cucumber *Holothuria hilla*. *Chemistry & Biodiversity*, **2006**, *3*, 1249–1254.

1055. Wu, P. P., Liu, K. C., Huang, W. W., Chueh, F. S., Ko, Y. C., Chiu, T. H., Lin, J. P., Kuo, J. H., Yang, J. S., & Chung, J. G. Diallyl trisulfide (DATS) inhibits mouse colon tumor in mouse CT-26 cells allograft model *in vivo*. *Phytomedicine*, **2011**, *18*, 672–676.

1056. Wu, P. P., Liu, K. C., Huang, W. W., Ma, C. Y., Lin, H., Yang, J. S., & Chung, J. G. Triptolide induces apoptosis in human adrenal cancer NCI-H295 cells through a mitochondrial-dependent pathway. *Oncology Reports*, **2011**, *25*, 551–557.

1057. Wu, S. H., Hang, L. W., Yang, J. S., Chen, H. Y., Lin, H. Y., Chiang, J. H., Lu, C. C., Yang, J. L., Lai, T. Y., Ko, Y. C., & Chung, J. G. Curcumin induces apoptosis in human non-small cell lung cancer NCI-H460 cells through ER stress and caspase cascade- and mitochondria-dependent pathways. *Anticancer Research*, **2010**, *30*, 2125–2133.

1058. Wu, Y., Chen, Y., Xu, Y. J., & Lu, L. Anticancer activities of curcumin on human Burkitt's lymphoma. *Zhonghua Zhong Liu Za Zhi*, **2002**, *24*, 348–352.

1059. Xiang, Y. Z., Shang, H. C., Gao, X. M., & Zhang, B. L. A comparison of the ancient use of ginseng in traditional Chinese medicine with modern pharmacological experiments and clinical trials. *Phytotherapy Research*, **2008**, *22*, 851–858.

1060. Xiao, D., Choi, S., Johnson, D. E., Vogel, V. G., Johnson, C. S., Trump, D. L., Lee, Y. J., & Singh, S. V. Diallyl trisulfide-induced apoptosis in human prostate cancer cells involves c-Jun N-terminal kinase and extracellular-signal regulated kinase-mediated phosphorylation of Bcl-2. *Oncogenesis*, **2004**, *23*, 5594–5606.

1061. Xiao, D., Herman-Antosiewicz, A., Antosiewicz, J., Xiao, H., Brisson, M., Lazo., J. S., & Singh, S. V. Diallyl trisulfide-induced G(2)-M phase cell cycle arrest in human prostate cancer cells is caused by reactive oxygen species-946] dependent destruction and hyperphosphorylation of Cdc25C. *Oncogenesis*, **2005**, *24*, 6256–6268.

1062. Xiao, D., & Singh, S. V. Diallyl trisulfide, a constituent of processed garlic, inactivates Akt to trigger mitochondrial translocation of BAD and caspase-mediated apoptosis in human prostate cancer cells. *Carcinogenesis*, **2006**, *27*, 533–540.

1063. Xiao, D., Zeng, Y., Hahm, E. R., Kim, Y. A., Ramalingam, S., & Singh, S. V. Diallyl trisulfide selectively causes Bax- and Bak-mediated apoptosis in human lung cancer cells. *Environmental and Molecular Mutagenesis*, **2009**, *50*, 201–212.

1064. Xiao, P. *A Pictorial Encyclopaedia of Chinese Medicine* (Vol. 1, pp. 88–90). Commercial Press, Hong Kong, **1988**.

1065. Xiao, W. L., Li, X. L., & Yang, L. M. Triterpenoids from *Schisandra rubiflora*. *Journal of Natural Products*, **2001**, *70*, 1056–1059.

1066. Xiao, W. L., Tian, R. R., Pu, J. X., & Wu, L. Triterpenoids from *Schisandra lancifolia* with anti HIV activity. *Journal of Natural Products*, **2006**, *69*, 277–279.

1067. Xu, B. J., & Chang, S. K. C. Comparative study on antiproliferation properties and cellular antioxidant activities of commonly consumed food legumes against nine human cancer cell lines. *Food Chemistry*, **2012**, *134*, 1287–1296.

1068. Xu, H. X., Zeng, F. Q., Wan, M., & Sim, K. Y. Anti-HIV triterpene acids from *Geum japonicum*. *Journal of Natural Products*, **1996**, *59*, 643–645.

1069. Xu, K., Liang, X., Gao, F., Zhong, J., & Liu, J. Antimetastatic effect of ganoderic acid T *in vitro* through inhibition of cancer cell invasion. *Process Biochemistry*, **2010**, *45*, 1261–1267.

1070. Xu, Z. Q., Flavin, M. T., & Jenta, T. R. Calanolides, the naturally occurring anti-HIV agents. *Current Opinion in Drug Discovery and Development*, **2000**, *3*, 155–166.

1071. Yadav, B., Bajaj, A., Saxena, M., & Saxena, A. K. *In Vitro* anticancer activity of the root, stem and leaves of *Withania Somnifera* against various human cancer cell lines. *Indian Journal of Pharmaceutical Sciences*, **2010**, *72*, 659–663.

1072. Yadav, P. N., Liu, Z., & Rafi, M. M. A diarylheptanoid from lesser galangal (*Alpinia officinarum*) inhibits proinflammatory mediators via inhibition of mitogen-activated protein kinase, p44/42, and transcription factor nuclear factor-kappa B. *Journal of Pharmacology and Experimental Therapeutics*, **2003**, *305*, 925–931.

1073. Yamahara, J., Kubomura, Y., Miki, K., & Fujimura, H. Anti-ulcer action of *Panax japonicus* rhizome. *Journal of Ethnopharmacology*, **1987**, *19*, 95–101.

1074. Yamamoto, K., Ishikawa, C., Katano, H., Yasumoto, T., & Mori, N. Fucoxanthin and its deacetylated product, fucoxanthinol, induce apoptosis of primary effusion lymphomas. *Cancer Letters*, **2011**, *300*, 225–234.

1075. Yan, M. H., Cheng, P., Ziang, Y., & Ma, Y. B. Periglaucines A-D anti HIV-1 alkaloids from *Pericampylus glaucus*. *Journal of Natural Products*, **2008**, *71*, 760–763.

1076. Yang, G. X., Li, Y. Z., & Hu, C. Q. Anti HIV bioactive stilbenes dimmers of *Caragana rosea*. *Planta Medica*, **2005**, *71*, 569–571.

1077. Yang, G. Y., Li, Y. K., Wang, R. R., & Xiao, W. L. Dibenzocyclooctadiene lignans from the fruits of *Schisandra wilsoniana* and their anti HIV activities. *Journal of Asian Natural Products Research*, **2010**, *12*, 470–476.

1078. Yang, H. C., Zeng, M. Y., Dong, S. Y., Liu, Z. Y., & Li, R. Antiproliferative activity of phlorotannin extracts from brown algae *Laminaria japonica* Aresch. *Chinese Journal of Oceanology and Limnology*, **2010**, *28*, 122–130.

1079. Yang, J. H., Han, S. J., Ryu, J. H., Jang, I. S., & Kim, D. H. Ginsenoside Rh2 ameliorates scopolamine-induced learning deficit in mice. *Biological and Pharmaceutical Bulletin*, **2009**, *32*, 1710–1715.
1080. Yang, J. S., Chen, G. W., Hsia, T. C., Ho, H. C., Ho, C. C., Lin, M. W., Lin, S. S., Yeh, R. D., Ip, S. W., Lu, H. F., & Chung, J. G. Diallyl disulfide induces apoptosis in human colon cancer cell line (COLO 205) through the induction of reactive oxygen species, endoplasmic reticulum stress, caspases casade and mitochondrial-dependent pathways. *Food and Chemical Toxicology*, **2009**, *47*, 171–179.
1081. Yang, J. S., Hour, M. J., Huang, W. W., Lin, K. L., Kuo, S. C., & Chung, J. G. MJ-29 inhibits tubulin polymerization, induces mitotic arrest, and triggers apoptosis via cyclin-dependent kinase 1-mediated Bcl-2 phosphorylation in human leukemia U937 cells. *Journal of Pharmacology and Experimental Therapeutics*, **2010**, *334*, 477–488.
1082. Yang, S., Chen, J., Guo, Z., Xu, X. M., Wang, L., Pei, X. F., Yang, J., Underhill, C. B., & Zhang, L. Triptolide inhibits the growth and metastasis of solid tumors. *Molecular Cancer Therapeutics*, **2003**, *2*, 65–72.
1083. Yao, X. J., Wainberg, M. A., & Parniak, M. A. Mechanism of inhibition of HIV-1 infection *in vitro* by purified extract of *Prunella vulgaris*. *Journal of Virology*, **1992**, *187*, 56–62.
1084. Yashphe. J., Feuerstein. I., Barel. S., & Segal, R. The antibacterial and antispasmodic activity of *Artemisia herba alba* Asso. II. Examination of essential oils from various chemotypes. *Pharmaceutical Biology*, **1987**, *25*, 89–96.
1085. Ye, W. C., Ji, N. N., Zhao, S. X., Liu, J. H., Ye, T., McKervey, M. A., & Stevenson, P. Triterpenoids from *Pulsatilla chinensis. Phytochemistry*, **1996**, *42*, 799–802.
1086. Yeh, C. T., Rao, Y. K., Yao, C. J., Yeh, C. F., Li, C. H., Chuang, S. E., Luong, J. H., Lai, G. M., & Tzeng, Y. M. Cytotoxic triterpenes from *Antrodia camphorata* and their mode of action in HT-29 human colon cancer cells. *Cancer Letters*, **2009**, *285*, 73–79.
1087. Yeh, F. T., Wu, C. H., & Lee, H. Z. Aloe-emodin novel anticancer herbal drug. *International Journal of Cancer*, **2003**, *3*, 27–31.
1088. Yin, X., Zhou, J., Jie, C., Xing, D., & Zhang, Y. Anticancer activity and mechanism of *Scutellaria barbata* extract on human lung cancer cell line A549. *Life Sciences*, **2004**, *75*, 2233–2244.
1089. Yodkeeree, S., Chaiwangyen, W., Garbisa, S., & Limtrakul, P. Curcumin, demethoxycurcumin and bisdemethoxycurcumin differentially inhibit cancer cell invasion through the down-regulation of MMPs and uPA. *Journal of Nutrition and Biochemistry*, **2009**, *20*, 87–95.
1090. Yoshiko, S., & Hoyoko, N. Fucoxanthin, a natural carotenoid, induces G1 arrest and GADD45 gene expression in human cancer cells. *In Vivo*, **2007**, 21, 305–309.
1091. Yoshioka, K., Kataoka, T., Hayashi, T., Hasegawa, M., Ishi, Y., & Hibasami, H. Induction of apoptosis by gallic acid in human stomach cancer KATO III and colon adenocarcinoma COLO 205 cell lines. *Oncology Reports*, **2000**, *7*, 1221–1223.
1092. Yosie, A., Effendy, M. A. W., Sifizul, T. M. T., & Habsah, M. Antibacterial, radical-scavenging activities and cytotoxicity properties of *Phaleria macrocarpa* (Scheff.) Boerl leaves in HEPG2 cell lines. *International Journal of Pharmaceutical Sciences and Research*, **2011**, *2*, 1700–1706.

1093. Yu, C. S., Huang, A. C., Lai, K. C., Huang, Y. P., Lin, M. W., Yang, J. S., & Chung, J. G. Diallyl trisulfide induces apoptosis in human primary colorectal cancer cells. *Oncology Reports*, **2012**, *28*, 949–954.

1094. Yu, D., Suzuki, M., Xie, L., Morris-Natschke, S. L., & Lee, K. H. Recent progress in the development of coumarin derivatives as potent anti-HIV agents. *Medicinal Research Reviews*, **2003**, *23*, 322–345.

1095. Yu, D., Morris-Natschke, S. L., & Lee, K. H. New developments in natural products-based anti-AIDS research. *Medicinal Research Reviews*, **2007**, *27*, 108–132.

1096. Yu, Y. B., Miyashiro, H., Nakamura, N., Hattori, M., & Park, J. C. Effects of triterpenoids and flavonoids isolated from *Alnus firma* on HIV-1 viral enzymes. *Archives of Pharmacal Research*, **2007**, *30*, 820–826.

1097. Yu, D., Wild, C. T., Martin, D. E., Morris-Natschke, S. L., Chen, C. H., Allaway, G. P., & Lee, K. H. The discovery of a class of novel HIV-1 maturation inhibitors and their potential in the therapy of HIV. *Expert Opinion on Investigational Drugs*, **2005**, *14*, 681–693.

1098. Yu, R. X., Hu, X. M., Xu, S. Q., Jiang, Z. J., & Yang, W. Effects of fucoxanthin on proliferation and apoptosis in human gastric adenocarcinoma MGC-803 cells via JAK/STAT signal pathway. *European Journal of Pharmacology*, **2011**, *657*, 10–19.

1099. Yuen, J. W., & Gohel, M. D. Anticancer effects of *Ganoderma lucidum*: A review of scientific evidence. *Nutrition and Cancer*, **2005**, *53*, 11–17.

1100. Yun, J. M., Afaq, F., Khan, N., & Mukhtar, H. Delphinidin, an anthocyanidin in pigmented fruits and veges, induces apoptosis and cell cycle arrest in human colon cancer HCT116 cells. *Molecular Carcinogenesis*, **2009**, *48*, 260–270.

1101. Zargari, A. *Iranian Medicinal Plants* (p. 421). Tehran University Press, Tehran, IR Iran, **1987**.

1102. Zeng, S., Wang, D., Cao, Y., An, N., Zeng, F., Han, C., Song, Y., & Deng, X. Immunopotentiation of caffeoyl glycoside from *Picrorhiza scrophulariiflora* on activation and cytokines secretion of immunocyte *in vitro*. *International Immunopharmacology*, **2008**, *8*, 1707–1712.

1103. Zhang, C. F., Nakamura, N., & Tewtrakul, S. Sesquiterpenes and alkaloids from *Lindera chunii* and their inhibitory activities against HIV-1 integrase. *Chemical and Pharmaceutical Bulletin*, **2002**, *50*, 1195–1200.

1104. Zhang, G. H., Wang, Q., Chen, J. J., Zhang, X. M., Tam, S. C., & Zheng, Y. T. The anti-HIV-1 effect of scutellarin. *Biochemical and Biophysical Research Communications*, **2005**, *334*, 812–816.

1105. Zhang, H. J., Tan, G. T., Hoang, V. D., Hung, N. V., Cuong, N. M., Soejarto, D. D., Pezzuto, J. M., & Fong, H. H. Natural anti HIV agents. Part IV. Anti-HIV constituents from *Vatica cinerea*. *Journal of Natural Products*, **2003**, *66*, 263–268.

1106. Zhang, J., Tang, Q., Zimmerman-Kordman, M., Reutter, W., & Fan, H. Activation of B lymphocytes by GLIS, a bioactive proteoglycan from *Ganoderma lucidum*. *Life Sciences*, **2002**, *71*, 623–638.

1107. Zhang, J., Tang, Q., Zhou, C., Jia, W., Da Silva, L., Nguyen, L. D., Reutter, W., & Fan, H. GLIS, a bioactive proteoglycan fraction from *Ganoderma lucidum*, displays anti-tumor activity by increasing both humoral and cellular immune response. *Life Sciences*, **2010**, *87*, 628–637.

1108. Zhang, L., Luo, R. H., Wang, F., & Jiang, M. Y. Highly functionalized daphnane diterpenoids from *Trigonostemon thyrsordeum*. *Organic Letters*, **2010**, *12*, 152–155.

1109. Zhang, M., Chen, H., Huang, J., Zhong, L., Zhu, C. P., & Zhang, S. H. Effect of *Lycium barbarum* polysaccharide on human hepatoma QGY7703 cells: Inhibition of proliferation and induction of apoptosis. *Life Sciences*, **2005**, *76*, 2115–2124.

1110. Zhang, S. Y., Yi, Y. H., & Tang, H. F. Bioactive triterpene glycosides from the sea cucumber *Holothuria fuscocinerea*. *Journal of Natural Products*, **2006**, *69*, 1492–1495.

1111. Zhang, W., Guo, Y. W., & Gu, Y. Secondary metabolites from the South China Sea invertebrates: Chemistry and biological activity. *Current Medicinal Chemistry*, **2006**, *13*, 2041–2090.

1112. Zhang, Y., Seeram, N. P., Lee, R., Feng, L., & Heber, D. Isolation and identification of strawberry phenolics with antioxidant and human cancer cell antiproliferative properties. *Journal of Agricultural and Food Chemistry*, **2008**, *56*, 670–675.

1113. Zhang, Y., Tang, L., & Gonzalez, V. Selected isothiocyanates rapidly induce growth inhibition of cancer cells. *Molecular Cancer Therapeutics*, **2003**, *2*, 1045–1052.

1114. Zhang, Z. F., Peng, Z. G., Gao, L., & Dong, B. Three new derivatives of anti HIV-1 polyphenols isolated from *Salvia yunnanensis*. *Journal of Asian Natural Products Research*, **2008**, *10*, 391–396.

1115. Zhou, M., Gu, L., Yeager, A. M., & Findley, H. W. Expression and regulation of Bcl-2, Bcl-xl and Bax correlate with p53 status and sensitivity to apoptosis in childhood acute lymphoblastic leukemia. *Blood*, **1997**, *89*, 2986–2993.

1116. Zhou, P., Takaishi, Y., & Duan, H. Coumarins and biocoumarins from *Ferula sumbul*: Anti HIV activity and inhibition of cytokinase release. *Phytochemistry*, **2000**, *53*, 689–697.

1117. Zhu, J., Jiang, Y., Wu, L., Lu, T., Xu, G., & Liu, X. Suppression of local inflammation contributes to the neuroprotective effect of ginsenoside Rb1 in rats with cerebral ischemia. *Neuroscience*, **2011**, *202*, 342–351.

1118. Zhu, K., Cordeiro, M. L., Atienza, J., Robinson, Jr. E. W., & Chow, S. Irreversible inhibition of human immunodeficiency virus type integrase by dicaffeoylquinic acids. *Journal of Virology*, **1999**, *73*, 3309–3316.

1119. Zibbu, G., & Batra, A. A review on chemistry and pharmacological activity of *Nerium oleander* I. *Journal of Chemical and Pharmaceutical Research*, **2010**, *2*, 351–358.

1120. Zou, Z., Yi, Y., Wu, H., Yao, L., Jiuhong, L. D., Liaw, C. C., & Lee, K. H. Intercedensides D-I, cytotoxic triterpene glycosides from the sea cucumber *Mensamaria intercedens* Lampert. *Journal of Natural Products*, **2005**, *68*, 540–546.

1121. Zubia, M., Fabre, M. S., Kerjean, V., Lann, K. L., Stiger-Pouvreau, V., Fauchon, M., & Deslandes, E. Antioxidant and antitumoral activities of some Phaeophyta from Brittany coasts. *Food Chemistry*, **2009**, *116*, 693–701.

1122. Zurier, R. B., Weissmann, G., Hoffstein, S., Kammerman, S., & Tai, H. H. Mechanisms of lysosomal enzyme release from human leukocytes, II: Effects of cAMP and cGMP, autonomic agonists, and agents, which affect microtubule function. *Journal of Clinical Investigation*, **1974**, *53*, 297–309.

APPENDICES

APPENDIX 1.1   General Bioactivities of Various Herbal Medicinal Plants

Plant Name	Treatment/Cure/Uses/Properties/Activity	References
Astragalus membranaceus	Anti-inflammatory, anti-oxidant, cardio protective, free radicals, immunotimulating, decrease peroxidation.	[484, 802]
Actaea dahurica	Cardiovascular diseases, coronary heart disease, atherosclerosis, pulmonary heart disease, thrombosis.	[1006]
Azadirachta indica	Leprosy, helminthiasis, respiratory disorders, constipation, skin infections.	[265]
Alisma plantago	Lower plasma cholesterol levels.	[376]
Allium sativum	Anti-bacterial, lower blood cholesterol levels.	[534]
Alpinia officinarum	Lower serum triglyceride, cholesterol.	[865, 866]
Alpinia oxyphylla	Hypertension, cerebrovascular disorders.	[492]
Andrographis paniculata	Upper-respiratory tract infections, common cold.	[119]
Angelica keiskei	Coronary heart disease, hypertension.	[915]
Aquilegia pubescens	Anti-inflammatory, anti-rheumatic activity.	[158, 182, 494]
Argemone mexicana	Diuretic, purgative, destroys worms, leprosy, skin-diseases, inflammations, wounds, bilious fever, ulcers, anti-helmintic, anti-bacterial, anti-malarial, anti-fungal, cough.	[23, 459]
Allium sativum	Anti-microbial, expectorant, cough, anti-septic, anti-histamine, cold, anti-fungal, asthma, anti-viral, lower blood cholesterol level, immunostimlant, anti-oxidant, anti-platelet aggregation, anti-inflammation, anti-cancer.	[333, 871]
Allium sativum	Cold, cough, asthma, immunostimulant, anti-bacterial, lower blood cholesterol, inhibit RNA synthesis.	[333, 871]
Agapanthus campanulatus	Piles, abdominal pain, tumors, spleen, asthma, rheumatism, anti-protease, anti-fungal, anti-bacterial.	[478, 481, 768]
Adhatoda vasica	Bronchodilatory, thrombopoeitic, antihistaminic, hypotensive inflammatory, uterotonic.	[175]
Artemisia herba-alba	Anti-malarial, cytotoxicity, diabetes, hypertension, antiseptic, vermifuge, enteritis, anti-helminthic, anti-malaria, antispasmodic, anti-bacterial, anti-spasmodic activity.	[476, 768, 1084]

APPENDIX 1.1 *(Continued)*

Plant Name	Treatment/Cure/Uses/Properties/Activity	References
Adhatoda vasica	Expectorant, antispasmodic, bronchodilator, anti-histaminic, menstrual disorders, uterine stimulant, eye infection, skin diseases, sore throat, bleeding, diarrhea.	[182]
Berries	Diabetes, liver, stomatitis, modulate enzyme activity, dysentery, bladder, kidney, cellular pathways, anti-inflammation, diarrhea, urinary tract, cancer, sore mouths, aphtha, nausea, sore throats, diarrhea, antioxidant, gene expression, antiatherosclerotic, cardiovascular diseases.	[456, 630, 842, 1039]
Bacopa monnieri	Antistress, promote nerve cells, action of kinase enzyme, tonic for brain, transmission of neural impulse, damaged nerve cell repair, increase memory, Improve synaptic impulse transmission.	[30]
Bauhinia variegata	Ulcers, leprosy, dysentery, tuberculosis, skin ailments, malaria, snakebite.	[508, 975]
Camellia sinensis	Cardiovascular, obesity, anti-oxidant, anti-bacterial, anti-inflammatory, anti-diabetic, anti-hypertensive activity, neuroprotection.	[131, 174, 276, 610, 726, 760]
Cassia fistula	Hypoglycemic, anti-cancer, abortifacient, anti-olic, anti-ertility, anti-microbial, estrogenic, laxative, antipyretic, anti-arthritic, purgative, analgesic, anti-viral, anti-inflammatory, anti-tussive, hepatoprotective, anti-implantation, anti-oxidant.	[77, 1024]
Cassia tora	Cardiac ataxia, constipation, anti-periodic, anti-helmintic, tonic for liver, heart, and ocular, abdomen pain, coughs, Hansen's disease, intestinal gas, digestion problem, ringworm, bronchitis.	[219]
Catharanthus roseus	Sore throat, eye irritation, stop bleeding, leukemia, diuretic, and expectorant, anti-viral, anti-bacterial, anti-microbial, anti-fungal, anti-cancer, anti-malaria activity, diabetes, astringent.	[83]
Curcumin	Anti-inflammatory, anti-mutagenic, anti-angiogenic, immunomodulatory, wound healing, neuroprotective.	[244, 504, 604, 704, 729, 895]
Emblica officinalis	Jaundice, diarrhea, anti-oxidant, anti-bacterial, anti-diabetic, anti-viral, purgative, anti-inflammatory, antimutagenic, cytoprotective, gastro protective, healing wounds, anti-tumor, hepatoprotective laxative, hypolipidemic, cough, chemo-protective, reducing fever, diarrhea, skin diseases, oral thrush, laxation, cleansing of bowels.	[8, 24, 80, 118, 425, 756, 758, 792, 891, 904]

APPENDIX 1.1 *(Continued)*

Plant Name	Treatment/Cure/Uses/Properties/Activity	References
Eucalyptus camaldulensis	Analgesic, inflammation, anti-pyretic, flu, analgesic, respiratory infections, anti-oxidant, cold, sinus congestion, anti-bacterial, anti-fungal.	[160]
Euphorbia hirta	Asthma, worm infestation, inflammation, respiratory, tract infection, diuretic, purgative, scavenging activity, sore, coughs, wound healing.	[67, 117, 393]
Emblica officinalis	Hypolipidaemic, anti-viral, hepatoprotective, anti-oxidant, anti-inflammatory, anti-diabetic, anti-clastogenic.	[246, 942]
Ficus sycomorus	Anti-diarrheal, anti-fungal, anti-tumor, anti-bacterial.	[656, 706]
F. racemosa, M. oleifera,	Bronchitis, breast tissue inflammation, skin, worm, HIV, anathema.	[444]
Ganoderma lucidum	Immuno-modulating, anti-inflammatory, anti-aging, tumoricidal.	[472, 565, 831, 979]
Ganoderma latifolia	Immuno-modulating, anti-inflammatory, anti-aging property.	[654]
Garcinia xanthochymus	Inhibit of cyclooxygenase and prostaglandin E2.	[671]
Gardenia fructus	Hemodynamic, choleretic, hepatoprotective activity.	[136, 181, 643]
Gymnema sylvestre	Pungent, reduce body heat, anti-inflammatory, bitter, tonic for digestion and liver, anti-diabetes, astringent, inflammations, hepatosplenomegaly, dyspepsia, constipation, hemorrhoids, dental, helminthiasis, cough, asthma, bronchitis, cardiopathy, jaundice, conjunctivitis, intermittent fever, piles, amenorrhea, leucoderma, urinary disorders, snakebite, anti-microbial, anti-oxidant, hepatoprotective, lipid lowering, uterine, stomachic, anti-hypercholesterolemic anti-allergic, cardiotonic, diuretic, laxative, anti-viral, immunostimulant, anti-allergic, hypoglycemic, hypolipidemic, anti-obesity, diuretic.	[10, 11, 302, 486, 664, 759, 797, 943]
Hornungia procumbens	Arthritis, back pain, neuralgia, headaches, digestion, fever, anti-inflammation	[105]
Hoslundia opposita	Sore throats, colds, venereal diseases, herpes, skin diseases, epilepsy, fever, malaria, anti-bacterial, anti-inflammation, purgative, diuretic, febrifuge, antibiotic, and antiseptic.	[32, 657]
Lycium barbarum	Diabetes, hyperlipidemia, cancer, hepatitis, hypo-immunity, thrombosis, male infertility, anti-aging.	[291, 560, 742, 776, 1010, 1109]

APPENDIX 1.1 *(Continued)*

Plant Name	Treatment/Cure/Uses/Properties/Activity	References
Mangifera indica	Anti-oxidative, anti-inflammatory, cardioprotective, anti-atherogenic.	[258, 289, 482, 642, 743, 803, 832, 851]
Murraya koenigii	Amoebiasis, swollen hemorrhoids, hepatitis, intestinal inflammation, pruritus, fresh cuts, nausea, burses, and oedema, diabetes, body pain, snakebite.	[83]
Mimusops elengi	Cardiotonic, alexipharmic, stomachic, antihelmintic, astringent, diarrhea, dysentery, constipation, headache.	[298, 407, 479]
Magnifera indica	Toothache, malaria, anti-diabetic, anti-oxidant, anti-allergic, anti-inflammatory, anti-bacterial, anti-pyretic, anti-diarrheal, atherogenic, anti-microbial activity, hypolipidemic, immunomodulation, gastro-protective, cardiovascular disorders.	[163, 709, 710, 855, 961]
Nerium oleander	Cardiotonic, diaphoretic, diuretic, anti-cancer, anti-bacterial, corns, warts, hard tumors, cancerous, and carcinoma, ulcer, aphrodisiac, chronic pain, abdomen pain, joints pain, snake-venom, anti-fungal, anti-bacterial, ophthalmia with copious lachrymation, anti-oxidant, anti-hyperglycemic.	[220, 333, 1119]
Ocimum sanctum	Anti-oxidant, scavenging, expectorant, analgesic, anti-cancer, anti-asthmatic, diaphoretic, fever, hepatoprotective, anti-fertility, hypotensive, antistress, SOD, anti-inflammatory, bronchitis, anti-diarrheal, arthritis, hypolipidemic, convulsions, anti-oxidants, cytotoxic, scavenging, anti-diabetic.	[308, 940, 983, 984, 986, 1043, 1048]
Picrorhiza kurroa	Immunomodulatory, hepatoprotective, anti-oxidant, anti-neoplastic, hypolipidemic, anti-neuropathic, neuroprotective, acute viral hepatitis.	[202, 537, 782, 875, 973, 974, 999, 1102]
Panax ginseng	Immunomodulatory, DNA damage inhibitory, anti-apoptotic, anti-obesity, inflammation, hepatoprotective, anti-hypertensive, anti-amnestic, anti-ulcer, neuroprotective.	[364, 439, 586, 687, 729, 730, 836, 1073, 1079, 1117]
Psidium guajava	Dysentery, diarrhea, upper respiratory tract infections, leucorrhea, cholera, skin diseases, external ulcers.	[670]
Punica granatum	Anti-peroxidative, hypoglycemic, anti-cancer.	[40, 127, 144, 370, 409, 447, 449, 662, 848, 990]
Polygonum bistorta	Astringent, demulcent, diuretic, febrifuge, laxative, styptic, bleeding, diarrhea, cholera, dysentery, bowel syndrome, peptic ulcers, ulcer colitis, menstruation.	[6, 9, 101, 162, 958]

APPENDIX 1.1   *(Continued)*

Plant Name	Treatment/Cure/Uses/Properties/Activity	References
Pseudotsuga macrocarpa	Impotency, hemorrhoids, diabetes mellitus, allergies, liver, lung, heart diseases, acne, kidney disorders, blood related diseases, stroke, migraine, skin ailments, diabetes, tumors, blood diseases, hypertension, anti-inflammation, cytotoxic, anti-bacterial, anti-malarial, anti-fungal, anti-viral, anti-inflammatory, fertility, gallbladder, leprosy, gonorrhea, fevers, hepatoprotective.	[25, 27, 187, 264, 341, 351, 352, 401, 473, 752, 956]
Parthenium hysterophorus	Antinociceptive, anti-inflammation.	[718]
Premna integrifolia	Fever, colic, diarrhea, dysentery, urine retention, flatulence, dyspepsia, RA.	[64, 68]
Phyllanthus emblica	Anti-oxidant.	[597]
Pueraria tuberose	Anti-aging, aphrodisiac, demulcent, lactagogue, purgative, cholagogue.	[426, 739, 764, 787]
Pongamia pinnata	Piles; rheumatic pains, hypertension, bronchitis, whooping cough, RA, skin diseases, teeth, and ulcers, inflammatory, anti-noneceptive, anti-plasmodial, anti-hyperglycemic, gonorrhea, anti-oxidant, anti-lipodoxidative, diarrheal, ulcer, anti-hyper ammonic.	[60, 143, 793, 816, 925, 926]
Panax quinquefolius	Herpes type 2 problem, psychomotor problems, psychiatric, and metabolic disorder, and immunomodulation.	[1059]
Pseuderanthemum palatiferum	Wounds, trauma, stomachache, colitis, high blood pressure, nephritis, diarrhea.	[201, 460, 698]
Randia dumetorum	Demulcent, diuretic, piles, antidysenteric, asthma, jaundice, diarrhea, emetic, gonorrhea.	[477]
Selaginella willdenowii	Wounds, fever, backache, gastric pains, urinary tracts, menstrual, skin diseases.	[258, 326, 334, 335, 462, 535, 854]
Selaginella doederleinii	Anti-cancer.	[535]
Selaginella lepidophylla	Kidney stone, gastric ulcer, rheumatism, anti-cancer, anti-oxidant, anti-microbial, anti-inflammatory, viral activity.	[214, 270, 746, 853, 976]
Saponin	Antiallergic, cytotoxic antitumor, anti-viral, immunomodulating, anti-hepatotoxic, antiphlogostic, anti-fungal.	[839]
Salacia chinensis	Thermogenic, urinary astringent, skin diseases, wounds, anti-inflammatory, diabetics, inflammations, leprosy.	[786]

APPENDIX 1.1 *(Continued)*

Plant Name	Treatment/Cure/Uses/Properties/Activity	References
Solanum xanthocarpum	Anthelmintic, cough, asthma, chest pain, flatulence, sore throat	[282, 967]
Solanum trilobatum	Anti-inflammation, anti-bacterial, anti-microbial, bronchial asthma.	[230, 250, 310, 719–721, 767, 903]
Syzygium cumini	Carminative, diuretic, digestive, antihelminthic, febrifuge, diabetes, pharyngitis, fever, constipating, stomachic, anti-bacterial, spleenopathy, urethrorrhea, ringworm, constipation, leucorrhoea, stomachalgia, gastropathy, strangury, dermopathy.	[78, 1023]
Silybum marianum	Liver, immune disorders.	[268, 916]
Tridax sp.	Anticoagulant, hair tonic, insect repellent, anti-diarrhea, dysentery.	[2, 124, 446, 834, 941]
Terminalia chebula	Anti-bacterial, anti-diabetic, renoprotective, anti-cancer, radioprotective, adaptogenic, antioxidant, antiviral	[16, 467]
Tridax procumbens	Pressure, cough, lung acute rhinitis, anti-malaria, anti-diarrhea, anti-dysentery, anti-oxidant, anti-septic, anti-hepatotoxic, anti-diabetic, anti-microbial, anti-inflammatory, anti-analgesic activity, airways disorder, neurological disorder, stomach, and headache, hair fall, immunomodulatory, gastritis, enteric fever, chest pain, wound healing, insecticidal, and parasiticidal properties.	[408, 692, 913]
Tilia argentea	Diuretic, dyspepsia, diaphoretic, antispasmodic, stomachic, cough, ingestion problems, liver, and gall bladder disorders, muscle cramp, neuromuscular problems, headache.	[70, 957]
Tripterygium wilfordii	RA, anti-inflammatory, immunosuppressive.	[934]
Thuja occidentalis	Oxi-stress, abdominal pain, small intestine obstruction, dermatitis, asthma, bleeding.	[702]
Tripterygium wilfordii	Autoimmune disease, anti-inflammation, kidney problem, arthritis, psoriasis, immunosuppression, Hansen's disease.	[157, 932, 933]
Terminalia bellirica	Cough, asthma, colic, diarrhea, dyspepsia, anemia, cancer, fever, inflammation, ulcers, rejuvenation, antioxidants, immunity, longevity, jaundice, constipation.	[115, 236, 405, 830]

APPENDIX 1.1   *(Continued)*

Plant Name	Treatment/Cure/Uses/Properties/Activity	References
Terminalia bellirica	Anti-microbial, anti-diabetic, anti-malarial, anti-parasitic, anti-viral, anti-fungal, anti-HIV activity, anti-mutagenic, hepatoprotection, lower blood pressure, myocardial necrosis, atherosclerosis.	[245, 304, 363, 525, 712, 753, 876, 978]
Terminalia chebula	Bleeding, piles, and eye disorders inhibits local anaphylaxis activity of the brain and its nerves, muscular, cataract, and clotting, immunomodulation, prevents blood clot, improves stomach functioning, maintains nutritional, imbalance, cardioprotective, anti-mutagenic, anti-oxidant, hypolipidemic, anti-cancer, anti-diabetic, anti-bacterial.	[56, 59, 207, 292, 451, 606, 611, 641, 658, 765, 824, 867]
Tamarindus indica	Alleviation of sunstroke, sore throats, wounds, paralysis, malaria, fever.	[142, 316, 952]
Uncaria tomentosa, Uncaria guianensis	Arthritis, bursitis, lupus, chronic fatigue syndrome, stomach, intestines disorders.	[272]
Urtica diocia	Alopecia, eczema, gout, urticaria, allergic rhinitis, RA, prostatic hypertrophy	[91, 260, 948]
Uncaria tomentosa	Asthma, cancer, cirrhosis, fevers, gastritis, diabetes.	[261, 350, 457]
Withania somnifera	Anti-oxidant, anti-tumor, anti-stress, anti-inflammatory, immunomodulatory, dropsy, hematopoietic, anti-aging, anxiolytic, tuberculosis, arthritis, bronchitis, disability, rheumatism, stomach, and lung inflammations, skin diseases, hypotensive, aphrodisiac, male sexual, eye diseases, bradycardiac, anti-cancer, anti-peroxidative, cardiotonic, radiosensitizing, thyro-regulatory, anti-amnestic.	[242, 249, 339, 410, 461, 488]
Xanthium strumarium	Diaphoretic, diuretic, sudorific, CNS depressant, styptic, gleet, leucorrhoea, smallpox, urinary, and renal complaints, menorrhagia, scrofulous tumors and cancer, herpes, inflammatory swellings, scrofula bladder infections.	[287]
Zingiber officinale	Giddiness, vomiting, motion disorder, autopsy, and pregnancy vomiting.	[522]
Ziziphora tenuior	Dysentery, fever, uterus infection, analgesic, gastrointestinal, vomiting, diarrhea.	[665, 667, 1101]

APPENDIX 1.2   Different Solvent Extracts of Various Parts of Herbal Plant, Enhanced Human Immune-Related Cells and Antibody Production

Plant Name	Parts	Solvents	Bioactivity	References
Allium sativum	Whole plant	NA	Mitogenic activity of lymphocytes.	[183]
Andrographis paniculata	Leaves/Aerial	Ethanol	Increased antibody production.	[500]
Azadirachta indica	Leaf	Acetone	Immunomodulation of PBMCs.	[50]
Achillea species	NA	NA	Stimulate erythrocytes, leucocytes.	[491]
Baliospermum montanum	Root	Aqueous	Phagocytosis, chemotactic, neutrophils intracellular killing potency.	[736]
Bauhinia vareigata	Stem bark	Ethanol, acetone	Enhance neutrophils, antibody.	[297, 734]
Boswellia carterii	Bark	Methylene	Immunomodulation of lymphocyte.	[55]
Calendula arvensis	Aerial	Ether	Immunomodulation of T-lymphocytes.	[45]
Inula crithmoides	Aerial	Ether	Immunomodulation of T-lymphocytes	[45]
Cinnamomum cassia	Fruit	Aqueous	Protect HIV-1 and HIV-2.	[770]
Citrus aurantifolia	Fruits	NA	Activation of mononuclear cells.	[615]
Couroupita guinensis	Flowers	Methanol	Increased antibody, neutrophils.	[762]
Glebionis coronaria	Aerial	Ether	Cytotoxicity.	[45]
Gracilaria verrucosa	Whole	Aqueous	Increased antibody.	[499]
Gymnema sylvestre	Leaf	Aqueous	Phagocytosis, neutrophils, chemotaxis.	[320, 607, 879]
Rhinacanthus nasutus	Whole	Methanol	Increased antibody.	[773]
Tanacetum parthenium	Leaves	Chloroform	Immunomodulation. Fibroblasts.	[905]
Terminalia chebula	Fruit	Aqueous	Immunomodulation.	[870]
Tripterygium wilfordii	NA	Methanol, Chloroform	Immunosuppression of T-cell proliferation.	[935]

Note: NA: Not available; PBMCs: human peripheral blood mononuclear cells; Host: Human.

APPENDIX 1.3　Various Medicinal Herbal Plants: Bioactivity for Inhibition and Control of HIV

Plant Name	Parts	Solvents	Bioactivity: HIV	References
Acer okamotoanum	Leaf	Ethyl acetate	Anti-HIV-1 integrase activity.	[468]
Ancistrocladus kor	NA	NA	Inhibits reverse transcriptase.	[102]
Agardhiella tenera	NA	NA	Inhibits HIV cytopathic effect.	[1050]
Agastache rugosa	Root	Methanol	Inhibit HIV replication.	[465]
Agrimonia pilosa	NA	Methanol	Inhibits HIV type-1.	[635]
Achyrocline satureioides	NA	NA	Irreversible inhibition of HIV-1	[1118]
Arctium lappa	NA	NA	Inhibits HIV-1 replication; blocks cell-to-cell, transmission.	[1052, 1083]
Arnebia euchroma	NA	NA	Inhibits HIV replication.	[442]
Acacia mellifera	Seed, root bark	Aqueous	Shyness of HIV-1 RT.	[813]
Azadirachta indica	NA	Aqueous	Shyness of HIV-1 RT.	[813]
Azadirachta indica	Leaf	NA	Inhibit dengue virus type-2, HIV-I.	[724]
Azadirachta indica	Leaf	Aqueous, Chloroform	Boost CD$^+$ cell count in HIV patients.	[621, 966]
Azadirachta indica	Leaf	NA	Block HIV-1 replica in C8166 CD$^+$ cells.	[49]
Albizia gummifera	Seed, root bark	Aqueous	Inhibition of HIV-1 reverse transcriptase.	[813]
Aloe barbadensis	NA	NA	Reduce symptoms associated with HIV	[622]
Aleurites moluccana	Leaf	Butanol	Anti HIV activity.	[322]
Artocarpus gomezianus	NA	NA	Anti HIV activity.	[406]
Artocarpus reticulates	NA	NA	Anti HIV activity.	[406]
Artocarpus heterophyllus	Seed	Hexane	Inhibition of HIV-1 RT activity.	[502]
Allium sativum	NA	NA	Inhibit HIV virus and replication HIV-1 RT.	[340, 502]

APPENDIX 1.3 *(Continued)*

Plant Name	Parts	Solvents	Bioactivity: HIV	References
Acorus calamus	NA	NA	Inhibitory effect against HIV-1 RT.	[502]
Asparagus officinalis	Leaf	Aqueous	Anti HIV activity, interference with the gp120/CD4 interaction.	[36]
Andrographis paniculata	Leaf	Methanol	Inhibit cell-to-cell transmission, viral replication, Inhibits HIV protease & RT.	[120, 708, 921, 1052]
Areca catechu	Bark	Methanol	Inhibition of HIV protease.	[510]
Alternanthera philoxeroides	NA	NA	Inhibitory activity against HIV.	[1052]
Agastache rugosa	Leaf	Methanol	Inhibit HIV-1 protease.	[636]
Alpinia galangal	NA	NA	Anti HIV activity.	[921]
Anisomeles indica	NA	NA	Anti HIV activity.	[921]
Ardisia japonica	NA	NA	Anti HIV activity.	[921]
Artemisia capallaris	NA	NA	Anti HIV activity.	[921]
Banksia micrantha	Leaf	Methanol	Inhibit RDDP & RNAse H Pascal activity HIV-1 RT.	[76]
Baccharis trinervis	Fruit	Aqueous, Ethanol	Block virus replica, attachment of virus cell, virus-or-cell-to-cell fusion.	[715]
Boesenbergia pandurata	NA	NA	Anti HIV activity.	[921]
Bulbine natalensis	Leaf	Aqueous, Chloroform	Anti HIV activity.	[511]
Clusia torresii	NA	NA	Anti HIV-1 activity derivatives-Clusianone, 7-epiclusianone.	[749]
Calophyllum lanigerum	NA	NA	Inhibit HIV-1 RT activity.	[41, 493, 601]
Calophyllum braziliense	Leaf	Acetone, methanol	Inhibit HIV-1 RT activity.	[129, 384, 921]

APPENDIX 1.3 *(Continued)*

Plant Name	Parts	Solvents	Bioactivity: HIV	References
Celastrus hindsii	NA	NA	Anti-HIV replication activity, Anti HIV activity.	[507, 1076]
Callophyllum cordato	NA	NA	Inhibits HIV-1 replication and cytopathic effects.	[224, 1070]
Combretum molle	NA	NA	Inhibits HIV-1 RT, Anti-HIV activity.	[75, 94]
Crataegus pinatifida	Leaf	Methanol	Inhibits HIV-1 protease.	[636]
Cinnamomum loureiroi	Seed, bark	Hexane	Inhibitory effect against HIV-1 RT.	[502]
Curcuma longa	NA	NA	Inhibition of HIV integrase.	[130]
Ceriops decandra	NA	NA	Inhibits virus adsorption	[769]
Cinnamomum aromiticum	Fruit	NA	Inhibit virus induced cytopathogenicity	[771]
Coptis chinensis	NA	NA	Inhibitory activity against HIV.	[1052]
Crataegus pinnatifida	NA	NA	Inhibits HIV-1 protease, Anti HIV activity.	[94, 636]
Charysanthemum morifolium	NA	NA	Inhibits integrase.	[542]
Castanopsis hystrix	NA	NA	HIV-RT inhibition.	[135]
Calophyllum inophyllum	Leaf	NA	Anti-HIV activity.	[527]
Calophyllum teysmannii	Leaf	NA	Anti-HIV activity.	[527]
Caranga rosea	NA	NA	Anti-HIV activity.	[1076]
Clausena excavate	NA	NA	Anti-HIV activity.	[542]
Coleus forskohlii	Aerial parts	Chloroform, ethyl acetate	Anti-HIV activity.	[94]
Coleus parvifolius	NA	NA	Anti-HIV activity.	[94]

APPENDIX 1.3 *(Continued)*

Plant Name	Parts	Solvents	Bioactivity: HIV	References
Cratoxylum arborescens	Leaf	NA	Anti-HIV activity.	[799]
Croton tiglium	Leaf	NA	Anti-HIV activity.	[799]
Capparis spinosa	NA	NA	Inhibit the replication of HSV-2.	[38]
Cordyceps militaris	NA	NA	Inhibits HIV-1 protease activities.	[417]
Cordyceps sinensis	NA	NA	Inhibitory HIV-1 replication in C8166 cells.	[317]
Detarium microcarpum	NA	NA	Anti-HIV activity, inhibit HIV-1 RT activity.	[624, 652]
Drymaria diandra	NA	NA	Anti-HIV activity.	[1032]
Dryopteris crassirhizoma	Root	NA	Ant-HIV-1 protease activity.	[543]
Dracaena loureiri	Leaf	Hexane	Inhibitory effect against HIV-1 RT.	[502]
Diospyros lotus	NA	NA	Inhibits of HIV 1 replication.	[595]
Erigeron breviscapus	NA	NA	Inhibits HIV-1 replication.	[1012, 1104]
Epicoccum nigrum	NA	NA	Inhibitory HIV-1 replication in C8166 cells.	[318]
Epimedium grandiflorum	NA	NA	Inhibitory activity against HIV.	[1052]
Eugenia jambolona	Bark	Aqueous, methanol	Inhibition of HIV protease.	[510]
Erythrina abyssinica	NA	NA	Inhibition of HIV-1 reverse transcriptase.	[813]
Erigeron breviscapus	NA	NA	Inhibits HIV-1 replication.	[1012, 1104]
Eclipta prostrate	NA	NA	Anti-HIV activity.	[543]
Epinetrum villous	NA	NA	Anti-HIV activity.	[543]
Erythrina senegalensis	NA	NA	Anti-HIV activity.	[541]
Euchresta formosana	NA	NA	Anti-HIV activity.	[585]
Eugenia caryophyllata	NA	NA	Anti-HIV activity.	[585]

APPENDIX 1.3 *(Continued)*

Plant Name	Parts	Solvents	Bioactivity: HIV	References
Euphorbia poisonii	NA	NA	Anti-HIV activity	[585]
Flammulina velutipes	Fruit	NA	Inhibits HIV-1 RT.	[1009]
Ferula sumbul	Root	Methanol	Anti-HIV activity.	[1116]
Garcinia speciosa	Seed, bark	Methanol	Inhibits HIV-1 reverse transcriptase.	[812]
Glycyrrhiza glabra	NA	NA	Inhibit giant cell formation, viral adsorption, inhibit infectivity, cytopathic activity, replication.	[402, 404]
Gossypiumher baceum	NA	NA	Inhibit viral replication.	[571]
Garcinia kingaensis	See, bark	Ethanol	Anti-HIV-1 protease activity.	[605]
Garcinia livingstoneii	NA	NA	Anti-HIV activity.	[322]
Garcinia volkensis	Seed, bark	Ethanol	Anti-HIV-1 protease activity.	[605]
Ganoderma sinense	NA	NA	Inhibition of HIV-1 protease.	[243, 838]
Garcinia semseii	Seed, bark	Ethanol	Anti-HIV activity.	[605]
Garcinia edulis	Seed, bark	Ethanol	Anti-HIV-1 protease activity.	[605]
Garcinia Ferrea	Seed, bark	Ethanol	Anti-HIV-1 protease activity.	[605]
Grewia mollis	NA	NA	Inhibition of HIV-1 RT.	[813]
Geum japonicum	Whole plant	Methanol	Inhibits HIV-1 protease.	[1068]
Garcinia hanburyi	NA	NA	Anti-HIV activity.	[800]
Gingko bilobas	NA	NA	Anti-HIV activity.	[800]
Garciniaman gostana	NA	NA	HIV-1 protease inhibition.	[155]
Glycosmis montana	NA	NA	Anti HIV activity.	[450]
Galanthus nivalis	NA	NA	Stops spread of HIV among lymphocytes GNA.	[817]

APPENDIX 1.3 *(Continued)*

Plant Name	Parts	Solvents	Bioactivity: HIV	References
Humulus lupulus	NA	NA	Inhibits HIV-1 induced cytopathic effect.	[1012, 1104]
Helictrus isora	NA	NA	Inhibition of HIV protease.	[708]
Hypoxis hemerocallidea	NA	NA	Inhibition of HIV protease.	[12, 88, 634]
Hippeastrum hybrids	NA	NA	Stops spread of HIV among lymphocytes, anti-HIV activity is found among MBLs.	[817]
Homolanthus nutans	NA	NA	Anti-HIV activity.	[322]
Hypericum perforatum	NA	NA	Inhibits HIV-1 replication.	[87, 528]
Hemsleya endecaphylla	NA	NA	Anti-HIV activity.	[403]
Illicium verum	NA	NA	Anti-HIV activity.	[886]
Inonotus obliquus	NA	NA	Inhibits HIV-1 protease.	[390]
Juglans mandshurica	Seed, bark	NA	Inhibits cytopathic activity.	[618, 638]
Jatropha curcus	Whole plant	Aqueous, methanol	Anti-HIV activity.	[618]
Kadsura angustifolia	NA	NA	Anti-HIV activity.	[290]
Kadsura longipedunculata	NA	NA	Anti-HIV activity.	[907]
Kaempferia parvifolia	Leaf	Aqueous, methanol	Anti-HIV activity.	[294]
Lentinus edodes	NA	NA	Prevents HIV-induced cytopathic effect.	[911]
Lomatium suksdorfii	NA	NA	Suppresses HIV viral 1 replication.	[1095, 1103]
Lycopodium japonicum	NA	NA	Anti-HIV activity.	[345]
Leptotrichilia sp.	NA	NA	Inhibition of HIV-1 RT.	[813]
Lawsonia inermis	NA	NA	Inhibitory effect against HIV-1 RT.	[502]
Lonicera japonica	NA	NA	Blocking of HIV RT.	[135, 1052]

APPENDIX 1.3 *(Continued)*

Plant Name	Parts	Solvents	Bioactivity: HIV	References
Lithospermum erythrorhizon	NA	NA	Blocking of HIV RT.	[1052]
Leucojum vernum	NA	NA	Anti-HIV activity.	[294]
Ligularia kanaitizensis	NA	NA	Anti-HIV activity.	[294]
Lindera chunii	Root	NA	Anti-HIV activity.	[1103]
Litsea verticillata	Root	Na	Anti-HIV activity.	[1103]
Melia azedarach	NA	NA	Blocking of HIV-1 RT.	[813]
Myrica salicifolia	NA	NA	Blocking of HIV-1 RT.	[813]
Maytenus buchnanii	NA	NA	Blocking of HIV-1 RT.	[813]
Maytenus senegalensis	NA	NA	Blocking of HIV-1 RT.	[813]
Myrothamnus flabellifolius	NA	NA	Polyphenols protect cell membranes against free radical anti-HIV RT activity.	[651]
Musa acuminate	NA	NA	Blocks HIV entry, inhibit of HIV-1 replication.	[912]
Marilalaxi flora	NA	NA	Inhibits HIV-1 RT.	[96]
Marila pluricostata	NA	NA	Anti-HIV activity.	[345]
Millettia erythrocalyx	NA	NA	Anti-HIV activity.	[345]
Momordica charantia	NA	NA	Inhibit integrase, Inhibition of HIV-1 RT.	[345, 1019]
Monotes africanus	NA	NA	Anti-HIV activity.	[627]
Murraya siamensis	NA	NA	Anti-HIV activity.	[628]
Ochna integerrimar	NA	NA	Anti-HIV activity.	[800]
Ocimum sanctum	NA	NA	Highest inhibition of recombinant HIV RT.	[36, 502]
Ocimum gratissimum	NA	NA	Reduction in HIV-1 RT activity.	[52]

APPENDIX 1.3 *(Continued)*

Plant Name	Parts	Solvents	Bioactivity: HIV	References
Polyalthia suberosa	Seed, leaf	Ethanol	Anti-HIV replication activity.	[556]
Peltophorum africanum	Seed, bark	Ethyl acetate	Inhibits HIV-1 RT, Anti-HIV activity.	[75, 944]
Phytolacca americana	Root bark	NA	Broad spectrum microbicide.	[211]
Pipturus albidus	NA	NA	Inhibit HIV growth.	[588, 625]
Prunus africana	NA	NA	Inhibition of HIV-1 reverse transcriptase.	[813]
Portula caoleracea	NA	NA	Interfere with viral growth.	[860]
Papaver somniferum	NA	NA	Interfere with HIV protein gp 120.	[964]
Psidium guajava	NA	NA	Inhibit HIV RT.	[431]
Phyllanthus emblica	Fruit	Methanol	Inhibit HIV RT.	[247]
Peltophorum africanum	NA	NA	Blocking of HIV-1 RT and binding of GP120.	[75, 277, 748, 951]
Prunella vulgaris	NA	NA	Inhibitory activity against HIV.	[137, 685]
Pericampylus glaucas	NA	NA	Anti-HIV activity.	[1075]
Phenax angustifolius	NA	NA	Anti-HIV activity.	[624, 750]
Phyllanthus amarus	NA	NA	Anti-HIV activity.	
Phyllanthus myritifolius	NA	NA	Anti-HIV activity.	
Pragnos tschimganica	NA	NA	Anti-HIV activity.	[864]
Plumbago indica	NA	NA	Inhibitory effect against HIV-1 RT.	[502]
Psiadia dentate	NA	NA	Anti-HIV activity.	[864]
Psychotria leptothyrsa	NA	NA	Anti-HIV activity.	[864]
Quercus coccifera	Leaf	NA	Inhibits HIV-1 protease.	[880]
Quercus myrsinifolia	NA	NA	Arrest HIV cell growth.	[995]

APPENDIX 1.3 *(Continued)*

Plant Name	Parts	Solvents	Bioactivity: HIV	References
Quercus infectoria	NA	NA	Arrest HIV cell growth.	[995]
Quercus pedunculata	NA	NA	Arrest HIV cell growth.	[995]
Rhus succedanea	NA	NA	Inhibits HIV-1 reverse transcriptase.	[573]
Rhamnus staddo	NA	NA	Inhibition of HIV-1 reverse transcriptase.	[813]
Rhus chinensis	NA	NA	Anti-HIV activity.	[313]
Ridolfia segetum	NA	NA	Anti-HIV activity.	[84]
Rosa damascene	NA	NA	Anti-HIV activity.	[84]
Rosa woodsii	Leaf	NA	Anti-HIV activity.	[443]
Rhizophora mucronata	NA	NA	Inhibits of RNA dependent DNA polymerase activity.	[36]
Russula paludosa	NA	NA	Inhibits HIV-1.	[708]
Scutellariai baicalensi	NA	NA	Anti-HIV RT activity, inhibit infectivity & replication.	[195, 267, 480, 493, 515, 554, 757, 1029]
Senecio scandens	NA	NA	Inhibitory activity against HIV.	[1052]
Sapium indicum	NA	NA	Inhibitory effect against HIV-1 RT.	[502]
Sideritis akmanii	NA	NA	Anti-HIV replication.	[110]
Sutherlandia frutescens	NA	NA	Anti-viral activity Treat HIV infection.	[204, 295, 312, 707]
Swertia franchetiana	NA	NA	Inhibits HIV-1 RT.	[1011]
Symphonia globulifera	NA	NA	Inhibits cytopathic effects of HIV.	[322]
Salvia miltiorrhyza	NA	NA	Anti-HIV activity.	[443]

APPENDIX 1.3 *(Continued)*

Plant Name	Parts	Solvents	Bioactivity: HIV	References
Salvia officinalis	NA	NA	Anti-HIV activity.	[57]
Salvia yunnanensis	Roct	Aqueous	Anti-HIV activity.	[1114]
Saururus chinenisis	NA	NA	Anti-HIV activity.	[540]
Schisandra lancifolia	Leaf, seed	NA	Anti-HIV activity.	[1066]
Schisandra propinqua	Arial part	NA	Anti-HIV activity.	[561]
Schisandra rubiflora	NA	NA	Anti-HIV activity.	[1065]
Schisandra sphenanthera	NA	NA	Anti-HIV activity.	[1065]
Schisandra wilsoniana	Fruit	NA	Anti-HIV activity.	[1077]
Stauntonia obovatifoliola	Seed	NA	Anti-HIV activity.	[1028]
Stephania cepharantha	Root	Methanol	Inhibits HIV replication, Anti-HIV activity.	[600]
Strychnos vanprukii	NA	NA	Anti-HIV activity.	[600]
Thalassia testudunium	NA	NA	Anti-HIV activity.	[810]
Tieghemella heckelii	Seed	NA	Anti-HIV activity.	[1108]
Trigonostemon thyrsoideum	Seed	NA	Anti-HIV activity.	[1108]
Trigonos temonlii	NA	NA	Anti-HIV activity.	[923]
Tripterigiumhy poglaucam	NA	NA	Anti-HIV activity.	[232]
Tripterigium wilfordii	NA	NA	Inhibits HIV replication, Anti-HIV activity.	[232, 366]
Terminalia chebula	NA	NA	Inhibits HIV integrase RT.	[17, 860]
Toddalia asiatica	NA	NA	Inhibition of HIV-1 reverse transcriptase.	[813]
Terminalia sericea	NA	NA	Inhibition against HIV-1 RT RDDP and RNase function.	[76]

APPENDIX 1.3 *(Continued)*

Plant Name	Parts	Solvents	Bioactivity: HIV	References
Tinospora cordifolia	NA	NA	Inhibition of gp 120 binding.	[36]
Terminalia bellirica	Fruit	NA	Inhibit HIV-1 activity.	[892, 977]
Terminalia arjuna	NA	NA	Inhibition of HIV protease.	[510]
Terminalia bellerica	NA	NA	Inhibit HIV RT, viral punicalagin, punicacortin adsorption.	[247]
Terminalia. horrida	NA	NA	Inhibit HIV RT.	[247]
Vatica astrotricha	NA	NA	Inhibits HIV-1 entry replication.	[727]
Vitex negundo	NA	NA	Inhibitory effect against HIV-1 RT.	[502]
Viola yedoensis	NA	NA	*In vitro* inhibitory activity against HIV.	[138, 684]
Vatica cinerea	Leaf, seed	NA	Inhibits replication.	[1105]
Woodwardia unigemmata	NA	NA	*In vitro* inhibitory activity against HIV.	[1052]
Warburgia ugandensis	NA	NA	Inhibition of HIV-1 reverse transcriptase.	[813]
White birch tree	NA	NA	Arrest HIV-1 Gag precursor maturation.	[441, 555, 1095]
Xanthoxylum chalybeum	NA	NA	Inhibition of HIV-1 reverse transcriptase.	[1095]
Triphala	NA	Multi-herbal	Improvement in HIV/AIDS positive people.	[747]
Chinese herbal	—		Inhibits HIV-1 gp120	[186]

Note: NA: not available; **Host:** Human.

APPENDIX 1.4 Anti-Cancer Bioactivity of Herbal Plant Extracts

Plant Name	Parts	Solvents	Cell Line Name	Anti-Cancer Bioactivity	References
Astragalus membranaceus	Root	NA	Myeloid (MBL-2)	Anti-tumor immune mechanism.	[1046]
Achillea millefolium	NA	Hexane, chloroform, Methanol	Cervical (HeLa), MCF-7, A431	Anti-proliferative activity.	[198]
Acorus tatarinowii	Crude	Ethanol-water	Hepatocellular (HepG2)	Cytotoxicity and apoptosis induction.	[602]
Adina rubella	NA	NA	NA	Anti-cancer activity.	[378]
Aegle marmelos	NA	Methanol	Erythroleukemic (HEL), Leukemic (K562), breast (MCF7, MDA-MB-231), melanoma (Colo38)	Anti-proliferative activity.	[517, 518]
Allium sativum	NA	NA	Human cancer cell lines	Anti-cancer activity.	[612]
Alpinia officinarum	NA	NA	Lung (COR L23), breast (MCF7)	Cytotoxic activity.	[354, 531, 532]
Alstonia macrophylla	Root	Methanol	Carcinoma (COR-L23, MOR-P)	Cytotoxic activity.	[454, 455]
Alstonia scholaris	Bark	Petroleum ether, Hydroalcoholic	Cervix (ME180, SiHa); leukemia (HL60, K562); lung (A549); breast (MCF7, MDA-MB-468); prostate (PC3, DU145); hepatoma (HEP G2), colon (HT29, Colo205); ovarian A2780, Ovkar-3); Oral (AW13516)	Anti-cancer activity.	[733]
Ampelopsis brevipedunculata	NA	NA	Human cell lines	Anticancer activity.	[378]
Amomum villosum	Curde	Ethanol-water	Hepatocellular (HepG2)	Cytotoxic activity.	[602]

APPENDIX 1.4 *(Continued)*

Plant Name	Parts	Solvents	Cell Line Name	Anti-Cancer Bioactivity	References
Amoora rohituka	Stem bark	NA	Breast (MDA-468; adenocarcinoma (MCF-7)	Growth arrest, apoptosis.	[778]
Angelica sinensis	NA	Acetone	Brain (DBTRG-05MG); colorectal (HT29); lung (A549); liver (J5)	Induction of apoptosis.	[161]
Angelica sinensis	Root	Aqueous	Cervical (HeLa)	Induction of apoptosis.	[125]
Anemarrhena asphodeloides	Crude	Aqueous	Breast, lung, pancreas, prostate, gastric (MKN45, KATO-III)	Anti-tumor activity, induction of apoptosis, cell death increases in caspase-3-like activity.	[873, 919]
Apple peels	Peel, pulp,	NA	Liver (HepG2); breast (MCF-7).	Inhibit the cell proliferation.	[170]
Argemone mexicana	Leaf	Methanol	Cervical (HeLa), breast (Mcf-75).	Anti-cancer activity.	[284]
Argemone mexicana	Seed	Ethanol	Liver (Hep-2, lung (A-549), neuroblastoma (IMR-32), colon (502713 HT-29).	Cytotoxic activity.	[733, 994]
Antrodia camphorata	Fruiting bodies	NA	Skin fibroblast (HS68); colon (HT-29, HCT-116, SW-480); mammary epithelial (MCF10A); breast (MDA-MB-231).	Cytotoxic activity.	[1086]
Brucea javanica	Fruit	Aqueous	Non-small cell lung (A549); hepatocellular(Hep3B); breast cancer (MDA-MB231); oesophageal (SLMT-1).	Anti-cancer activity, caspase 3 activations.	[526]
Calotropis procera	Root bark	NA	SF295 and MDA-MB-435.	Cytotoxicity, effects on viability/cell morphology.	[881]
Cannabis sativa	NA	NA	Breast, brain.	Inhibit the growth.	[286]

APPENDIX 1.4 *(Continued)*

Plant Name	Parts	Solvents	Cell Line Name	Anti-Cancer Bioactivity	References
Calotropis procera	Whole plant	Ethanol	Liver (Hep-2); neuroblastoma (IMR-32); lung (A-549); colon (502713, HT-29)	Cytotoxic activity.	[837]
Catimbium speciosum	Crude	Ethanol-water	Hepatocellular (HepG2)	Cytotoxic activity.	[602]
Centella asiatica	Leaf	Aqueous	Breast (MDA MB-231.	Cytotoxic activity.	[274]
Camellia reticulata	Fruit	NA	Gastric (SNU-668).	Anti-tumor activity, chromatin condensation, apoptotic body formation, reduced BCL-2 expression; increase BAX CASP-3, caspase-3 activity.	[471]
Citrus natsudaidai	NA	NA	Lung cancer cells.	Anti-tumor activity.	[429]
Cladogynos orientalis	Crude	Ethanol-water	Hepatocellular (HepG2).	Cytotoxic activity.	[602]
Coconut peel	Pulp, fruit	Hexane, methanol	KB cells.	Anti-oxidant activity, cytotoxic activity.	[463]
Cortex mori	Root bark	Aqueous	Leukemia (K562, B380).	Cytotoxic activity, apoptosis.	[672]
Cynometra cauliflora	Whole plant	Methanol	Promyelocytic (HL-60).	Cytotoxic activity.	[917]
Cynara scolymus	Edible part	Petroleum ether	Hepatoma (HepG2)	Induced apoptosis, activation of caspase-3 activity.	[631]

APPENDIX 1.4 *(Continued)*

Plant Name	Parts	Solvents	Cell Line Name	Anti-Cancer Bioactivity	References
Diospyros kaki	Pulp	Methanol	Squamous (HSC-2), submandibular gland (HSG).	Cytotoxic activity	[452]
Dillenia suffruticosa	NA	NA	Breast (MCF-7, MDA-MB-231).	Inhibits proliferation via induction of G2/M arrest.	[39]
Euchresta formosana	Crude	NA	Hepatocellular (HCC, Hep3B).	Induce apoptosis, overexpression of ROS, GADD153, Bax, and caspase-3, decline of RhoA, ROCK1, FAK, matrix metalloproteinase-1, -2, -9, -10.	[371]
Ebenus boissieri	Rot, Arial part	Hexane	Breast (293T, MDA-MB231).	Induced apoptosis, caspase-3, -9, increased TNF-α, IFN-γ.	[395, 396]
Emblica officinalis	NA	NA	Colon (HCCSC).	Suppresses cell proliferation, induces apoptosis via suppression of c-Myc, cyclin D1	[788]
Euphorbia hirta	NA	NA	Hep-G2.	Anti-proliferative activity.	[106]
Flueggea leucopyrus	NA	NA	Endometrial (AN3CA).	DNA fragmentation, nuclear condensation, enhanced caspase 3, 9.	[827]
Flammulina velutipes	NA	NA	HepG2, MCF-7, SGC7901, A549.	Anti-oxidant activity	1018

APPENDIX 1.4 *(Continued)*

Plant Name	Parts	Solvents	Cell Line Name	Anti-Cancer Bioactivity	References
Fomitopsis nigra	NA	NA	Oral squamous (YD-10B).	Induced apoptosis via the ROS-dependent mitochondrial dysfunction, increased expression ratio of Bax/Bcl-2.	[81, 539]
Ganoderma lucidum	NA	NA	Cervical (HeLa).	Inhibit the growth	[159, 1069]
Ganoderma lucidum	Fruit	Ethanol	Breast (MCF-7, MDA-MB-231).	DNA damage, impairment of G1 cell cycle.	[1053]
Ganoderma lucidum	NA	NA	Breast (MCF-7)	Apoptosis, enhance pro-apoptotic Bax, p21/Waf1, suppress cyclin D1, cell cycle arrest.	[374]
Glochidion daltonii	Crude	Ethanol-water	Hepatocellular (HepG2)	Cytotoxic activity.	[602]
Grapeseed	NA	NA	Prostate, breast (MDA-MB468), colon (CaCo2 cells)	G1 cell cycle and cell-growth arrest, apoptosis, MAPK activation.	[7, 253]
Heliotropium indicum	NA	NA	Breast (SKBR3).	Anti-proliferative activity.	[649]
Hydrangea angustipetala	Leaf	Aqueous	Gastric (AGS, SNU-1).	Decreases in PARP and pro-caspase 3.	[369]
Lawsonia inermis	Leaf	Chloroform	Liver (HepG2); breast (MCF-7; MDA-MB-231).	Cytotoxic activity.	[252]

APPENDIX 1.4 *(Continued)*

Plant Name	Parts	Solvents	Cell Line Name	Anti-Cancer Bioactivity	References
Melissa officinalis	NA	NA	Colon (HCT-116).	Anti-proliferative, cytotoxic activity.	[251]
Morus alba	Root bark	NA	Colorectal (SW480); leukemia (HL60).	Induced cell growth arrest, reduced procaspases-3, 8, 9 levels by activating ATF3 expression	[255, 464]
Moringa oleifera	Seed	Methanol	Liver (Hep-2); colon (HT-29); neuroblastoma (IMR-32); lung (A-549).	Cytotoxic activity.	[640]
Ochradenus baccatus	Arial part	Aqueous	Hepatocellular (HepG2).	Antiproliferative Effect.	[947]
Origanum dayi	Arial part	Aqueous	Hepatocellular (HepG2).	Anti-proliferative effect.	[947]
Parthenium hysterophorus	NA	NA	Lung (A-549); breast (MCF-7; CNS SF-295); ovary (IGROVI); prostate (PC-3).	Cytotoxic activity.	[336]
Pseudotsuga macrocarpa	NA	NA	Breast (MDA-MB-231; MCF-7).	Anti-cancer & anti-proliferative, induction of apoptosis.	[955]
Psychotria macrocarpa	Crude	NA	Hepatoma (HepG2).	Cytotoxic, antioxidant activity.	[1092]
Pinus kesiya	Crude	Ethanol-water	Hepatocellular (HepG2).	Cytotoxic activity.	[602]
Phellinus igniarius	NA	NA	Lung (A549); liver (Bel7402)	Cytotoxic activity.	[645]
Phlomis platystegia	Arial part	Aqueous	Hepatocellular (HepG2)	Anti-proliferative effect.	[947]

APPENDIX 1.4 *(Continued)*

Plant Name	Parts	Solvents	Cell Line Name	Anti-Cancer Bioactivity	References
Phyllanthus amarus	NA	NA	Lung (A549); breast (MCF-7), hepatocellular (HepG2, Huh-7); Prostate (PC-3); melanoma (MeWo).	Induction of apoptosis, cell cycle via caspases activity.	[546, 930]
Phyllanthus niruri	NA	NA	Lung; leukemia; Caucasian lung large cell melanoma (MeWo); prostate (PC-3); lung (A549); carcinoma (COR-L23); T-lymphoblastic (MOLT-4); breast (MCF-7; hepatocellular HepG2; Huh-7).	Induction of apoptosis, cell cycle via caspases activity.	[212, 775, 798, 930]
Phyllanthus urinaria	NA	NA	Melanoma (MeWo); lung (A549); hepatocellular (HepG2; Huh-7); prostate (PC-3); breast (MCF-7).	Induction of apoptosis, cell cycle via caspases activity.	[546, 930]
Phyllanthus watsonii	NA	NA	Melanoma (MeWo); lung (A549); hepatocellular (HepG2, Huh-7); prostate (PC-3); breast (MCF-7).	Induction of apoptosis, cell cycle via caspases activity.	[546, 930]
Phyllanthus emblica	Fruit, leaf, seed	Methanol	Lung (A549); liver (HeGP2); cervix (Hela); breast (MDA-MB-231); ovarian (SKOV3); colorectal (HCT116); colorectal (SW620).	DNA fragmentation, increased caspase-3/7, -8 activity, enhancement of *Fas* protein.	[62, 214, 241, 683]
Polyherbal drug	NA	NA	Mucinous (MGC-803); cachectic (LAMK1).	Inhibit cell proliferation via induction of apoptosis.	[576, 584, 777]
Psidium cathianum	NA	NA	Gastric (SNU-16)	Inhibits the proliferation via induction of apoptosis.	[650]
Psidium guajava	Leaf	Hexane	Prostate (PC-3).	Anticancer activity by suppressing AKT; Inhibits AKT/mTOR/S6K1	[815]

APPENDIX 1.4 *(Continued)*

Plant Name	Parts	Solvents	Cell Line Name	Anti-Cancer Bioactivity	References
				signaling pathway; induced apoptosis.	
Pulsatilla chinensis	Root	NA	Lung (P-388).	Cytotoxic activity.	[1085]
Rubia cordifolia	Root	Methanol	Carcinoma (HT-29); breast (MCF-7); liver (HepG2).	Inhibit the growth.	[884]
Streptocaulon juventas	Root	Methanol	Fibrosarcoma (HT-1080).	Anti-proliferative activity.	[968]
Salvia miltiorrhiza	NA	NA	Breast cancer (MDA-MB-231).	Expression of ICAM-1, VCAM-1, TNF-α.	[696]
Sarris cernuss	Whole plant	Methanol	Colon (CL-187); Breast (MDA-MB-231).	Cytotoxic activity.	[54]
Scutellaria barbata	NA	NA	Lung (A549).	DNA damage, arrest cell cycle, cytotoxic activity.	[353, 809, 1088]
Smilax kraussiana	Whole plant	Ethanol	Cervical (HeLa, SiHa); fibroblast (CCL-110).	Cytotoxic activity.	[794]
Strobilanrhes crispus	Leaf	Chloroform	Liver (HepG2); breast (MCF-7; MDA-MB-231).	Cytotoxic activity.	[252]
Tomato (*Solanum lycopersicum*)	Peel	NA	Intestinal cells (Caco-2).	Anti-cancer activity.	[796]
Triphala	Polyherbal	NA	Breast (MCF-7).	Cytotoxic effects.	[829]
Trigonella foenum	Whole plant	Ethanol	Liver (Hep-2), colon (502713, HT-29); neuroblastoma (IMR-32); lung (A-549).	Cytotoxic activity.	[837]

APPENDIX 1.4 *(Continued)*

Plant Name	Parts	Solvents	Cell Line Name	Anti-Cancer Bioactivity	References
Varthemia iphionoides	Arial part	Aqueous	Hepatocellular (HepG2).	Anti-proliferative effect.	[947]
Vaccinium stamineum	Fruit	NA	Lung (A549): leukemia (HL-60).	Proliferation, induction of apoptotic, inhibition of UVB-induced AP-1 and NF-kappaB activity	[1014]
Withania somnifera	Leaf	NA	Colon (HCT-116); CNS (SF-268); lung (NCI-H460), breast (CNS, MCF-7).	Anti-proliferative.	[412]
Withania somnifera	Root, leaf, seed	Ethanol	Colon (HCT-15); neuroblastoma (IMR-32); prostate (PC-3, DU-145); lung (A-549).	Cytotoxicity activity.	[1071]

Note: NA: not available; **Host:** Human.

APPENDIX 1.5 Anti-Cancer Mechanisms of Herbal Bioactive Compounds

Compound Name	Cell Line Name	Functions/Bioactivity/Properties/Mechanisms	References
5-Fluorouracil	Colon	Bad, Bcl-X(L), Bcl-2, Bax, Bak, p53 protein induced expression.	[695]
6-gingerol promyelocytic	Leukemic (HL-60)	Induced cell death, DNA fragmentation, Inhibits Bcl-2.	[1007]
6-gingerol, 6-paradol	Promyelocytic (HL-60)	Suppressed proliferation through the induction of apoptosis.	[533]
7, 8-Dihydroxyflavanone	Lung (A549), leukemia (K562).	Jun-Fos-DNA complex formation, cytotoxic effect.	[325]
10-paradol, 6-dehydroparadol, 6-paradol	Oral squamous (KB).	Induced apoptosis, caused proteolytic cleavage of pro-caspase-3.	[458]
Alisol B acetate (triterpene)	Prostate (PC-3)	Induced apoptotic, loss of $\Delta\Psi m$, activation caspase-8, -9, executor caspase-3.	[381]
Alisol B acetate/triterpene	Lymphoblastic (CEM cells)	Decreased cell viability. $\Delta\Psi m$, increased ratio of Bax/Bcl-2.	[148]
Alisol B acetate	Prostate (PC-3)	Induces Bax translocation, apoptosis via Bcl-2 pathways.	[1051]
Allitridi	Gastric (BGC823)	Down regulate cyclin D1, upregulate p27Kip1 protein level.	[521]
Allitridi	Gastric (BGC823).	Down-regulation of Bcl-2 and caspase-3 activity, DNA dissolution.	[520]
Allyl isothiocyanate (AITC)	Prostate (PC-3, LNCaP).	Reduce Bcl-2 expression, control G2/M, cyclin B1, cell division cycle (Cdc)-25B and Cdc25C).	[896]
Alkaloids/schischkinnin, montamine	Colon (Caco-2)	Anti-cancer activity.	[872]

APPENDIX 1.5 *(Continued)*

Compound Name	Cell Line Name	Functions/Bioactivity/Properties/Mechanisms	References
Alkaloid	Leukemia (Jurkat J6).	Inhibit the growth.	[105, 218]
Alkaloids	Promyelotic HL-60.	Inhibition of protein biosynthesis, microtubule formation, induces apoptosis.	[805]
Aloe-emodin	Lung (CH27).	Induced apoptosis, DNA fragmentation, Bcl-X(L), Bag-1, Bak expression; translocation of Bak, Bax, activated caspase-3, -8, -9	[538]
Aloe-emodin	Neuroblastoma (IMR-5, IMR-32, SJ-N-KP, AF8).	Induced apoptosis, DNA fragmentation	[1087]
Aloe-emodin	Liver (Hep G2, Hep 3B).	Inhibits cell proliferation, induced apoptosis, p53, and p21 expression, increase in Fas/APO1 receptor and Bax expression.	[506]
Amino-polysaccharide (G009)	Promyelocytic (HL-60).	Decrease DNA strand breaks.	[538]
Angelicin	Neuroblastoma (SH-SY5Y).	DNA fragmentation, increased cellular cytotoxicity, down-regulate Bcl-xL, Mcl-1, Bcl-2, induced caspase-8, and MAP kinases, PI3K/AKT/GSK-3β activity, up-regulation of caspase-9, -3 activity.	[780]
Anthocyanins T	Tumor cells.	Induction of apoptosis	[368]
Anthocyanins.	Leukemia (HL60), colon (HCT116).	DNA fragmentation, induced apoptotic cell bodies	[448]
Anthocyanin	Colon, breast, lung, gastric.	Inhibit cancer cell growth.	[100]
Anthraquinones	Breast, CNS, colon, lung.	Cytotoxic effects.	[180]
Anthraquinone	HT-29.	Cytotoxicity effects.	[885]
Anthraquinones) chrysophanol, rhein, physcion, aloe-emodin)	Breast (MCF-7; MDA-MB-231).	Cytotoxicity effects.	[437]

APPENDIX 1.5 *(Continued)*

Compound Name	Cell Line Name	Functions/Bioactivity/Properties/Mechanisms	References
Anthraquinones (diketopiperazines, isocoumarins, ihydrobisfuran)	Promyelotic (HL-60)	Induction apoptosis and DNA fragmentation.	[969]
Apigenin	Hapatoma (Hep G2, Hep 3B, PLC/PRF/5).	Induce apoptosis, increased p53 accumulation, enhanced p21/WAF1 level through the p53-dependent pathway, induction of p21 expression, G2/M phase arrest.	[167]
Apigenin	Lung (A549).	Inhibits VEGF transcriptional, decreased HIF-1α, activation of AKT, and p70S6K1.	[579]
Apigenin	Colon (SW480; HT-29; Caco-2).	G2/M arrest, block p34(cdc2) and cyclin B1 protein.	[1016]
Astaxanthin	Colon cancer.	Cell proliferation.	[761]
Baicalein, a flavonoid	HL-60.	Induction apoptosis, DNA fragmentation.	[811]
Baicalein, geniposide, alisol B acetate, saikosaponin-d	Hepatoma (Hep3B).	DNA fragmentation assays, induced G2/M arrest, H_2O_2 generation, NF-kB, increased caspase-3 activity.	[176]
Berberine	Oral (HSC-3).	Induces apoptosis, ROS, Ca^{2+} production, inhibition cell growth in G0/G1-phase, cellular DNA synthesis, activation caspase-3, loss of ΔΨm.	[567]
Berberine	Tongue (SCC-4).	Induced apoptosis, caspase-8, -9, -3 activities, loss of ΔΨm, reduction Bax/Bcl-2 ratio.	[361]
Betulinic acid	Betulinic acid.	Induce cell death, activation of p38, MAPKs, ROS.	[924]
Betulinic acid	Melanoma cell lines.	Cytotoxic activity.	[61, 755]
Bryostatin	Acute lymphoblastic leukemia (ALL).	Increased 26S proteolytic activity, Bcl-2 protein.	[1004]

APPENDIX 1.5 *(Continued)*

Compound Name	Cell Line Name	Functions/Bioactivity/Properties/Mechanisms	References
Caffeic acid/hydroxycinnamic acid	Breast (MCF-7).	Inhibit NFκB, induced apoptosis, p53-regulated Bax, activated Fas-L, caspases, JNK. SB203580, p38 MAPK.	[1026]
Calotropine/glycoside	Epidermoid.	Anti-tumor activity.	[1045]
Capsaicin	Colon (colo 205).	Increased ROS, Ca^{2+}, decreased $\Delta\Psi m$, increase Fas, cytochrome c, Bax, activations of caspase-8, -9, -3.	[592]
C-Phycocyanin (C-PC)	Myeloid (K562).	Induced apoptosis, cell shrinkage, membrane blebbing, nuclear condensation, down-regulation of Bcl-2.	[901]
Chrysophanol	Liver (J5).	Induces necrosis through the production of ROS.	[591]
Chrysophanol	Liver (Hep3B).	Induction of apoptosis, morphological changes; S phase arrest, DNA damage, ROS, Ca^{2+}; decreased $\Delta\Psi m$ and ATP levels, regulators Bax and Bcl-2, caspase-8, -9.	[686]
Curcumin	Nasopharyngeal (NPC-TW 076).	G2/M phase arrest, decrease cyclin A, B, Cdk1, up-regulate Bax, down-regulation of Bcl-2, activation caspase-9, -3.	[503]
Curcumin	Lung (NCI-H460).	Induces morphologic, up-regulate of BAX, BAD, DNA damage down-regulate of BCL-2, BCL-X(L), XIAP, CDK1; increased ROS, Ca2+, ER stress, loss of $\Delta\Psi m$, activation of caspase-3 through the FAS/caspase-8, growth arrest.	[1057]
Curcumin	Hepatoma (G2).	Induced cell death, cytochrome c and caspase-3 activation, cleavage of PARP, ROS generation, stimulated intracellular oxidative stress	[132]

APPENDIX 1.5 *(Continued)*

Compound Name	Cell Line Name	Functions/Bioactivity/Properties/Mechanisms	References
Curcumin	Fibrosarcoma cells.	Decrease MMP-2, -9; MT1-MMP, uPA expression, block TIMP-2.	[1089]
Curcumin	Tumor cells (AK-5).	Inducing apoptosis, activation of ROS, loss of ΔΨm, cytochrome c release.	[82]
Curcumin	Leukemic (K562).	Decrease WT1 protein and WT1 mRNA.	[34]
Curcumin	Colon (HCT-116).	p21-mediated cell cycle arrest, reduce pro-caspase-3, and PARP-1 cleavage.	[1027]
Curcumin	Cervical (HeLa).	Apoptotic cell death, arrest cell cycle, DNA disintegration, reduce ROS, Ca²⁺, Bcl-2, Bcl-xL; increase ΔΨm, Bax.	[178]
Curcumin	Burkitt's lymphoma (CA46).	DNA fragmentation, down-regulate of c-myc, bcl-2, mutant-type p53, up-regulate of Fas expression.	[1058]
Curcumin	Head & neck squamous.	Induced apoptosis.	[343, 1044]
Gypenosides	Tongue (SCC4).	Down-regulation of NF-kB, COX2, ERK1/2, MMP-9, matrix metalloproteinase-9, ΔΨm.	[594]
Gypenosides	Tongue (SCC4).	Apoptosis via ER stress, DNA disintegration, arrest G0/G1 via CHK2, up-regulate sub-G1, ROS, Ca²⁺, ΔΨm; down-regulate RAS, NF-kB.	[151]
Gypenosides (Gyp)	Hepatoma (Huh-7, Hep3B, HA22T).	Morphological changes, up-regulate Bax, Bak, Bcl-X(L), down-regulate Bcl-2, Bad, activation of caspase-1, -9, -3	[1013]
Gypenosides (Gyp)	Cervix (Ca Ski).	Inhibits NAT mRNA expression.	[172]
Gypenosides (Gyp)	Colon (colo 205).	DNA scrap, enhance ROS, Ca2+, ΔΨm production, down-regulation of Bcl-2, Bcl-xl; up-regulate Bax; activate caspase-3 activity.	[150]

APPENDIX 1.5 *(Continued)*

Compound Name	Cell Line Name	Functions/Bioactivity/Properties/Mechanisms	References
Cyanidin 3-glucoside	HS578T	Apoptosis induction, arrest G2/M, chromatin abbreviation, decreased CDK-1, -2, 3-glucoside, cyclin B1, E expression, activate caspase-3.	[153]
Glycosides, cervicosides, prostanoid sclaviridenones	Cancer cell lines.	Anti-tumor activity.	[1110]
Danthron	Gastric (SNU-1).	Enhance Bax-triggered pathways, DNA damage, caspase-3, -8, -9 activity, ROS, mitochondrial interruption, Bcl-2.	[165]
Delphinidin/anthocyanidin	Colon (HCT116).	Enhance apoptosis, PARP cleavage, Bax, caspases-3, -8, -9 activity, down-regulate NF-κB/p65 at Ser(536), Bcl-2, protein, IKKalpha.	[1100]
Demethoxycurcumin (DMC).	Lung (NCI-H460).	Enhance apoptosis, ROS, Ca^{2+}, caspase-3, -8, -9 activity; down-regulate ΔΨm, ER stress proteins (GADD153, GRP78, ATF-6α, ATF-6β, caspase-4, IRE1β).	[485]
Demethoxycurcumin (DMC)/curcumin	Colon (HCT116).	Inhibiting proliferation and inducing apoptosis	[922]
Demethoxycurcumin (DMC)/curcumin	Renal.	Enhanced apoptosis, supper ROS synthesis, caspase-3 activity, Cyt c release.	[544]
Demethoxycurcumin (DMC)/curcumin	Glioma U87.	Activation of Bcl-2 mediated G2 checkpoint, induction of G2/M arrest and apoptosis.	[599]
Demethoxycurcumin (DMC)/curcumin Bisdemethoxycurcumin	Leukemic (K562, HL60, U937, Molt4).	Decreased WT1 mRNA and protein level.	[35]
Diallyl sulfide (DAS), diallyl disulfide	Colon colo 205 humans.	Decrease MEKK3, MKK7JNK1/2, PI3K, Ras, p38, ERK1/2, COX-2, NF-κB; Enhance apoptosis.	[512]

APPENDIX 1.5 *(Continued)*

Compound Name	Cell Line Name	Functions/Bioactivity/Properties/Mechanisms	References
Diallyl trisulfide (DATS), Diallyl disulfide (DADS)	Lung (A549, MRC-5).	Increase DNA fragmentation in intracellular Ca^{2+}	[822]
Diallyl trisulfide (DATS)	Gastric (BGC823).	Induce apoptosis.	[559]
Diallyl trisulfide (DATS)	Colon (HCT-15, DLD-1)	Increased caspase-3, Cys-12β, Cys-354β.	[367]
Diallyl trisulfide (DATS)	Prostate (LNCaP), breast (MCF-7).	Enhance apoptosis, control cell cycle growth, alteration transduction pathways signal.	[754]
DAS, DADS, DATS	Hepatoma (HepG2).	Modulation of CYP1-mediated bioactivation of B[a]P	[179]
Diallyl trisulfide (DATS)	Lung (H358, H460).	Arrest G2-M, DNA disintegration, decreased Cdk1, Bcl-2, Bcl-x; increase Bax, BID, Bak.	[1063]
Diallyl trisulfide (DATS).	Colorectal cancer (CRC).	DNA condensation, increased ROS production, cytochrome c, Apaf-1. AIF, caspase-3, -9 levels	[1093]
Diallyl trisulfide (DATS)	Depatoma (HepG2).	Inhibits cell proliferation, induction of caspase 3 activity.	[391]
Diallyl trisulfide (DATS), diallyl sulfide (DAS)	Prostate (PC-3, DU145).	Activation of JNK1/2, reduced Bcl-2, Bax interaction	[1060]
Diallyl trisulfide (DATS)	CNE2 cells.	Induce cell death, activate p38MAPK, and caspase-8.	[416]
Diallyl trisulfide (DATS)	Colon (HT29, colo 205).	Enhance apoptosis.	[1055]
Diallyl trisulfide (DATS)	Prostate (PC-3 and DU145).	Involve Ser(216) phosphorylation and Cdc25C; arrest G(2)-M phase, DNA break.	[356]
Diallyl trisulfide (DATS)	Cancer cell lines.	Enhance apoptosis and interferer cell cycle.	[357]
Diallyl trisulfide (DATS)	Prostate (PC-3, DU 145).	Arrest G(2)-M phase, enhance ROS production, Cdk inhibitor p21, Ser(216) phosphorylation, down Cdc 25 C activity.	[1061]

APPENDIX 1.5 *(Continued)*

Compound Name	Cell Line Name	Functions/Bioactivity/Properties/Mechanisms	References
Diallyl trisulfide (DATS)	Prostate (DU145, PC-3).	Enhance ROS synthesis, arrest G(2)-M phase, involve Cdc25C.	[33]
Diallyl trisulfide (DATS)	Colon (Colo 205).	Decrease Bcl-xL, Bcl-2, ΔΨm; increased Bak, Bax, cdc25c-ser-216-9, cyclin B, caspase-3, -8, -9 activity, Fas, JNK, p53 level, phospho-Ask1.	[1080]
Diallyl trisulfide (DATS)	Cervical (Ca Ski).	Production of ROS, Ca²⁺, induce ΔΨm, increased p53, p21, Bax level, decrease Bcl-2 level, induced caspase-3 activity.	[574]
Diallyl trisulfide (DATS)	Prostate (DU145, LNCaP).	Inhibition IL-6 nuclear translocation (STAT3/pSTAT3), phosphorylation of Janus-kinase 2.	[134]
Diallyl trisulfide (DATS)	Prostate (LNCaP-C4-2, LNCaP-C81, LNCaP).	Collapse ΔΨm, increase Bak, decrease Bcl-xL, Bcl-2 expression, and inhibit ROS production.	[474]
Diallyl trisulfide (DATS)	Prostate (DU145, PC3, LNCaP)	Growth inhibition and apoptosis stimulate IL-6 induced.	[653]
Diallyl trisulfide (DATS)	Prostate (DU145, NRP-154).	STAT3 activation, inhibition of STAT3 triggers apoptosis.	[66]
Diallyl trisulfide (DATS)	Breast (MDA-MB-231).	Cell-death signaling pathway, increase ROS, activate of ASK1, downstream signal JNK pathway, Inhibits Bim phosphorylation.	[531, 1062]
Diallyl trisulfide (DATS)	Prostate (PC-3, DU145).	Enhance c-Jun N-terminal kinase, kinase-phosphorylation of Bcl-2, inhibit Akt, Ser(473), Thr(308).	[228]
Diallyl trisulfide (DATS)	Colon (HCT-15).	Enhance cyclin B1 protein expression, arrest G2 -M phase, inhibit phosphatase activity I.	[483]
Diallyl trisulfide (DATS)	Lymphoblastic leukemia cell lines.	Apoptosis of Bcl-2, Bcl-xl, Bax.	[283, 1115]

APPENDIX 1.5 *(Continued)*

Compound Name	Cell Line Name	Functions/Bioactivity/Properties/Mechanisms	References
Diallyl trisulfide (DATS)	Neural cell.	Bcl-2 inhibits the necrosis induced by glutathione depletion.	[436]
Diallyl trisulfide (DATS)	LNCaP, HCT-116.	Down-regulation of Cdk1, Cdc, Cdc25C, DNA fragmentation, and caspase-3 activation.	[229]
Diallyl trisulfide (DATS)	Prostate (PC-3, DU145).	Inhibit Ser(473/155/136), Thr(308)phosphorylation, Ak, insulin-like growth factor receptor 1.	[228]
Diallyl trisulfide (DATS)	Breast.	Inhibit ROS generation, arrest cell-cycle, and enhance apoptosis.	[962]
Diosgenin/steroidal saponin	Colon (HT-29, HCT-116).	Increase 5-LOX expression, enhance leukotriene B4 production.	[552]
Diosgenin/steroidal saponin	Colon (HCT-116).	Inhibit apoptosis, 3-hydroxy-3-methylglutaryl CoA reductase expression.	[783]
Diosgenin/steroidal saponin	Colon (HT-29).	Enhance apoptosis and cell growth, inhibit Bcl-2, caspase-3 activity.	[784]
Diosgenin/steroidal saponin	Breast (HER2).	Supper mTOR and Akt phosphorylation, FAS; increase JNK phosphorylation.	[164]
Diosgenin/steroidal saponin	Breast (BCa).	Decrease cyclin D1, pAkt, Bcl-2, Akt kinase, cdk-2, -4, NF-kappaB activity, arrest G1 cell cycle.	[894]
Diosgenin/steroidal saponin	Prostate (PC-3).	Suppressed phosphorylation of PI3K, Akt, ERK, JNK, NF-κB.	[154]
Diosgenin and hecogenin	Cervical (CaSki).	Increased caspase-3 activity, DNA fragmentation, apoptosis.	[269]
Diosgenin/steroidal saponin	Osteosarcoma (1547).	Induced NF-kB binding to DNA caused a cell cycle arrest apoptosis, increase of p53 protein expression.	[191]

APPENDIX 1.5 (Continued)

Compound Name	Cell Line Name	Functions/Bioactivity/Properties/Mechanisms	References
Diosgenin/steroidal saponin	Osteosarcoma (1547).	Enhance apoptosis, NF-κB, p53, p21 mRNA expression; arrest G1 phase.	[646]
Diosgenin/steroidal saponin	Leukemia (K562).	Caused G2/M arrest and apoptosis, decreased cyclin B1, p21Cip1/Waf1, Ca^{2+}, activation of caspase-3, downregulate Bcl-2, Bcl-xL, up-regulate Bax.	[550, 580]
Diosgenin	Erythroleukemia.	Induced apoptosis	[549]
Diosgenin	Tumor cells.	Enhance phosphorylation of NF-κB, p65, decrease TNF, disturb cyclin D1, c-myc, COX-2 proliferation, Akt, IAP1, Bcl-X(L), TRAF1, cFLIP, MMP-9, Bcl-2, Bfl-1/A1.	[869]
Diosgenin	Erythroleukemia (TIB-180/HEL).	Arrest G2/M phase, decrease p53-pathway, enhance PARP cleavage, Bax/Bcl-2 ratio; DNA disintegration.	[550]
Diosgenin	HEL cells.	Decrease ERK activity, NF-kB, p38 MAPK, Bcl-xL, Akt pathways; enhance caspase-3 and PARP cleavage.	[548]
Diosgenin/steroidal saponin	Erythroleukemia (K562, HEL).	Induce apoptosis associated with COX-2, activate p38 MAPK, JNKs, DNA fragmentation.	[563]
Diterpenoid quinone, agastaquinone	Cell lines (A549, SK-OV-3, SK-MEL-2, XF498, and HCT15).	Cytotoxic activity.	[536]
Diterpenoids	Lung (H460), liver (HepG2).	Cytotoxic activity.	[898]
DLBS1425/Phaleria	Breast (MDA-MB-231, MCF-7).	Enhance apoptosis, caspase 9; DNA disintegration, control Bax and Bcl-2; decrease cPLA2, COX-2, c-fos, HER-2, VEGF-C.	[954]

APPENDIX 1.5 *(Continued)*

Compound Name	Cell Line Name	Functions/Bioactivity/Properties/Mechanisms	References
Ellagitannins (ETs)	Cancer cell lines.	Antioxidant activity.	[128]
Ellagitannins	Cervical (HeLa).	Reduced cell proliferation.	[806]
Emodin	Liver, lung (A549).	Induces apoptosis of caspase-2, -3, -9, generation of ROS, disruption of MMP, decrease mitochondrial Bcl-2, inactivation of ERK and AKT.	[900]
Emodin	Hepatoma (HepG2/C3A, PLC/PRF/5, SK–HEP-1).	Enhance apoptosis, p53/21, caspase-3, Fas/APO-1; DNA disintegration, arrest cell cycle.	[863]
Emodin	Cervical (Bu 25TK),	Enhance apoptosis, nuclear abbreviation, caspases-3, -9 activity, poly(ADP-ribose) polymerase cleavage, DNA disintegration.	[894]
Enterolactone	Colon (colo 201).	Induces apoptosis, inhibits growth down-regulate Bcl-2, PCNA, up-regulate Caspase-3.	[205]
Epigallocatechin-3-gallate (EGCG)	Lung.	Upregulation of miR-210, induced stabilization of HIF-1α.	[1008]
Epigallocatechin-3-gallate (EGCG)	Epidermoid (A431), keratinocyte (HaCaT), prostate (DU145).	DNA fragments, characteristic of apoptosis.	[15]
Epigallocatechin-3-gallate (EGCG)	Prostate (DU145).	Arrest cell cycle growth, enhance apoptosis, and modulate p53.	[319]
Epigallocatechin-3-gallate (EGCG)	Oral (OC2).	Inhibit MMP-2, -9, uPA expression.	[360]
Epigallocatechin-3-gallate (EGCG)	Hepato (HLE, PLC/PRF/5, HepG2, HuH-7).	Decrease Bcl-2α, Bcl-xl.	[695]

APPENDIX 1.5 *(Continued)*

Compound Name	Cell Line Name	Functions/Bioactivity/Properties/Mechanisms	References
Epigallocatechin-3-gallate (EGCG)	Ovarian (OVCA 433).	Activation of COX-1, COX-2-dependent pathways by ET-1, Inhibits ET-1/ET(A)R, ET(A)R-mediated COX-1/2 mRNA level.	[889, 890]
Epigallocatechin-3-gallate (EGCG)	Colon.	Inhibit COX-2, AMPK, VEGF, Glut-1.	[387]
Epigallocatechin-3-gallate (EGCG)	Colon, colorectal (HT-29, HCA-7).	Inhibit COX-2 activity via NF-kappaB activation, decrease Akt, and ERK1/2.	[375, 741]
EGCG, EGC, ECG	Colon mucosa, colon.	Supper COX-2, TBX/HHT formation, enhance PGE(2) production.	[365]
Flavonoids	Liver (HepG2), breast (MCF-7).	Antioxidant and antiproliferative activities.	[346]
Ferulic acid	Keratinocyte (HaCaT).	Enhance apoptosis control genes (c-fos, RPA, IL-6, p53-p21, TNF-α) and formation of CPDs.	[572]
Flavonoid glycoside	Breast (A375, HL60).	Cytotoxicity and anti-proliferative effect, DNA fragmentation, cell cycle arrest at G(1) phase.	[562]
Flavonoids	Colon (Caco2).	Anti-oxidant and anti-proliferative effects, inhibition of cellular growth.	[826]
Fucoidans	Melanoma, Colon.	Anti-tumor activity.	[256]
Fucoxanthin/fucoxanthinol	Leukemia (HTLV-1).	Enhance GADD45α, apoptosis; arrest cell cycle, decrease cyclin D1/2, CDK4/6, Bcl-2, cIAP2, XIAP.	[400]
Fucoxanthin (FX)/fucoxanthinol (FXOH)	Lymphoma (PEL).	Arrest G₁ phase and caspase-dependent apoptosis NF-κB, Akt, AP-1.	[1074]

APPENDIX 1.5 *(Continued)*

Compound Name	Cell Line Name	Functions/Bioactivity/Properties/Mechanisms	References
Fucoxanthin (FX)	Gastric (MGC-803).	Induce apoptosis, decreased CyclinB1, surviving, STAT3, reduction STAT3, p-STAT3, CyclinB1, downregulate CyclinB1, arrest G2/M phase, decrease CyclinB1-JAK/STAT signal pathway.	[1098]
Fucoxanthin (FX)	Hepato (HepG2).	Arrest G0/G1 phase, inhibit Rb-Serine 780 phosphorylation.	[208]
Fucoxanthin (FX)	Hepato (HepG2, DU145)	Induced GADD45A, induction of Gi arrest.	[1090]
Fucoxanthin (FX)	Colon.	Arrest G0/G1 phase, decrease p21WAF1/Cip1, pRb-Ser780/807/811 phosphorylation.	[209]
Fucoxanthin (FX)	Leukemia (HL-60).	Enhance apoptosis, ROS production, caspases -3, -7 cleavage, PARP; inhibit Bcl-xL expression.	[470]
Fuscocineroside C/atriterpene glycoside	Cancer cells.	Cytotoxic activity, Inhibits proliferation.	[1111, 1120]
Gammalinolenic acid (GLA)	Monocytes	Suppresses release of IL-1β, reduces pro-IL-1 β mRNA.	[280]
Gammalinolenic acid	PBMC.	Secretion of IL-1β.	[217]
Gammalinolenic acid	Monocytes	Production of IL-1β.	[311]
Gammalinolenic acid	Neutrophils.	Reducing the secretion of IL-1β, increases cAMP, decline secretion of lysosomal products in neutrophils.	[1122]
Ganoderic acid Me (GA-Me)	Metastatic lung cancer.	Inhibits cell adherence to ECM, suprrer MMP2/9 activity.	[152]
Ganodermanontriol (GNDT)	Colorectal (HCT-116, HT-29).	Inhibits proliferation, Cdk-4, PCNA, ß-catenin expression.	[414]

APPENDIX 1.5 *(Continued)*

Compound Name	Cell Line Name	Functions/Bioactivity/Properties/Mechanisms	References
Ganoderic acids (GAs)	Cancer cell lines.	Cytotoxic and anti-angiogenesis mechanisms.	[862]
Ganoderic acid Mf (GA-Mf), S (GA-S).	Cervical (HeLa).	Enhance apoptosis, Bax/Bcl-2 ratio; arrest cell cycle, inhibit MMP.	[581]
Ganoderic acid T (GA-T)	Cervical (HeLa).	Inhibits cell proliferation, cell cycle arrest, induced apoptosis, decreased the MMP, enhanced caspase-3, -9 activities	[582]
Ganoderic acid T (GA-T)	Lung (95-D).	Enhance apoptosis, p53, Bax, caspase-3 activity; arrest cell cycle, inhibit MMP, cytochrome c	[929]
Gallic acid	Stomach (KATO III), colon (Colo 205).	Induced apoptosis, DNA fragmentation	[1091]
Gallic acid	Esophageal (TE-2), gastric (MKN-28), colon (HT-29 and Colo201); breast (MCF-7), cervix (CaSki).	Up-regulate Bax, induced caspase-cascade activity, down-regulate Bcl-2, Xiap, Akt/mTOR pathway.	[263]
Gallic acid	Glioma (U87 and U25).	Suppress of ADAM17, down-regulate PI3K/Akt, Ras/MAPK signaling pathways, inhibit cell viability, reduce cell-mediated angiogenesis.	[596]
Gallic acid	Lymphocytes (PBLs).	Inhibit lipid peroxidation, induced apoptosis, phosphatidylserine externalization, DNA fragmentation	[883]
Gallic acid	Prostate (PC-3).	Supper MMP-2, -9 level, GRB2, PKC, JNK, SOS1, ERK1/2, p38, NF-κB, p-AKT (Thr308), p-AKT (Ser473); enhance PI3K, AKT, TIMP1; decrease FAK, Rho A.	[577]
Gallic acid	Prostate (DU145).	Induce caspase-9, -3, poly (ADP-ribose) poly-merase (PARP) cleavages.	[992]

APPENDIX 1.5 *(Continued)*

Compound Name	Cell Line Name	Functions/Bioactivity/Properties/Mechanisms	References
Gallic acid	Colon (colo 205).	Enhance apoptosis, Bax, caspase-3, cytochrome c; inhibit Bcl-2 level.	[581]
Garlic and lasun	Fibroblasts (FS4), hamster kidney (BHK21), Burkitt lymphoma (BJA-B).	Cytotoxic activity.	[841]
Ginger/Phenolics	Prostate cancer.	Modulated cell-cycle, apoptosis, caspase activity, $\Delta\Psi m$	[440]
Ginseng saponin K	Astroglioma.	Supper MMP-9, ERK, JNK, p38 MAPK, AP-1, activation.	[427]
Ginseng saponins	Prostate (LNCaP).	Induced apoptotic morphology, interfered Bcl-2, caspase-3.	[583]
Ginseng neutral polysaccharides	Sarcoma (S180).	Proliferation of lymphocytes, increased NK cell, enhanced phagocytosis, NO production by macrophages, increase TNF-α.	[688]
Ginsenoside-Rb1	Breast (MCF-7).	Activated of estrogen-responsive luciferase reporter gene.	[547]
Gossypol	Glioma (HS 683, U373, U87, U138), adrenal (SW-13); breast (MCF-7, T47-D); cervical (HeLa); melanoma (SK-MEL-3); colon (Colo 201, BRW).	Cytotoxicity.	[193]
Gossypol and gossypolone	Breast (MCF-7, MCF-7 Adr; MDA-MB-231).	Antiproliferative activity, disorganization, and loss of cytoplasmic organelles, change cellular morphology.	[300]

APPENDIX 1.5 *(Continued)*

Compound Name	Cell Line Name	Functions/Bioactivity/Properties/Mechanisms	References
Gossypol	Malignant (MCF-7, MCF-7/adr); breast (HBL-100).	Antiproliferative activity, cytotoxic activity.	[564]
Halocynthiaxanthin/fucoxanthin	Malignant (GOTO).	Suppression of cell proliferation.	[694]
Halocynthiaxanthin, fucoxanthinol	Breast (HL-60, MCF-7); colon (Caco-2).	DNA fragmentations, induce apoptosis, suppressing Bcl-2 protein expression.	[490]
Hexahydrocurcumin	Colorectal (SW480).	Cell cycle arrest in G1/G0 phase.	[145]
Hillaside C/triterpene	Leukemia, breast, colon.	Inhibits the growth, induction of apoptosis, reduction of Bcl.	[1054]
HMJ-30/quinazoline	Osteogenic (U-2 OS, HOS, 143B).	Inhibit cell growth, enhance apoptosis, DNA break, caspase-8, -9, -3 pathways, Mcl-1, Bcl-2, BAD, Bax, t-Bid proteins.	[172]
Isoflavonoids, lignans	Cancer.	Influence intracellular enzymes, malignant cell proliferation, differentiation, protein synthesis, growth factor action.	[620]
Isoprenoids/terpenoids (terpenes)	Liver cancer cells.	Cytotoxic activity.	[337]
Isothiocyanate	Leukemia, myeloma, colon, breast, hepatoma.	Inhibits growth.	[1113]
Kaempferol (flavonoids)	Osteogenic (U-2 OS, HOS, 143B); fetal osteoblast (hFOB).	Induced apoptosis, reduces viabilities, increase Ca^{2+}, $\Delta\Psi m$.	[380]
Limonin	Leukemia (CEM/ADR5000); colon (Caco-2).	Decreased P-gp activity, enhanced doxorubicin cytotoxicity.	[248]
Limonoids	Pancreatic (Panc-28).	Induction of apoptosis, cytotoxicity, expression of Bax, Bcl-2, caspase-3, p53.	[735]

APPENDIX 1.5 *(Continued)*

Compound Name	Cell Line Name	Functions/Bioactivity/Properties/Mechanisms	References
MJ-29	Leukemia (U937, HL-60, K562, KG-1).	Arrest cell growth; enhance Bcl-2 phosphorylation, procaspase-9, cytochrome *c*, apoptotic protease-activating factor-1.	[1081]
Neoxanthin, fucoxanthin	Prostate (PC-3).	Induction of apoptosis, morphological changes, DNA fragmentation, increased hypodiploid cells (%), cleavages of caspase-3, PARP, reduced Bax and Bcl-2 proteins.	[495]
Niacin, butyrate	Breast (MDA-MB-435); colon (LSLiM6).	Inhibit production of cyclic-AMP, enhance apoptosis, prevent colony formation.	[668]
Paclitaxel	Ovarian, breast, small/non-small lung, head, neck.	Anti-cancer activity.	[1021]
Paclitaxel	Osteogenic sarcoma (U-2 0s).	Induced apoptosis, caspase-3 activity, G2/M-cycle arrest.	[593]
Paclitaxel	Esophageal; squamous (ESCC, TE-2, TE-13, TE-14).	Arrest G2/M cell-cycle, DNA fragmentation.	[262]
Parthenolide	Lymphoma (TK6).	Inhibits cell growth.	[807]
Parthenolide/sesquiterpene lactone	Carcinoma (A549), endo-thelial (HUVEC); medul-loblastoma (TE671); colon (HT-29).	Inhibits proliferation.	[722]
Parthenolide	Eagle's (9KB).	Cytotoxic activity.	[74]
Peridinin	Colorectal (DLD-1).	Decrease cell viability, activating caspase-8, -9.	[902]
Phenolics	Colon (HT29/HCT-116); Oral (CAL-27), prostate (LNCaP/DU145).	Anti-proliferative activity.	[1112]

APPENDIX 1.5 *(Continued)*

Compound Name	Cell Line Name	Functions/Bioactivity/Properties/Mechanisms	References
Phenolics	Lung fibroblasts (CCD-25LU).	Reduced cell damage, protection of oxidative induced stress.	[93]
Phenolics	Breast (MCF-7), osteosarcoma (HOS-1); prostate (PC-3, PNT1A).	Decreased viability. Inhibits proliferation, induced cell death.	[825]
Phenolics	Oral (KB/CAL-27), breast (MCF-7), prostate (LNCaP), colon (HT-29/HCT116).	Increasing inhibition of cell proliferation. Stimulate apoptosis.	[849]
Phenolics	Daudi, Jurkat, and K562.	Cytotoxic activity.	[1121]
Phenolics, procyanidin, saponin	Digestive (CAL27/AGS/); HepG2/SW480; Caco-2; breast (MCF-7), ovary (SK-OV-3).	Cellular antioxidant activities.	[1067]
Phloroglucinol	Breast MCF-7.	Enhance pro-apoptotic gene, caspase-3, -9 activity, DNA damage, PARP cleavage; inhibits anti-apoptotic gene, NF-kB.	[489]
Phlorotannin	Hepatocellular (BEL-7402).	Anti-proliferative activity.	[1078]
Polyoxygenated steroids	Tumor (CCRF-CEM, DLD1)	Cytotoxicity activity.	[235]
Pheophorbide	Hepatoma (HepG2).	Reduction of P-glycoprotein activity, arrest G2/M phase, cyclin-A1, cdc2.	[928]
Plumbagin, juglone	Lymphocytes.	Apoptotic pathway, DNA condensation, apoptotic body formation, membrane blebbing, generation of ROS.	[852]
Polysaccharides (PS)	Keratinocytes (NHK, HaCaT).	Increased proliferation, ATP-synthesis.	[221]

APPENDIX 1.5 *(Continued)*

Compound Name	Cell Line Name	Functions/Bioactivity/Properties/Mechanisms	References
Polysaccharides (PS)	Leukemic (HL-60, U937).	Enhance IL-1β, -6, IFN-γ, TNF-α production.	[1005, 1015]
Polysaccharides (PS)	Brain tumors.	Involve defense mechanisms via phagocytosis.	[53, 613]
Polysaccharides (PS)	Leukemia (U937).	Phagocytosis, producing cytoplasmic superoxide.	[566]
Pomegranate	Cancer cell lines.	Cell proliferation, cell cycle, invasion, and angiogenesis.	[524]
Proanthocyanidins (GSPs)	Epidermoid (A431).	Induces apoptosis, reactivation of MAP kinase phosphatases, decrease PI3K, Akt, NF-kappaB/p65-targeted of COX-2, PCNA, iNOS, MMP-9, cyclin D1, decrease cell proliferation.	[623]
Proteoglycan (GLIS)	Sarcoma (S180).	Increase phagocytosis, production of IL-1β, nitric oxide, activation of B lymphocytes, IL-1β, TNF-α, reactive nitrogen intermediates, like NO. GLIS.	[1107]
Protopanaxadiol ginsenosides (PPDGs) such as: Rb1 and Rb2	Macrophages (U937).	Suppressed TNF-α production, increased protein kinase A,C.	[173]
Pumpkin	Cancer.	Regulates metabolism, exert detoxifying, slightly dehydrating.	[42]
Pyrrolizidine alkaloids (PAs)	Mutagenic, carcinogenic.	Anti-tumor activity	[51, 231]
Quercetin	Colon, colorectal.	Protective effect.	[227]
Quercetin	Colon (HT29, Caco-2).	Affect cell viability/lactate release, decrease total cellular ATP.	[9]
Quercetin	Colon (HCT-116, HT29, mammary (MCF-7).	Decrease in cell proliferation.	[981]

APPENDIX 1.5 *(Continued)*

Compound Name	Cell Line Name	Functions/Bioactivity/Properties/Mechanisms	References
Quercetin	Colon (Caco-2).	Decrease CDC6, cyclin D1, CDK4, cell proliferation, arrest cell cycle, enhance sub-G1 phase, interfere TCF, and MAPK signal transduction.	[982]
Quercetin	Colon (Caco-2).	Decrease in cell differentiation	[225]
Quercetin	Colon (CO115).	Arrest G(1)/S, G(2)/M phases	[661]
Quercetin	Prostate (PC-3, DU-145).	Arrest G(1),S,G(2),M phases, induce tumor suppressor genes.	[669]
Quercetin	Breast (MDA-MB468).	Arrest G2-M phase, inhibit p53 protein expression.	[48]
Quercetin	Prostate (PC-3).	Arrest G2-M phase, increase Bax, Cdc2/Cdk-1, caspase-3activity; decrease cyclin B1 (Bcl-2, Bcl-X(L), pRb) phosphorylation, p21.	[996]
Resveratrol	Breast cancer.	Enhance apoptosis, nuclear, and DNA disintegration, decrease Bax, Noxa, Bim expression.	[18]
Resveratrol	Lymphoma U937.	Arrest S phase.	[728]
Resveratrol	Colon (CaCo-2).	Arrest S/G2 phase, reduced ODC activity, and intracellular putrescine content.	[846]
Resveratrol	Epidermoid (A431).	Inhibit WAF-1/p21-mediated G(1)-phase cell cycle, cyclin D1/D2-cdk6, -4, E-cdk2	[14]
Resveratrol	Prostate.	Arrest sub-G1 and S phases, inhibit proliferation-specific genes.	[423]
Rhein	Nasopharyngeal (NPC).	Induces apoptosis, increased nuclear condensation, DNA fragmentation, increase GRP 78, PERK, A TF6, and CCAA, activation of caspase-3, -8, -9, -12, increased ROS, Ca^{2+}, decrease MMP.	[569]

APPENDIX 1.5 *(Continued)*

Compound Name	Cell Line Name	Functions/Bioactivity/Properties/Mechanisms	References
Rhein	Colonic (CaCo-2).	Induced nitrate production, chemotaxis, and apoptosis.	[781]
Rhein	Hepatoblastoma G2 (Hep G2).	Arrest G1 phase, inhibit p53, p21/WAF1 or increase CD95 (mCD95L, sCD95L) expression.	[505]
Rhein	Cell lines.	Decrease SOD, phagocytic, chemotaxis, macrophage, neutrophils, activity.	[888]
Rhein	Colonic (Ca Ski).	Induced ΔΨm, increase Fas, p53, p21, Bar, caspase-3, -8, -9, DNA fragmentation.	[399]
Rhein	Colon (Caco-2).	Reduced cell proliferation, MAP, ERK phosphorylation, induced DNA damage, ROS.	[47]
S-allyl cysteine (SAC)	Colon (SW-480 and HT-29).	Enhance caspase-3 activity, arrest G1 to S phase.	[868]
Saikosaponin-A	Breast (MDA-MB-231, MCF-7).	Control proliferation; enhance Bax to Bcl-2, caspase-3, c-myc activity.	[149]
Saikosaponin D	Hepatoma (Hep G2, Hep 3B).	Inhibit cell proliferation, arrest G1 phase, and enhance p21/WAF1 Fas/APO-1, Bax expression.	[372]
Saponin	Lung (A549).	Cytotoxic activity, extrinsic apoptosis pathway.	[373]
Sesquiterpene	Cervical (HeLa).	Induced cellular shrinkage, chromatin condensation, the appearance of apoptotic bodies, G2/M phase cell cycle arrest, loss of ΔΨm.	[949]
Solamargine	Lung (H441, H520, H661, H69).	Induce TNFRs, TRADD/FADD signal cascades, TNF-α/β, caspase-3 activity, inhibit cytochrome c, Bcl-2, Bcl-xL, DNA disintegration.	[578, 818]
Steroid glycosides, spirostan, pregnane furostansteroid saponins	Leukemia.	Cytotoxicity activity.	[766]

APPENDIX 1.5 *(Continued)*

Compound Name	Cell Line Name	Functions/Bioactivity/Properties/Mechanisms	References
Tetraclinis articulata essential oil (TA-EO)	Breast, ovarian.	Cytotoxic effect, induction of apoptosis.	[114]
Tetrandrine, fangchinoline	MDR Caco-2, CEM/ADR5000.	Reduced P-gp expression.	[908]
TMS (2, 4, 3′,5′-tetra-methoxystilbene)	Breast Cancer.	Inducing apoptosis, nuclear condensation, DNA fragmentation, reduced for Bax, Bim, and Noxa.	[18]
Triterpenoids	Breast (HepG2, MCF-7), colon (Caco-2).	Antiproliferative activity.	[349]
Triterpenoid/saponins	Cervix (HeLa), hepatocyte (Hep-G2), fibrosarcoma (HT1080), promyelocytic (HL-60).	Cytotoxic activity.	[681]
Triptolide	Adrenal (NCI-H295).	Induced apoptosis in sub-G1 phase, production of ROS, Apaf-1, cytochrome c, AIF, Endo G, caspase-9, -3, decreased $\Delta\Psi m$.	[1056]
Triptolide	Leukemic, acute myeloid (AML).	Decrease in XIAP, Bcl-2, Bcl-X(L), Mcl-1 expression, loss $\Delta\Psi m$, caspase-9 resistant/caspase-8 sensitive activity.	[126]
Triptolide (PG490)	Prostatic.	Induced apoptosis by p53-independent mechanisms, reduced p21(WAF1/CIP1), hdm-2, bcl-2.	[481]
Triptolide (TPL)	Melanoma (B16), breast (MDA-435), bladder (TSU), gastric (MGC80-3).	Induced apoptosis, reduced cell cycle-regulated molecules expression.	[1082]
Triptolide (TPL)	Ovarian cancer cells.	Reduce PI3 kinase inhibitor LY294002, carboplatin	[1030, 1130]

APPENDIX 1.5 *(Continued)*

Compound Name	Cell Line Name	Functions/Bioactivity/Properties/Mechanisms	References
Triptolide	Proximal tubular (HK-2).	Enhance p17 cleaved, caspase 3 activity.	[874]
Triptolide (PG490)	Tumor cells.	Blocks NF-κB activation, TNF-α-induced apoptosis, increase p53 expression, decline p21(WAF1/CIP1)/p53-responsive gene.	[139]
Triptolide	Breast tumors	Cytotoxicity activity.	[857]
Triptolide	Gastric (AGS, MKN-, 28, 45, SGC-7901).	Suppressed NK-kappaB activation.	[418]
Triptolide (PG490)	Tumor cells.	Cytotoxicity of TNF-alpha, activation of NF-kappaB, enhance c-IAP1/2 (hiap-1/hiap-2) expression.	[545]
Triptolide	Fibroblasts.	Inhibits IL-1α, $\Delta\Psi m$ (MMP1 and MMP3).	[570]
Triterpenes	Hepatocellular (HuH-7).	Function of ERKs, JNK inhibitors, activates p38 MAPKs.	[422]
Triterpene	RAW 264.7	Suppress TNF-α, IL-6, NO, PGE(2), p65 phosphorylation, decrease iNOS, COX-2, AP-1 subunit c-Jun, cyclin D1, CDK4, and cyclin B1 expression, inhibit NF-kappaB.	[234]
Triterpene	Lung, hepatoma (HepG2).	Up-regulation VEGF, IL-6, TNF-α; Inhibits EGFR	[475, 568]
Triterpenes	Cancer cell lines.	Cytotoxic and anti-angiogenesis mechanisms.	[1099]
Withanolides, Withaferin A	Lung (NCI-H460), colon (HCT-116), CNS (SF-268), breast (MCF-7)	Inhibits cell growth.	[412]
Zhejiang saffron	Lung (A549, H446).	Cell proliferation, apoptosis, caspase-3, -8, -9 activity.	[575]

CHAPTER 2

HERBAL EXTRACTS AND THEIR BIOACTIVITIES: COMPARATIVE PHYTOCONSTITUENT ANALYSIS OF SELECTED MEDICINAL PLANTS USING GC-MS/FTIR TECHNIQUES

C. STANLEY OKEREKE, O. UCHE ARUNSI, E. MARTINA ILONDU, and S. CHIEME CHUKWUDORUO

ABSTRACT

The research study in this chapter was aimed to determine and characterize the bioactive compounds present in three medicinal plants (*Ageratum conyzoides, Chromolaena odorata,* and *Ficus exasperata*) commonly used in Nigeria, especially in the southeast Geopolitical Zone for the management of ailments. The biological activities of these plants were established using phytochemical detection schemes of GC-MS and FT-IR. From the results obtained, the methanolic leaf extract of *Chromolaena odorata* recorded the highest bioactive compounds followed by those of *Ageratum conyzoides* and *Ficus exasperata.* These confirm the wide use of *C. odorata* in herbal medicine. The blend of the different leaf extracts may be of health benefits owing to the distribution of the compounds within each plant sample. However, it is recommended that a more thorough investigation should embark to quantitatively isolate, purify, package, and market-specific bioactive compounds.

2.1 INTRODUCTION

Medicinal plants contain essential biochemicals, which confer health benefits to humans when properly prescribed. These biochemicals not only serve as drugs but also a defense against biological intruders. The involvement of herbal extracts to treat and manage infectious diseases is as old as antiquity and is attracting much attention by scientists all over the globe [32, 57], especially in rural areas, where folk medicine is well demonstrated. Today, plants remain one of the most untapped natural resources and have been heavily harvested to ease the health burdens, such as: cardiovascular diseases, obesity, hypertension, diabetes, peptic ulcers, cancers, rheumatoid arthritis, and neurological disorders. However, consensus has been reached recently; in which case, drugs derived from plants were observed to have a potential panacea for diseases and infections that constantly short-live the lifespan of man [69]. This breakthrough in traditional medicinal practice is connected to the following benchmarks: bioavailability, higher safety margin, efficacy, quality, affordability [26, 41], and accumulation of chemicals that target metabolic pathways [55].

These comparative advantages of herbal medicine over conventional drugs have led to a decline in the prescription of some drugs. For instance, non-steroidal anti-inflammatory drugs (NSAIDs) used to relieve pains have been marked as potential ulcerogens in recent times. This is because they competitively inhibit Cyclooxygenases (Prostaglandin E synthase), an enzyme that is saddled with the responsibility of prostaglandin's biosynthesis [56]. The legendary role played by traditional medicinal practitioners in disease prevention has also been applauded by the World Health Organization (WHO) [23, 80].

The integration of plant sources in the treatment of diseases has evolved the name herbal medicine that refers to the use of herbs for therapeutic or medicinal purposes [1]. Medicinal plants contain certain organic molecules called bioactive compounds or phytoconstituents, which are present in medicinal plants and when ingested as food or drugs, promote good health. They can act as precursors for the synthesis of endogenous molecules, which can either facilitate the running of a particular metabolic pathway or retard the progression of one that may trigger off uncontrolled production of 'nonsense' and malignant products (e.g., cancer, AIDS, and many ailments of clinical concerns). Owing to the functionality of these molecules as demonstrated *in vitro, in vivo,* and *in silico,* attempts have been made to define these compounds vis-à-vis their biological roles (bioactivities). A database nicknamed '*Ethnobotany and Phytochemistry Database*' was created to achieve this goal [25].

The need to screen, isolates, quantifies, and characterizes bioactive compounds in the field of ethnobotany, phytochemistry, ethnopharmacology, and ethnomedicine has long been advocated. Following this, a lot of techniques have been introduced. They include the preliminary qualitative and quantitative screening [33, 74]; which identify the presence and concentrations of phytochemicals (flavonoids, saponins, tannins, cardiac glycosides, alkaloids, terpenes, carotenoids, phenolics, hydrocinnamic acids) in plants.

Over the past decades, there exists a preponderance of claims by traditional medicinal practitioners that herbal products are a better elixir for ailments than their synthetic counterparts. This claim has kindled interests among scientists all over the globe in exploring the intrinsic properties of medicinal plants, thus leading to the invention of detection schemes for the characterization of phytoconstituents. These techniques or detection schemes are GC-MS (gas chromatography-mass spectrometry), FTIR (Fourier transform infrared), FID (flame ionization detector), NMR (nuclear magnetic resonance) and HPLC (high-performance liquid chromatography). In this chapter, the techniques of GC-MS and FT-IR were employed. While GC-MS identifies biological molecules in extracts at an infinitesimal level (< 1ng) [28], the FT-IR technique sorts out their functional groups within 400–4000 cm^{-1} of the infrared (IR) region [10].

The present study was aimed at determining and characterizing the bioactive compounds resident in three medicinal plants (*Ageratum conyzoides, Chromolaena odorata,* and *Ficus exasperata*) commonly used in Nigeria, especially in the southeast Geopolitical Zone for the management of ailments, and their established biological activities using more recent phytochemical detection schemes of GC-MS and FT-IR. These plants were selected because of their importance as a source of traditional medicines in Nigeria, especially in Uturu, Isiukwuata Local Government Area, Abia State in Nigeria.

2.2 LITERATURE REVIEW

2.2.1 *PHARMACOGNOSTIC PROFILE OF AGERATUM CONYZOIDES (LINN.)*

The plant weed, *Ageratum conyzoides* (Linn) or Billy goat in English, is a member of the *Asteraceae* indigenous to West Africa, Central

America, Southeast Asia, South China, and India. It is commonly nicknamed Ufuopioko, Imiesu, and Nriewu among the Igedes, Yorubas, and Igbos, respectively. In folk medicine, different parts of the plant are consumed for therapeutic purposes ranging from wounds, spasm, colic, fever, migraine, pneumonia, inflammation, asthma, leprosy, stomach disorders to skin diseases [7, 8, 18, 37, 66]. Specifically, *A. ageratum* has shown to possess wound healing, antifungal, and antibacterial activities [30, 36, 58].

2.2.2 PHARMACOGNOSTIC PROFILE OF CHROMOLAENA ODORATA (LINN.)

Chromolaena odorata (Linn.) King and Robinson is a flowering shrub indigenous to South and Central America and have spread throughout the tropics, including Nigeria [7]. It is commonly known as Siam weed, triffid weed, bitter bush orjuck in the bush (English), Awolowoakintolataku (Yoruba), Ana-afu-okuno, Obiarakara (Igbo), or independence weed (in all Nigeria). It is named independence weed because it was introduced during the time Nigeria gained her independence. *C. odorata* has a lot of ethnopharmaceutical benefits. In Nigeria, a fresh juice squeezed out from the leaves of the plant is used to stop bleeding [3]. The decoction of the leaves and stems of *C. odorata* has been reported to be effective in the treatment of propionibacterium acnes [16] while that of the flowers are used as tonic, antipyretic, and heat tonic [15]. Other documented ethnopharmaceutical evidences include management of spasms, infections due to protozoa, trypanosoma, bacteria, and fungi, hypertensive, and inflammation [6, 38, 49, 62].

2.2.3 PHARMACOGNOSTIC PROFILE OF FICUS EXASPERATA (VAHL.)

Ficus exasperata Vahl. is commonly distributed in West Africa [7]. It is known by many nicknames, such as: sandpaper tree, fig tree, white fig (in English), ewe ipin, oporo, opoto (in Yoruba), ogbu (in Igbo) and achedin-nini (in Hausa) [5]. Especially in rural areas, the leaves of the plant are used to scrub utensils, polish wooden plates and furniture [12], feed for animals such as goats, sheep, and chimpanzees [75] and treat coughs, high blood pressure, rheumatism, arthritis, cancer, and wounds [19]. The sap is used to stop bleeding and hasten the discharge of placenta in cows after calf delivery [39].

Several folkloric claims of the leaf extract of *F. Exasperata* have been added in the pharmacopeia. Included in the exhaustive list are hypotensive and hypoglycaemic [2, 14], heamostative opthalmia, coughs, and hemorrhoids [54]. Other ethnopharmacological activities are: antioxidant, anticonvulsant, antiarthritic, antibacterial, anti-inflammatory, antipyretic, antinociceptive, anticandidal, antidiabetic, hypotensive, antifungal, and lipid-lowering potentials [44, 47, 70, 79].

2.3 MATERIALS AND METHODS

2.3.1 COLLECTION AND IDENTIFICATION OF PLANT SAMPLES

The leaves of *Ageratum conyzoides*, *Chromolaena odorata,* and *Ficus exasperata* were collected from the courtyard of the Faculty of Biological and Physical Sciences, Abia State University, Uturuin August, 2016 and were identified by an experienced Taxonomist in Plant Science and Biotechnology Department.

2.3.2 PROCESSING AND EXTRACTION OF PLANT SAMPLES

The leaves of the plants were carefully sorted to remove irrelevant materials, washed in distilled water, and allowed to air-dry under shade for 12 days. After drying, the leaves were macerated separately into a fine powder, packaged in a polythene bag and stored at room temperature until use. Two grams (2 g) of each sample was dissolved in 50mL of methanol and stirred gently for 72 hours. After extraction, the mixtures were strained with muslin cloth and filtered carefully using Whatman filter paper No. 1. The resultant filtrates were concentrated on a water bath (BJE-750 Gallen Kamp England) until a dark green filtrate was formed. This was transferred to glass vials and kept at 4°C before use.

2.3.3   QUALITATIVE PHYTOCHEMICAL ANALYSIS

The test carried out was based on procedures outlined in the literature [33, 75].

2.3.4   DETERMINATION OF BIOACTIVE COMPOUND BY GC-MS

The phytoconstituents present in the methanolic leaf extracts of *Ageratum conyzoides, Chromolaena odorata,* and *Ficus exasperata* were analyzed using Clarus-500 Perkin-Elmer Gas Chromatograph (GC) coupled with a mass detector. The procedures outlined by Kalimuthu et al., [42] were adopted.

2.3.5   DETERMINATION OF FUNCTIONAL GROUPS OF BIOACTIVE COMPOUND BY FT-IR

The determination of functional groups of the bioactive compounds present in the different plant extracts was done using the Bruker Tensor-27 Spectrometer coupled with IR in the wavelength range of 400–400 cm^{-1}. The detailed methodology of Krishnaveni et al., [45] was used.

2.3.6   IDENTIFICATIONS OF BIOACTIVE COMPOUNDS AND THEIR FUNCTIONAL GROUPS

The National Institute of Standard and Technology (NIST) database was used to ascertain the chemical structures, formula, structures, names, % peak areas and retention time of unknown bioactive compounds present in the samples, while the functional groups were determined from the FT-IR spectrum and were identified based on the peak values in the region of IR radiation.

2.4   RESULTS

Table 2.1 reveals the ecological attributes of *Ageratum conyzoides, Chromolaena odorata,* and *Ficus exasperata. Ageratum conyzoides,* and *Ficus exasperata* are both annual plants, while *Chromolaena odorata* is a perennial plant. *A. conyzoides* grows as herb while *C. odorata* and *F. exasperate* grow as shrub. Table 2.2 shows the preliminary phytochemical screening of *Ageratum conyzoides, Chromolaena odorata,* and *Ficus exasperata.* Alkaloids, flavonoids, and steroids were present in the methanolic leaf extracts of the three plants.

TABLE 2.1 Ecological Attributes of the Plants Used in This Study

Botanical Name	Common Name	Local Name	Life Form	Habit of Growth
Ageratum conyzoides	Goat weed	Ufuopioko (Igede, Middle Belt) Akwukwo-nwaosinaka or Agadi-isi-awo-ocha (Igbo) Ime-esu (Yoruba)	A	H
Chromolaena odorata	Siam weed	Awolowo weed or Akintola-taku (Yoruba Ogididiri (Igbo)	P	S
Ficus exasperata	Sandpaper plant	Ewe-ipin, Oporo, opoto (Yoruba) Ogbu (Igbo) Achedinnini (Hausa)	A	S

Legend: P = Perennial, A = Annual, H = Herb, S = Shrub.

TABLE 2.2 Preliminary Phytochemical Screening of the Plant Extracts

Phytochemical	Ageratum conyzoides	Chromolaena Odorata	Ficus Exasperata
Alkaloids	++	++	++
Anthocyanin	–	–	+
Anthraquinones	+	–	–
Cardiac glycosides	–	–	+
Flavonoids	+	++	+
Phenols	+	+	–
Saponins	+	–	+
Steroids	+	+	+
Tannins	+	–	+
Terpenes	+	+	–

Key: + low concentration, ++ moderate concentration, +++ high concentration, – absent

The results of the GC-MS for *Ageratum conyzoides, Chromolaena odorata* and *Ficus exasperata* are presented in Figures 2.1–2.3 and Tables 2.3–2.6. Exactly 20, 69 and 18 bioactive molecules were identified in the methanolic

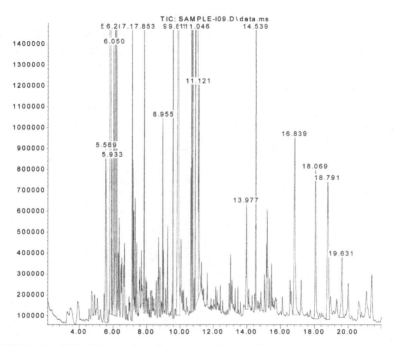

FIGURE 2.1   GC-MS for methanolic extract of leaves of *Ageratum conyzoides.*

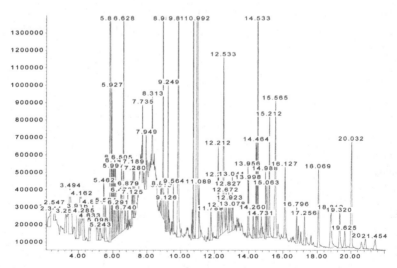

FIGURE 2.2   GC-MS for methanolic extract of leaves of *Chromolaena odorata.*

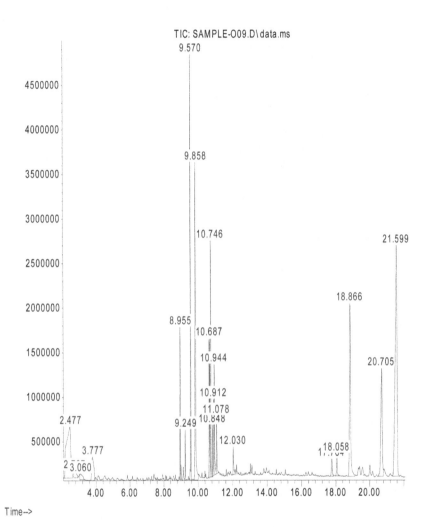

FIGURE 2.3 GC-MS for methanolic extract of leaves of *Ficus exasperate.*

leaf extracts of *A. conyzoides, C. odorata* and *F. exasperata*, respectively. The Figure 2.1 and Table 2.3 summarize the phytocompounds that were found in the leaf extract of *A. conyzoides*. In this list, Alpha-Linolenic acid was predominant followed by Precolene-1, Caryophyllene, n-hexadecanoic acid, squalene, and phytol.

- (-)-Spathulenol (1.39%);
- α-Linolenic acid (27.05%);
- 1-Heptatriacotanol (2.28%);
- 1-Octadedelyne (0.83%);
- 7-t-Butyl-3, 3-dimethyl-1-inlanone (1.25%);
- 8, 11, 14-Eicosatrienoic acid, (Z,Z,Z) (1.02%);
- Aromandendrene (2.02%);
- Caryophyllene (10.44%);
- Cyclopentaneundecanoic acid (1.51%);
- D-Germacrene (0.93%);
- Epi-β-Santalene (2.27%);
- Hexadecanoic acid, methyl ester (2.06%);
- Humulene (4.31%);
- Methyl-linolenate (2.75%);
- n-Hexadecanoic acid (9.01%);
- Phytol (5.39%);
- Precolene 1 (11.44%);
- Squalene (6.98%), Vitamin E (2.60);
- Stigmasterol (1.58%).

TABLE 2.3 Bioactive Compounds Identified from the Methanolic Leaf Extract of *Ageratum conyzoides* by GC-MS

S/N	RT	Compound Name	MW	Formula	Area (%)
1	5.569	Aromandendrene	204	$C_{10}H_{20}O$	2.02
2	5.852	Caryophyllene	204	$C_{10}H_{23}NO$	10.44
3	5.933	D-Germacrene	204	$C_{15}H_{24}$	0.93
4	6.050	Epi-β-Santalene	204	$C_{15}H_{24}$	2.27
5	6.157	Humulene	204	$C_{15}H_{24}$	4.31
6	6.205	Precolene 1	190	$C_{15}H_{24}$	11.44
7	7.131	(-)-Spathulenol	220	$C_{15}H_{24}O$	1.39
8	8.955	1-Octadedelyne	250	$C_{13}H_{20}O_2$	0.83
9	9.570	Hexadecanoic acid, methyl ester	270	$C_{14}H_{30}$	2.06
10	9.891	n-Hexadecanoic acid	256	$C_{18}H_{34}$	9.01
11	10.688	Methyl-linolenate	292	$C_{10}H_{18}O$	2.75
12	10.746	Phytol	296	$C_{30}H_{50}O_4$	5.39
13	10.912	7-t-Butyl-3, 3-dimethyl-1-inlanone	216	$C_{14}H_{30}$	1.25
14	11.046	α-Linolenic acid	273	$C_{17}H_{32}O_2$	27.05

TABLE 2.3 *(Continued)*

S/N	RT	Compound Name	MW	Formula	Area (%)
15	11.121	Cyclopentaneundecanoic acid	254	$C_{17}H_{34}O_2$	1.51
16	13.977	8, 11, 14-Eicosatrienoic acid, (Z,Z,Z)	306	$C_{16}H_{32}O_2$	1.02
17	14.539	Squalene	410	$C_{19}H_{40}$	6.98
18	16.839	Vitamin E	430	$C_{19}H_{32}O_2$	2.60
19	18.069	Stigmasterol	412	$C_{19}H_{32}O_2$	1.58
20	18.791	1-Heptatriacotanol	536	$C_{20}H_{40}O$	2.28

TABLE 2.4 Bioactive Compounds Identified from the Methanolic Leaf Extract of *Chromolaena odorata* by GC-MS

S/N	RT	Compound Name	MW	Formula	Area (%)
1	2.344	Erythritol	122	$C_4H_{10}O_4$	0.82
2	2.547	Tetrahydro-4H-pyran-4-ol	102	$C_5H_{10}O_2$	0.41
3	3.258	1-Pentene	70	C_5H_{10}	0.62
4	3.494	1, 3, 5-Hexatriene, 3-methyl-,(E)-	94	C_7H_{10}	2.45
5	3.916	Catechol	110	$C_6H_6O_2$	1.20
6	4.093	Cycloheptatrienylium, iodide	218	C_7H_7I	0.93
7	4.163	Isopropenyl bromide	120	C_3H_5Br	0.41
8	4.285	2-Propanone, dimethylhydrazone	100	$C_5H_{12}N_2$	0.68
9.	4.633	Hydroquinone	110	$C_6H_6O_2$	0.65
10.	4.831	2-Isopropoxyethylamine	103	$C_5H_{13}NO$	0.50
11.	5.098	DL-Cystine	240	$C_6H_{12}N_2O_4$	0.45
12.	5.243	Metaraminol	167	$C_9H_{13}NO_2$	0.43
13.	5.462	α-Cubebene	204	$C_{15}H_{24}$	1.09
14	5.564	2-Octyne	110	C_8H_{14}	0.85
15	5.842	Calarene	204	$C_{15}H_{24}$	7.64
16	6.927	β-Cubebene	204	$C_{15}H_{24}$	2.06
17	5.997	1, 5-Heptadien-3yne	92	C_7H_8	0.97
18	6.045	α-Farnesene	204	$C_{15}H_{24}$	1.18
19	6.104	Methyl-4, 7, 10, 13, 16, 19-docosahexaenoate	342	$C_{23}H_{34}O_2$	1.13
20	6.146	Humulene	204	$C_{15}H_{24}$	1.35
21	6.291	1, 2-Propanediol, 3-Chloro	110	$C_3H_7ClO_2$	0.57
22	6.473	Spiro[5.5]undec-2-ene, 3, 7, 7-trimethyl-11-methylene-, (-)-	204	$C_{15}H_{24}$	0.75

TABLE 2.4 *(Continued)*

S/N	RT	Compound Name	MW	Formula	Area (%)
23	6.505	Diazoadamantane	162	$C_{10}H_{14}N_2$	1.00
24	6.628	Benzene, 1, 2-diethyl-	134	$C_{10}H_{14}$	3.39
25	6.740	Arginine	174	$C_6H_{18}O_2$	0.34
26	6.879	7-Octen-2-ol, 2-methyl-6-methylene-	154	$C_{10}H_{18}O$	0.37
27	7.125	2-Propanone, dimethylhydrazone	100	$C_5H_{12}N_2$	0.42
28	7.189	Ethanamine, 2-phenoxy-	137	$C_8H_{11}NO$	0.76
29	7.280	Taurultam	136	$C_3H_8N_2O_2S$	0.86
30	7.730	Oxetane, 2, 2, 4-trimethyl-	100	$C_6H_{12}O$	0.82
31	7.949	3-Aminocrotononitrile	82	$C_4H_6N_2$	0.78
32	8.313	1-[-]-4-Hydroxy-1-methylproline	145	$C_6H_{11}NO_3$	0.86
33	8.815	1-Nitro-2-acetamido-1, 2-dideoxy-d-glucitol	252	$C_8H_{16}N_2O_7$	0.34
34	8.874	1, 2, 3, 4-Butanetetrol, [S-R*, R*)]-	122	$C_4H_{10}O_4$	0.54
35	8.955	1, 9-Nonanediol, dimethanesulfonate	316	$C_{11}H_{24}O_6S_2$	3.93
36	9.126	1, 2-Benzenediol, 4-[2-(methylamino) ethyl]-	167	$C_9H_{13}NO_2$	0.42
37	9.249	2-Cyclohexen-1-one, 4, 4-dimethyl-	124	$C_8H_{12}O$	1.35
38	9.564	Methyl-2-O-methyl-β-D-xylopyranoside	178	$C_7H_{14}O_5$	0.51
39	9.864	n-Hexadecanoic acid	256	$C_6H_{10}O_3$	5.90
40	10.741	Oxirane, decyl	184	$C_{12}H_{24}O$	3.29
41	10.992	9, 12, 15-Octadecanoic acid, (Z,Z,Z)-	278	$C_{18}H_{30}O_2$	11.18
42	11.789	Adipamide	144	$C_6H_{12}N_2O_2$	0.66
43	12.212	Bicyclo[3.1.1]heptane, 2, 66-trimethyl-[1R-(1α, 2α, 5α)]-	138	$C_{10}H_{18}$	1.62
44	12.244	Bicyclo[2.2.1]-Heptan-2-one, 4, 7, 7-trimethyl-,semicarbazone	209	$C_{11}H_{19}N_3O$	0.74
45	12.388	Acetic acid, 2-(1-methyl-2-oxohydrazino)-N'-[(E)-(2-hydroxyphenyl) methylidene]hydazide, N-oxide	252	$C_{10}H_{12}N_4O_4$	0.58
46	12.533	1H-3a, 7-Methanoazulene, octahydro1, 4, 9, 9-tetramethyl-	206	$C_{15}H_{26}$	2.85
47	12.672	Tricyclo[4.3.1.1(3, 8)undecane-1-carboxylic acid	194	$C_{12}H_{18}O_2$	0.49
48	12.827	Benzenemethanol, alpha-[(methylamino) methyl]-	151	$C_9H_{13}NO$	0.75
49	12.923	Phenylephrine	167	$C_9H_{13}NO_2$	0.98

TABLE 2.4   *(Continued)*

S/N	RT	Compound Name	MW	Formula	Area (%)
50	13.041	Cyclohexanol, 1R-4-trans-acetamido-2, 3-trans-epoxy-	171	$C_8H_{13}NO_3$	0.71
51	13.956	1-Eicosanol	298	$C_{20}H_{42}O$	0.84
52	13.998	5-Hydroxy-4,'7-dimethoxyflavanone	300	$C_{17}H_{16}O_5$	0.62
53	14.260	Pyridine, 2, 3-dimethyl-	107	C_7H_9N	0.46
54	14.421	4,'5-Dihydroxy-7-methoxyflavanone	286	$C_{16}H_{14}O_5$	1.15
55	14.464	Tricyclo[4.3.1.0(3, 8)]decan-10-ol	152	$C_{10}H_{16}O$	1.16
56	14.533	Squalene	410	$C_{30}H_{50}$	3.69
57	14.731	Benzaldehyde, 3-hydroxy-,oxime	137	$C_7H_7NO_2$	0.40
58	14.988	Phenol, 2, 6-dimethyl-5-methylphenyl)-	167	$C_8H_9NO_3$	0.89
59	15.212	Ethanone, 1-(2-hydroxy-5-methylphenyl)-	166	$C_9H_{10}O_3$	1.89
60	15.565	Leucopterin	195	$C_6H_5N_5O_3$	3.02
61	16.127	Benzaldehyde, 2-hydroxy-5-nitro-	167	$C_7H_5NO_4$	1.50
62	16.796	Vitamin E	430	$C_{29}H_{50}O_2$	0.62
63	17.256	Sarcosine, N-(3-cyclopentylpropionyl)-, tetradecylester	409	$C_{25}H_{28}O$	1.65
64	18.069	2-Pentadecyn-1-ol	224	$C_{15}H_{28}O$	1.55
65	18.812	Estran-3-one, 17-(aceyloxy)-2-methyl-, (2α, 5β, 17β)	332	$C_{21}H_{32}O_3$	1.41
66	19.320	2, 4-Dimethylamphetamine	163	$C_{11}H_{17}N$	0.99
67	19.625	Metaraminol	167	$C_9H_{13}NO_2$	0.46
68	20.032	Cyclohexane, 1-ethenyl-1-methyl-2, 4-bis(1-methylethenyl)-,[1S-(1α, 2β, 4β]-	204	$C_{15}H_{24}$	2.72
69	20.871	Acetamide, N-(aminocarbinyl)2-chloro-	136	$C_3H_2ClN_2O_2$	0.34

TABLE 2.5   Bioactive Compounds Identified from the Methanolic Leaf Extract of *Ficus exasperata* by GC-MS

S/N	RT	Compound Name	MW	Formula	Area (%)
1	2.477	Glycerin	92	$C_3H_8O_3$	12.20
2	2.707	Eucalyptol	154	$C_{10}H_{18}O$	0.82
3	3.060	3-(Prop-2-enolyoxy)dodecane	240	$C_{15}H_{28}O_2$	0.81
4	3.777	Levomenthol	156	$C_{20}H_{40}O$	2.52
5	8.955	3, 7, 11, 15-Tetramethyl-2-hexadecen-1-ol	296	$C_{20}H_{40}O$	2.61
6	9.249	Hexadecanoic acid, Methyl ester	270	$C_{17}H_{34}O_2$	0.93

TABLE 2.5   *(Continued)*

7	9.570	n-Hexadecanoic acid	256	$C_{16}H_{32}O_2$	7.48
8	9.858	Methyllinolelaidate	294	$C_{19}H_{34}O_2$	11.62
9.	10.650	Linoleoyl chloride	298	$C_{18}H_{31}ClO$	2.49
10.	10.687	Phytol	296	$C_{20}H_{40}O$	2.48
11.	10.746	Methyl stearate	298	$C_{19}H_{38}O_2$	4.25
12.	10.848	Linolenic acid	280	$C_{18}H_{32}O_2$	0.90
13.	10.944	Octadecanoic acid	284	$C_{18}H_{36}O_2$	2.83
14	11.078	Methyl arachidonate	318	$C_{21}H_{34}O_2$	1.42
15	17.764	Methylenecholestan-3-ol	400	$C_{28}H_{48}O$	0.86
16	18.058	β-Sitosterol	414	$C_{29}H_{50}O$	1.00
17	18.866	Urs-12-ene-24-oic acid, 3-oxo-, methylester, (+)-	468	$C_{31}H_{48}O_3$	10.32
18	20.705	Geranyl linalool	290	$C_{20}H_{34}O$	23.18

TABLE 2.6   Distribution of Major Phytoconstituents in Methanolic Leaf Extracts of *A. conyzoides, C. odorata,* and *F. exasperata*

S/N	Phytoconstituents	Distribution		
		A. conyzoides	C. odorata	F. exasperata
1	Phytol	+	−	+
2	Squalene	+	+	−
3	Alpha-linolenic acid	+	+	−
4	Precolene-1	+	−	−
5	Caryophyllene	+	−	−
6	n-Hexadecanoic acid	+	+	+
7	Calarene	−	+	−
8	Geranyl linalool	−	−	+
9	Glycerin	−	−	+
10	Methyllinolelaidate	−	−	+
11	Urs-12-ene-24-oic acid, 3-oxo-, methylester	−	−	+
12	9, 12, 15-Octadecanoic acid, (Z,Z,Z)-	−	+	−

The methanolic leaf extract of *C. odorata* showed the presence of phytochemicals that are listed in Figure 2.2 and Table 2.4. The 9, 12,

15-Octadecanoic acid, (Z,Z,Z) was predominant followed by calarene and n-Hexadecanoic acid. In this plant, biochemicals were:

- Erythritol (0.82%);
- 1-Pentene (0.62%);
- Cycloheptatrienylium, iodide (0.93);
- 2-Propanone, dimethylhydrazone (0.68%);
- Hydroquinone (0.65%);
- 2-Octyne (0.85%);
- Ethanamine, 2-phenoxy (0.76%);
- 3-Aminocrotononitrile (0.78%);
- 1-[-]-4-Hydroxy-1-methylproline (0.86%);
- Adipamide (0.66%);
- Acetic acid, 2-(1-methyl-2-oxohydrazino)-N'-[(E)-(2-hydroxyphenyl) methylidene]hydazide, N-oxide (0.58%);
- Benzenemethanol, alpha-[(methylamino)methyl] (0.75%);
- Phenylephrine (0.98%);
- Cyclohexanol, 1R-4-trans-acetamido-2, 3-trans-epoxy (0.71%);
- 1-Eicosanol (0.84%);
- 5-Hydroxy-4,'7-dimethoxyflavanone (0.62%);
- Vitamin E (0.62%);
- Calarene (7.64%);
- Sarcosine, N-(3-cyclopentylpropionyl)-,tetradecylester (1.65%);
- Squalene (3.69%);
- Ethanone, 1-(2-hydroxy-5-methylphenyl) (1.89%);
- n-Hexadecanoic acid (5.90%);
- 1, 9-Nonanediol, dimethanesulfonate (3.93%);
- Tricyclo[4.3.1.0(3, 8)]decan-10-ol (1.16%);
- Bicyclo[3.1.1]heptane, 2, 66-trimethyl-[1R-(1α, 2α, 5α)] (1.62%);
- Tricyclo[4.3.1.1(3, 8)undecane-1-carboxylic acid (0.49%);
- Taurultam (0.86%), Oxetane, 2, 2, 4-trimethyl (0.82%);
- 1, 2, 3, 4-Butanetetrol, [S-R*, R*)] (0.54%);
- 1, 3, 5-Hexatriene, 3-methyl-,(E) (2.45%);
- Spiro[5.5]undec-2-ene, 3, 7, 7-trimethyl-11-methylene-, (-) (0.75%);
- Bicyclo[2.2.1]-Heptan-2-one,4,7,7-trimethyl-,semicarbazone(0.74%);
- 1H-3a, 7-Methanoazulene, octahydro1, 4, 9, 9-tetramethyl (2.85%);
- 9, 12, 15-Octadecanoic acid, (Z,Z,Z) (11.18%);
- Cyclohexane, 1-ethenyl-1-methyl-2, 4-bis(1-methylethenyl)-,[1S-(1α, 2β, 4β] (2.72%);

- Arginine (0.34%);
- 1-Nitro-2-acetamido-1, 2-dideoxy-d-glucitol (0.34%);
- Acetamide, N-(aminocarbinyl)2-chloro (0.34%);
- 7-Octen-2-ol, 2-methyl-6-methylene (0.37%);
- Benzaldehyde, 3-hydroxy-, oxime (0.40%);
- Tetrahydro-4H-pyran-4-ol (0.41%);
- Isopropenylbromide (0.41%);
- 2-Propanone, dimethylhydrazone (0.42%);
- 1, 2-Benzenediol, 4-[2-(methylamino)ethyl] (0.42%);
- Metaraminol (0.43%);
- DL-Cystine (0.45%);
- Pyridine, 2, 3-dimethyl (0.46%);
- Metaraminol (0.46%);
- 2-Isopropoxyethylamine (0.50%);
- Methyl-2-O-methyl-β-D-xylopyranoside (0.51%);
- 1, 2-Propanediol, 3-Chloro (0.57%);
- 1, 5-Heptadien-3yne (0.97%);
- α-Cubebene (1.09%);
- 4,'5-Dihydroxy-7-methoxyflavanone (1.15%);
- α-Farnesene (1.18%);
- Catechol (1.20%);
- Humulene (1.35%);
- 2-Cyclohexen-1-one, 4, 4-dimethyl (1.35%);
- Estran-3-one, 17-(aceyloxy)-2-methyl-, (2α, 5β, 17β) (1.41%);
- Benzaldehyde, 2-hydroxy-5-nitro (1.50%);
- 2-Pentadecyn-1-ol (1.55%);
- 2, 4-Dimethylamphetamine (0.99%);
- β-Cubebene (2.06%);
- Phenol, 2, 6-dimethyl-5-methylphenyl) (0.89%);
- Leucopterin (3.02%);
- Oxirane, decyl (3.29%);
- Diazoadamantane (1.00%), Benzene, 1, 2-diethyl (3.39%);
- Methyl-4, 7, 10, 13, 16, 19-docosahexaenoate (1.13%).

Figure 2.3 and Table 2.5 indicate phytochemicals in the methanolic leaf extract of *Ficus exasperata* using GC-MS spectral analysis. Geranyl linalool was predominant followed by glycerin, Methyllinolelaidate, Urs-12-ene-24-oic acid, 3-oxo-, methylester, (+) and n-Hexadecanoic acid. Biochemicals were:

- β-Sitosterol (1.00%);
- 3-(Prop-2-enolyoxy) dodecane (0.81%);
- 3, 7, 11, 15-Tetramethyl-2-hex-adecen-1-ol (2.61%);
- Eucalyptol (0.82%);
- Geranyl linalool (23.18%);
- Glycerin (12.20%);
- Hexadecanoic acid, methylester (0.93%);
- Levomenthol (2.52%);
- Linolenic acid (0.90%);
- Linoleoyl chloride (2.49%);
- Methyl arachidonate (1.42%);
- Methyl stearate (4.25%);
- Methylenecholestan-3-ol (0.86%);
- Methyllinolelaidate (11.62%);
- n-Hexadecanoicacid (7.48%);
- Octadecanoicacid (2.83%);
- Phytol (2.48%);
- Urs-12-ene-24-oic acid, 3-oxo-, methylester, (+) (10.32%).

The distribution of the major phytochemicals in the leaf extract of *A. conyzoides, C. odorata,* and *F. exasperata* is presented in Figure 2.4 and Table 2.6. The Venn diagram summarizes the occurrence of phytochemicals in *A. conyzoides, C. odorata,* and *F. exasperata,* with n-Hexadecanoic acid as most predominant botanical among the major bioactive compounds isolated from the three plant samples. Squalene and alpha-linolenic acid were present in both *A. conyzoides* and *C. odorata;* while phytol was present in *A. conyzoides* and *F. exasperata.*

The outcome of the FT-IR spectral analysis of *A. conyzoides* showed the following functional groups (Table 2.7):

- Alcohols;
- Aliphatic amines;
- Alkanes, aromatic organophosphorus compounds;
- Arenes;
- Aromatic nitro compounds;
- Aromatics;
- Bonded phosphorus oxide;
- Bromoalkanes;
- Disulfides;

- Meta-substituted benzenes;
- Phenols;
- Saturated aldehydes;
- Terminal alkynes;
- Tri-substituted alkenes;
- Vinyl group.

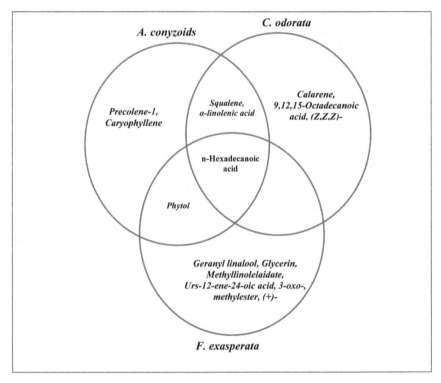

FIGURE 2.4 Venn diagram showing the distribution of major phytoconstituents in methanolic leaf extracts of *A. conyzoides*, *C. odorata*, and *F. exasperata*.

The absorption bands at 3274.20 cm⁻¹ revealed alcohols and phenols OH strong bond, 2919.68 to 2851.63 cm⁻¹ showed alkane CH_3, CH_2 and CH bends, 2116.41 cm⁻¹ revealed terminal alkyne bends, 1733.44 cm⁻¹ revealed saturated aldehyde bend, 1604.63 cm⁻¹ revealed are nestretching, 1418.67 cm⁻¹ revealed stretching due to aromatic organophosphorus compounds, 1317.13 cm⁻¹ indicated bend due aromatic nitro compounds, 1241.03 cm⁻¹ revealed phosphorus oxide bending, 1025.99 cm⁻¹ revealed aliphatic amines C-N bend, 893.80 and 765.22 cm^{-1} showed meta-substituted benzenes, aromatics stretching, 832.79 cm^{-1} indicated trisubstituted alkenes, vinyl group bend, 589.27 and 572.93 cm^{-1} revealed bromoalkane stretching and 536.20 cm^{-1} indicated disulfide S-S bend.

TABLE 2.7   Functional Groups of Methanolic Leaf Extract *Ageratum conyzoides* by FT-IR

S/N	Wave Number (cm^{-1})	Functional Group
1	536.20	S-S (disulfides)
2	572.93	C-X (Bromo alkanes, alkylhalides)
3	589.27	C-X (Bromo alkanes, alkylhalides)
4	663.13	O-H out of plane bend (Alcohols and phenols)
5	765.22	C-H (meta substituted benzene, aromatics)
6	832.79	C-H (trisubstituted alkenes, vinyl group)
7	893.80	C-H (meta substituted benzene, aromatics)
8	1025.99	C-N (Aliphatic amines)
9	1241.03	P-O (bonded phosphorus oxide)
10	1317.13	N-O (Aromatic nitro compounds)
11	1374.65	CH$_3$ deformation (Alkanes)
12	1418.67	P-C (Aromatic organophosphorus compounds)
13	1604.63	C=C (Arenes)
14	1733.44	C=O (Saturated aldehydes)
15	2116.41	C≡C (Terminal alkynes)
16	2851.63	CH$_3$, CH$_2$, and CH (alkanes)
17	2919.68	CH$_3$, CH$_2$, and CH (alkanes)
18	3274.20	O-H (Strong bond) (Alcohols and phenols)

The possible functional groups (Table 2.8) assigned to the bioactive compounds of the methanolic leaf extract of *C. odorata* consist of:

- Alcohols;
- Alkynes;
- Arenes;
- Aromatic ethers;
- Aromatic organophosphate compounds;
- Fluoroalkanes;
- Phenols;

- Saturated aldehyde;
- Tri-substituted alkenes.

TABLE 2.8 Functional Groups of Methanolic Leaf Extract of *Chromolaena odorata* by FT-IR

S/N	Wave Number (cm⁻¹)	Functional Group
1	609.40	C-H deformation (Alkynes)
2	767.83	O-H bend (Alcohols and Phenols)
3	816.50	C-H (tri substituted Alkenes, vinyl group)
4	1027.02	C-X (Fluoroalkanes, alkyl halides)
5	1156.15	C-O (Aromatic ethers)
6	1245.85	C-H (Arenes)
7	1412.70	P-C (Aromatic organophosphorus compounds)
8	1601.55	C=C (Aromatics, with benzene ring)
9	1729.69	C=O (Saturated aldehyde)
10	2850.27	CH_3, CH_2, and CH (Alkanes)
11	2917.87	CH_3, CH_2, and CH (Alkanes)
12	3271.98	O-H bend (Alcohols and Phenols)

The absorption bands at 3271.98 cm⁻¹ revealed O-H stretching of strong bond, 2917.87 to 2850.27 cm⁻¹ are attributed to $C-H_3$, $C-H_2$ and C-H bends of alkanes, 1729.69 cm⁻¹ indicated saturated aldehydes, 1601.55 cm⁻¹ is as a result of aromatic and benzene ring stretching, 1412.70 cm⁻¹ is due to aromatic organophosphorus compounds bend, 1245.85 showed arenes stretching, 1156.15 cm⁻¹ indicated aromatic ethers bend, 1027.02 cm⁻¹ is due to fluoro-alkanes bend, 816.50 cm⁻¹ indicated tri-substituted alkenes, vinyl group C-H bending while 609.40 cm⁻¹ showed C-H deformation of alkynes present in the extract.

The bioactive compounds of the methanolic leaf extract of *Ficus exasperata* showed the presence of the following functional groups (Table 2.9):

- Alcohols;
- Aldehydes;
- Aliphatic;
- Alkanes;
- Amides;

- Amines;
- Bonded phosphorus oxide;
- Carboxylic acids derivatives;
- Ketones;
- Phenols;
- Primary amines and amides.

TABLE 2.9 Functional Groups of Methanolic Leaf Extract of *Ficus exasperata* by FT-IR

S/N	Wave Number (cm⁻¹)	Functional Group
1	766.15	O-H bend (Alcohols and phenols)
2	1027.05	C-N (Aliphatic amines)
3	1243.92	P-O (bonded phosphorus oxide)
4	1319.28	N-O (Aliphatic nitro compounds)
5	1374.44	CH₃ deformation (Alkanes)
6	1421.55	O-H in plane bend (Alcohols and Phenols)
7	1612.73	-NH₂ bend (1° Amines and amides)
8	1733.05	C=O (Amides, ketones, aldehydes, carboxylic acids/derivatives)
9	2850.10	O-H (carboxylic acids)
10	2917.81	O-H (carboxylic acids/derivatives)
11	3271.27	O-H (Alcohols and phenols)

The absorption bands at 3271.27 cm⁻¹ indicated the O-H stretching of strong bond, 2917.81 to 2850.10 cm⁻¹ are attributed to O-H bend of carboxylic acids/derivatives, 1733.05 cm⁻¹ revealed bends due to amides, ketones, aldehyde, and carboxylic acids/derivatives, 1612.73 cm⁻¹ indicated primary amines and amides –NH₂ stretching, 1421.55 cm⁻¹ showed O-H in plane bend of alcohols and phenols, 1374.44 cm⁻¹ showed alkanes CH₃ deformation, 1319.28 cm⁻¹ is attributed to aliphatic nitro-compounds bend, 1243.92 cm⁻¹ revealed bonded phosphorus oxide P-O stretching, 1027.05 cm⁻¹ is due to aliphatic amines C-N bending while that of 766.15 cm⁻¹ indicated O-H bending of alcohols and phenols.

From the bioactive compounds found in the methanolic leaf extracts of the plants, the biological activities of few have been elucidated in Table 2.10.

TABLE 2.10 Bioactivities of Selected Bioactive Compounds Identified in the Methanolic Leaf Extracts of the Plants Under Study

Bioactive Compounds	Bioactivities	References
(-)-Spathulenol	Precursor of various sesquiterpenes	[73]
3, 7, 11, 15-Tetramethyl-2-hexadecen-1-ol	Antimicrobial and anti-inflammatory properties	[71]
9, 12, 15-Octadecanoic acid	Anti-inflammatory, hypocholesterolemic, cancer preventive, hepatoprotective, nematicide, insect fuge, antihistaminic, and antiarthritic activities.	[13, 63]
Arginine	Wound healing, immune function	[72]
Aromadendrene	Bactericidal activity	[22]
Caryophyllene	Anti-inflammatory, anticarcinogenic, skin penetration-enhancing, antinociceptive, neuroprotective, anxiolytic, anti-depressant, and anti-alcoholism effects.	[11, 31, 43, 68]
Cystine	Precursor to the antioxidant glutathione and iron-sulfur, maintenance of protein conformation.	[46, 51]
Epi-β-Santalene	Antidepressant, anti-inflammatory, antifungal, astringent, sedative, insecticide, and antiseptic agents	[21]
Farnesene	Antimicrobial, anti-inflammatory, anti-leishmanial, antiproliferative, antimutagenic, anti-inflammatory, analgesic, and gastroprotective effects	[61, 64, 76]
Geranyl linalool	Flavor enhancer	[17, 78]
Germacrene-D	Anti-inflammatory effects	[27]
Glycerin	Antiretinopathic, antimicrobial, and anti-inflammatory activities	[29, 40]
Hexadecanoic acid,	Antioxidant, anti-inflammatory anti-arthritic and hypocholesterolemic	[24, 48]
Hexadecanoic acid, methyl ester	Immunomodulatory, mosquito repellant, antimicrobial, and anti-inflammatory activities	[34, 81]
Humulene	Anti-inflammatory effects on TNF-α and IL1β generation in carrageenan-injected rats	[27]
Hydroquinone	Skin treatment	[59]
Methyl Linolenate	Antioxidant and catalase activator activity	[77]
Methyl Stearate	Anti-diarrheal, cytotoxic, and anti-proliferative activities	[53]
Methylarachidonate	Hypotension, anti-diabetes, antitumor, and anti-arthritis effects	[67]

TABLE 2.10 *(Continued)*

Bioactive Compounds	Bioactivities	References
Methylenechlestan-3-ol	Hypocholesterolemic effect	[82]
Methyllinolelaidate	Antioxidant and catalase activator properties	[77]
Octadecenoic acid	Antifungal potential	[71]
Phytol	Antimicrobial, anticancer, anti-inflammatory, antidiuretic, immunostimulatory, and anti-diabetic activities	[20, 52, 77]
Squalene	Anti-inflammatory, antioxidant, antibacterial, cancer-preventive, immunostimulant, and antitumor effects	[63, 77]
Urs-12-ene-24-oic acid, 3-oxo-, methyl ester, (+)-	Antibacterial activity	[50]
Vitamin E	Anti-aging, analgesic, antidiabetic, anti-inflammatory, antioxidant, anti-coronary, antibronchitic, hepatoprotective, hypocholesterolemic, antidermatitic, antitumor, anticancer, antiulcerogenic, antispasmodic, antileukemic properties	[63, 65]
β-Sitosterol	Antihyperlipoproteinaemic, antibacterial, antitumor, and antimycotic activities	[9, 84]

2.5 DISCUSSION

The pharmacological properties of many plants used in herbal medicine are due to the combined actions of many phytoconstituents. The techniques for the identification of phytochemicals have made phytochemistry a relevant field in ethnomedicine. The present study was aimed at determining and characterizing the bioactive compounds resident in three medicinal plants (*Ageratum conyzoides, Chromolaena odorata,* and *Ficus exasperata*) and their established biological activities using more recent phytochemical detection schemes of GC-MS and FT-IR.

The phytochemicals detected in the methanolic leaf extracts of *Ageratum conyzoides, Chromolaena odorata,* and *Ficus exasperata* (Table 2.2) have a lot of pharmacological properties [33, 60]. Consistent intake of foods rich in secondary metabolites (phytochemicals) has been known over the years to offer substantive health benefits. Alkaloids are basic nitrogenous plant products and have been known to possess analgesic, antispasmodic, and bactericidal functions [4]. Flavonoids are a large family of hydroxylated polyphenolic compounds and have been associated with anti-inflammatory, antitumor, antivirus, antibacterial, and antioxidant activities [83].

GC-MS and FT-IR are sophisticated techniques used in the determination and characterization of bioactive compounds residents in plants [55]. Some of the bioactive compounds identified in the methanolic leaf extracts of *A. conyzoides, C. odorata,* and *F. exasperate* possess many biological properties (Table 2.10) [25]. Eighteen (18) bioactive compounds, including some of those recorded in this study, have been reported in *A. conyzoides* [36]. The FT-IR spectral analysis is an exquisite technique for the determination of the functional groups presents in medicinal plants [10].

The identified functional groups in these three plants were: disulfides, bromoalkanes, alcohols, phenols, meta substituted benzenes, aromatics, trisubstituted alkenes, vinyl group, aliphatic amines, bonded phosphorus oxide, aromatic nitro compounds, alkanes, alkynes, aromatic organophosphorus compounds, arenes, saturated aldehydes and terminal alkynes, fluoroalkanes, aromatic ethers, primary amines and amides, amides, ketones, aldehydes, and carboxylic acids derivatives. These groups may be responsible for various pharmacological activities of the plants considered in this study.

2.6 SUMMARY

The current study was aimed at determining and characterizing the different bioactive compounds with their functional groups contained in the methanolic leaf extracts of *Ageratum conyzoides, Chromolaena odorata,* and *Ficus exasperata* using sophisticated detection schemes of GC-MS and FT-IR. The preliminary phytochemical analysis revealed the existence of alkaloids, anthocyanins, anthraquinones, cardiac glycosides, flavonoids, phenols, saponins, steroids, tannins, and terpenes, with alkaloids, flavonoids, and steroids evenly distributed in all three plant samples. A total of 107 bioactive compounds were identified, out of which 20 were from *A. conyzoides,* 69 from *C. odorata,* and 18 from *F. exasperata,* respectively.

The major compounds identified in the methanolic leaf extract of *A. conyzoides* include α-Linolenic acid (27.05%), Precolene-1 (11.44%), Caryophyllene (10.44%), n-Hexadecanoic acid (9.01%), Squalene(6.98%) and Phytol (5.39%).

The results also revealed that 9, 12, 15-Octadecanoic acid, (Z,Z,Z) (11.18%), Calarene (7.64%) and n-Hexadecanoic acid (5.90%) occurred abundantly in *C. odorata* while Geranyl linalool (23.18%), Methyl-linolelaidate (11.62%), Urs-12-ene-24-oic acid, 3-oxo-, methyl ester, (+) (10.32%) and n-Hexadecanoic acid (7.48%) were predominant in *F. exasperata*.

The Venn diagram showing the distribution of the compounds revealed that n-Hexadecanoic acid was present in all three samples; squalene and α-linolenic acid in *A. conyzoides* and *C. odorata*; and phytol in *A. conyzoides* and *F. exasperata*. The degree of bioaccumulation of phyto-constituents in the order of *C. odorata* > *Ageratum conyzoides* > *Ficus exasperata*.

These biological molecules responsible for their various pharma-cological activities possess the functional groups, such as: disulfides, bromoalkanes, alcohols, phenols, meta-substituted benzenes, aromatics, trisubstituted alkenes, vinyl group, aliphatic amines, bonded phosphorus oxide, aromatic nitro compounds, alkanes, alkynes, aromatic organophos-phorus compounds, arenes, saturated aldehydes and terminal alkynes, fluo-roalkanes, aromatic ethers, primary amines and amides, amides, ketones, aldehydes, and carboxylic acids derivatives. These findings conclude that the leaf extracts of these three plants and their blends in the right dosage are potential pharmacological regiments for ameliorating ailments.

KEYWORDS

- *Ageratum conyzoides*
- aldehydes
- alkaloids
- alkenes
- alkynes
- arenes
- botanicals
- carboxylic acid
- *Chromolaena odorata*

- **ethnopharmacology**
- *Ficus exasperata*
- **fluoroalkanes**
- **Fourier-transform infrared spectroscopy (FTIR)**
- **gas chromatography-mass spectrometry (GC-MS)**
- **ketones**
- **organophosphorus**
- **phenols**
- **steroids**
- **tannins**
- **terpenoids**

REFERENCES

1. Acharya, N. A., Deepak, M., & Shrivastava, A. *Indigenous Herbal Medicines: Tribal Formulations and Traditional Herbal Practices* (pp. 440–520). India, Aavishkar Publishers Distributor, **2008**.

2. Adewole, S. O., Adenowo, T. K., Thajasvarie, N., & Ojewole, J. A. O. Hypoglycaemic and hypotensive effects of *Ficus exasperata* Vahl (Moraceae) leaf aqueous extract in rats. *African Journal of Traditional, Complementary and Alternative Medicine,* **2011**, *8*(11), 275–283.

3. Afolabi, C. Phytochemical constituents and antioxidant properties of extracts from the leaves of *Chromolaena odorata. Scientific Research and Essay,* **2007**, *2*(6), 191–198.

4. Ahmad, A., Husain, A., Mujeeb, M., Khan, S. A., Najini, A. K., & Siddique, N. A. A review on therapeutic potential of *Nigellesarivas* a miracle herb. *Asian Pacific Journal of Tropical Biomedicine,* **2003**, *3*(5), 337–352.

5. Aiyeloja, A. A., & Bello, O. A. Ethnobotanical potentials of common herbs in Nigeria: A case study of Enugu state. *Educational Research and Review,* **2006**, *1*(1), 16–22.

6. Akinmoladun, A. C., Ibukun, E. O., & Dan-Ologe, I. A. Phytochemical constituents and antioxidant properties of extracts from the leaves of *Chromolaena odorata. Science Research Essays,* **2007**, *2*, 191–194.

7. Akobundu, I. O., & Agyakwa, C. V. *Handbook of West African Weeds* (2nd edn., p. 564). International Institute for Tropical Agriculture, Ibadan, Nigeria, **1998**.

8. Amadi, B. A., Duru, M. K. C., & Agomuo, E. N. Chemical profiles of leaf, stem, root and flower of *Ageratum conyzoides. Asian Journal of Plant Science and Research,* **2012**, *2*(4), 428–432.

9. Awad, A. B., Roy, R., & Fink, C. S. ß-sitosterol, a plant sterol, induces apoptosis and activates key caspases in MDA-MB-231 human breast cancer cells. *Oncology Reports*, **2003**, *10*, 497–500.

10. Aysal, P., Ambrus, A. D., Lehotay, S. J., & Cannavan, A. Validation of an efficient method for the determination of pesticide residues in fruits and vegetables using ethyl acetate for extraction. *Journal of Environmental Science Health*, **2007**, *42*, 481–490.

11. Smith, B. P. C., Hayasaka, Y., Tyler, M. J., & Williams, B. D. ß-caryophyllene in the skin secretion of the Australian green tree frog, *Litoriacaerulea*: An investigation of dietary sources. *Australian Journal of Zoology*, **2004**, *52*, 521–530.

12. Blench, R. *Nupe Plants and Trees: Their Names and Uses.* Cambridge Press, United Kingdom, **2008**. https://www.academia.edu/3776854/Nupe_plant_names_and_uses (Accessed on 30 September 2019).

13. Blondeau, N., Lipsky, R. H., Bourourou, M., Duncan, M. W., Gorelick, P. B., & Marini, A. M. Alpha-linolenic acid: An omega-3 fatty acid with neuroprotective properties: Ready for use in the stroke clinic. *Biomedical Research International*, **2014**, *2015*, 1–8.

14. Buniyamin, A. A., Eric, K. I. Q., & Fabian, C. A. Pharmacognosy and Hypotensive evaluation of *Ficus exasperata* Vahl (Moraceae) leave. *Acta Poloniae Pharmaceutica-Drug Research*, **2007**, *64*(6), 543–546.

15. Bunyapraphatsara, N., & Chokechaijaroenporm, O. Thai medicinal plants. *National Center for Genetic Engineering and Biotechnology, Bangkok*, **2000**, *4*, 622–626.

16. Chakraborty, A. K., Sujit, R., & Umesh, K. P. *Chromolaena odorata*: An overview. *Journal of Pharmacy Research*, **2011**, *43*, 573–576.

17. Chen, W., & Viljoen, A. M. A review of a commercially important fragrance material. *South African Journal of Botany*, **2010**, *76, 643–651.*

18. Chopra, R. N., Nayar, S. L., & Chopra, I. C. *Glossary of Indian Medicinal Plants* (p. 9). NISCIR, New Delhi, **2002**.

19. Cousins, O. N., & Micheal, A. H. Medicinal properties in the diet of Gorillas: An Ethnopharmacological Evaluation. *African Study Monograph*, **2002**, *23*, 65–89.

20. Daines, A., Payne, R., Humphries, M., & Abell, A. The synthesis of naturally occurring vitamin k and vitamin k analogs. *Current Organic Chemistry*, **2003**, *7*(16), 1625–1634.

21. Daramwar, P. P., Srivastava, P. L., Priyadarshini, B., & Thulasiram, H. V. Preparative separation of α- and β-santalenes and (Z)-α- and (Z)-β-santalols using silver nitrate-impregnated silica gel medium pressure liquid chromatography and analysis of sandalwood oil. *Analyst*, **2012**, *137*, 4564–4570.

22. Deng, G. B., Zhang, H. B., Xue, H. F., Chen, S. N., & Chen, X. L. Chemical composition and biological activities of essential oil from the rhizomes of *Iris bulleyana. Agricultural Science, China*, **2009**, *8*(6), 691–669.

23. Derwich, E., Benziane, Z., & Boukir, A. Chemical composition and antibacterial activity of leaves of essential oil of *Laurus nobilis* from Morocco. *Australian Journal of Basic Applied Sciences*, **2009**, *3*, 3818–3824.

24. Dhanalakshmi, R., & Manavalan, R. Bioactive compounds in leaves of *Corchorus trilocularis* L. by GC-MS analysis. *International Journal of Pharmtech. Research*, **2014**, *6*(7), 1991–1998.

25. Duke, J. Duke's phytochemical and ethnobotanical databases [online], **1998**. www.ars-grin.gov/duke/ (Accessed on 30 September 2019).

26. Ezekwesili, C. N., Ghasi, S., Adindu, C. S., & Mefoh, N. C. Evaluation of the anti-ulcer property of aqueous extract of unripe *Musa paradisiacal* Linn. peel in Wistar rats. *African Journal of Pharmacy and Pharmacology*, **2014**, *89*(39), 1006–1011.

27. Fernandes, E. S., Passos, G. F., Medeiros R., Da Cunha, F. M., Ferreira, J., Campos, M. M., Pianowski, L. F., & Calixto, J. B. Anti-inflammatory effects of compounds alpha-humulene and (−)-trans-caryophyllene isolated from the essential oil of *Cordiaverbenacea*. *European Journal of Pharmacology*, **2007**, *569*(3), 228–236.

28. Florence, A. R., & Jeeva, S. FTIR and GC-MS spectral analysis of *Gmelina asiatica* L. leaves. *Science Research Reporter*, **2015**, *5*(2), 125–136.

29. Frank, M. S., Nahata, M. C., & Hitty, M. D. Glycerol: A review of its pharmacology, pharmacokinetics, adverse reactions and clinical use. *Pharmacotherapy*, **1981**, *1*, 47–60.

30. Gbadamosi, I. T. Evaluation of antibacterial activity of six ethnobotanicals used in the treatment of infectious diseases in Nigeria. *Botany Research International*, **2012**, *5*(4), 83–89.

31. Gertsch, J., Leonti, M., & Raduner, S. Beta-caryophyllene is a dietary cannabinoid. *Proceedings of the National Academy of Sciences of the United States of America*, **2008**, *105*(26), 9099–9104.

32. Gilani, A. H. Role of medicinal plants in modern medicine. *Malaysian Journal of Science*, **2005**, *24*(1), 1–5.

33. Harbone, J. B. *Phytochemical Methods: A Guide to Modern Technology of Plant Analysis* (3rd edn., pp. 203–214). Springer, New York, **2008**.

34. Hema, R., Kumaravel, S., & Alagusundaram, K. GC-MS determination of bioactive components of *Murrayakoenigii*. *Journal of American Science*, **2011**, *7*(1), 80–83.

35. Igwenyi, I. O., Isiguzo, O. E., Aja, P. M., Ugwu-Okechukwu, P. C., Ezeani, N. N., & Uraku, A. J. Proximate composition, mineral content and phytochemical analysis of the African oil bean (*Pentaclethra macrophylla*) seed. *American-Eurasian Journal of Agriculture and Environmental Science*, **2015**, *15*(9), 1873–1875.

36. Ilondu, E. M., Ojeifo, I. M., & Emosairue, S. O. Evaluation of antifungal properties of *Ageratum conyzoides, Spilanthesfilicaulis* and *Tithoniadiversifolia* leaf extracts and search for their compounds using gas chromatography-mass spectrum. *ARPN Journals of Agricultural and Biological Sciences*, **2014**, *9*(11), 375–384.

37. Ilondu, E. M., & Okoegwale, E. E. Some medical plants used in the management of Dermatophytic diseases in Nigeria. *African Journal of Environmental Studies*, **2002**, *3* (1&2), 146–151.

38. Inyang, C. U., & Adegoke, A. A. Antimicrobial properties and preliminary phytochemical screening of *Chromolaena odorata* (siam or sapysa weed) leaf. *Nigerian Journal of Microbiology*, **2008**, *22*(1), 1652–1659.

39. Irene, I. I., & Iheanacho, U. A. Acute effect of ethanol extracts of *Ficus exasperata* Vahl on kidney function in albino rats. *Journal of Medical Plant Research*, **2007**, *1*, 27–29.

40. Jananie, R. K., Priya, V., & Vijaylakshmi, K. Determination of bioactive component of cynodondactylon by GC-MS analysis. *New York Science Journal*, **2011**, *4*, 4–6.

41. Jini, J., Sindhu, T. J., Girly, V., David, P., Kumar, B. D., Bhat, A. R., & Krishakumar, K. Review on *in vivo* and *in vitro* studies on the pharmacological activities of *Musa* species. *World Journal of Pharmacy and Pharmaceutical Sciences*, **2014**, *3*(2), 1133–1142.

42. Kalimuthu, K., & Prabakaran, R. Preliminary phytochemical screening and GC-MS analysis of methanol extract of *Ceropegiapusilla*. *International Journal of Research in Applied Natural and Social Sciences*, **2013**, *1*(3), 49–58.

43. Katsuyama, S., Mizoguchi, H., & Kuwahata, H. Involvement of peripheral cannabinoid and opioid receptors in β-caryophyllene-induced antinociception. *European Journal of Pain*, **2013**, *17*(5), 664–675.

44. Kazeem, M. I., Oyedapo, B. F., Raimi, O. G., & Adu, O. B. Evaluation of *Ficus exasperata* Vahl leaf extracts in the management of diabetes mellitus in vitro. *Journal of Medical Science*, **2013**, *13*(4), 269–275.

45. Krishnaveni, M., & Saranya, S. Phytoconstituent analysis of *Nigella Sativa* seeds using analytical techniques. *Bulletin of Environmental, Pharmacological and Life Sciences*, **2016**, *5*(3), 25–38.

46. Lill, R., & Mühlenhoff, U. Iron-sulfur protein biogenesis in eukaryotes: Components and mechanisms. *Annual Review of Cell Biology*, **2006**, *22*, 457–486.

47. Lourens, A. C., Reddy, D., Baser, K. H., & Viljoen, V. V. S. F. *In vitro* biological activity and essential oil composition of four indigenous South African *Helichrysum Species*. *Journal of Ethnopharmacology*, **2004**, *95*(2/3), 253–258.

48. Maruthupandian, A., & Mohan, V. R. GC-MS analysis of ethanol extract of *Wattakaka volubilis* (L.f) Stapf. *leaf. International Journal of Phytomedicine*, **2011**, *3*, 59–62.

49. Mbajiuka, C. S., Obeagu, E. I., Chude, C. N., & Ihezie, O. E. Antimicrobial effects of *Chromolaena odorata* on some human pathogens. *International Journal of Current Microbiology and Applied Sciences*, **2014**, *3*(3), 1006–1012.

50. Natarajan, D., Gomathi, M., & Yuvarajan, A. Phytochemical and antibacterial evaluation of *Barleria montana*. *Asian Journal of Pharmaceutical and Clinical Research*, **2012**, *5*(3), 44–46.

51. Nelson, D. L., & Cox, M. M. *Lehninger, Principles of Biochemistry* (3rd edn., p. 1119). Worth Publishing: New York, **2000**.

52. Netscher, T. Synthesis of vitamin E. *Vitamins & Hormones*, **2007**, *76*, 155–202.

53. Noryawati, M., Bibiana, W. L., Laora, O., & Sri, R. Antidiarrheal activity of apus Bamboo (*Gigantochloa apus*) leaf extract and its bioactive compounds. *American Journal of Microbiology*, **2013**, *4*(1), 1–8,

54. Odunbaku, O. A., Ilusanya, O. A., & Akasoro, K. S. Antibacterial activity of ethanolic leaf extract of *Ficus exasperata* on *Escherichia coli* and *Staphylococcus albus*. *Science Research Essay*, **2008**, *3*(11), 562–564.

55. Okereke, S. C., Arunsi, U. O., Nosiri, C. I., & Nwadike, C. Gas chromatography mass spectrometry/Fourier transform infrared (GC-MS/FTIR) spectral analyzes of *Tithonia diversifolia* (Hemsl.) leaves. *Journal of Medicinal Plants Research*, **2017c**, *11*(19), 345–350.

56. Okereke, S. C., Arunsi, U. O., Ngwogu, A. C., Onyike, J. O., & Chukwudoruo, S. C. Identification of the active ingredients in methanolic leaf extract of *Aspilia africana* (Pers.): Adams and their effects on ibuprofen induced ulcer model in Wistar rats. *International Journal of Science and Research*, **2018**, *7*(1), 310–319.

57. Okunrobo, L., Usifoh, C., Ching, P., & Bariweni, M. Anti-inflammatory evaluation of methanol extract and aqueous fraction of the leaves of *Anthocleista djalonensis*. *The International Journal of Pharmacology*, **2008**, *7*, 28–38.

58. Oladejo, O. W., Imosemi, I. O., Osuagwu, F. C., Oluwadara, O. O., Aiku, A., Adewoyin, O., Ekpo, O. E., Oyedele, O. O., & Akang, E. E. U. Enhancement of cutaneous wound healing by methanolic extracts of *Ageratum conyzoides* in Wistar rat. *African Journal of Biomedical Research,* **2003**, *6*(1), 27–31.

59. Olumide, Y. M., Akinkugbe, A. O., Altraide, D., Mohammed, T., Ahamefule, N., Ayanlowo, S., Onyekonwu, C., & Essen, N. Complications of chronic use of skin lightening cosmetics. *International Journal of Dermatology,* **2008**, *47*(4), 344–353.

60. Oyeleke, G. O., Odedeji, J. O., Ishola, A. D., & Afolabi, O. Phytochemical screening and nutritional evaluation of african oil bean (*Pentaclethra macrophylla*) seeds. *Journal of Environmental Science, Toxicology and Food Technology (IOSR-JESTFT),* **2014**, *8*(2), 14–17.

61. Paiva, L. A. F., Gurgel, L. A., Campos, A. R., Silveira, E. R., & Rao, V. S. N. Attenuation of ischemia/reperfusion-induced intestinal injury by oleo-resin from *Copaifera langsdorffii* in rats. *Life Science,* **2004**, *75*, 1979–1987.

62. Phan, T. T., Wang, L., See, P., Grayer, R. J., Chan, S. Y., & Lee, S. T. Phenolic compounds of *Chromolaena odorata* protect cultured skin cells from oxidative damage: Implication for cutaneous wound healing. *Biological and Pharmaceutical Bulletin,* **2001**, *24*, 1373–1379.

63. Rajeswari, G., Murugan, M., & Mohan, V. R. GC-MS analysis of bioactive components of *Hugonia mystax* L. *Research Journal of Pharmaceutical, Biological and Chemical Sciences,* **2012**, *3*(4), 301–308.

64. Santos, A. O., Ueda-Nakamura, T., Dias, F. B. P., Veiga, V. F., Pinto, A. C., & Nakamura, C. V. Antimicrobial activity of Brazilian copaiba oils obtained from different species of the *Copaifera*genus. *Memorias do Instituto Oswaldo Cruz, Rio de Janeiro,* **2008**, *103*(3), 277–281.

65. Sato, K., Gosho, M., Yamamoto, T., Kobayashi, Y., Ishii, N., Ohashi, T., Nakade, Y., Ito, K., Fukuzawa, Y., & Yoneda, M. Vitamin E has a beneficial effect on nonalcoholic fatty liver disease: A meta-analysis of randomized controlled trials. *Nutrition,* **2015**, *31*, 923.

66. Sharma, P. D., & Sharma, O. P. Natural products chemistry and biological properties of the *Ageratum* plant. *Toxicological and Environmental Chemistry,* **1995**, *50*, 213–217.

67. Simopoulos, A. P. Omega-3 fatty acids and antioxidants in edible wild plants. *Biological Research,* **2004**, *37*, 263–277.

68. Singh, G., Marimuthu, P., De Heluani, C. S., & Catalan, C. A. Antioxidation and biocidal activities of *carumnigrum* (seed) essential oil, oleoresin, their selected components. *Journal of Agricultural and Food Chemistry,* **2006**, *54*(1), 174–181.

69. Sofowora, A. *Medicinal Plant and Traditional Medicine in Africa* (3rd edn. pp. 172–188). Spectrum Books Limited: Ibadan, **1999**.

70. Sonibare, M. A., Ogunwande, I. A., Walker, T. M., Setzer, W. N., Soladoye, M. O., & Essien, E. Volatile constituents of *Ficus exasperata* leaves. *Natural Product Communications,* **2006**, *1*, 763–765.

71. Sudha, T., Chidambarampillai, S., & Mohan, V. R. GC-MS analysis of bioactive components of aerial parts of *Fluggealeucopyrus* Willd (Euphorbiaceae). *Journal of Applied Pharmaceutical Science,* **2013**, *3*(5), 126–130.

72. Tapiero, H., Mathé, G., Couvreur, P., & Tew, K. D. L-Arginine. *Biomedicine and Pharmacotherapy,* **2002**, *56*(9), 439–445.

73. Telascrea, M., De Araújo, C. C., Marques, M. O. M., Facanali, R., De Moraes, P. L. R., & Cavalheiro, A. J. Essential oil leaves of *Cryptocaryamandioccana* Meisner (Lauraceae): Composition and intraspecific chemical variability. *Biochemical and Systemic Ecology*, **2007**, *35*, 222–232.

74. Trease, G. E., & Evans, W. C. *Pharmacognosy* (15th edn., pp. 42–44, 221–229, 246–249, 304–306, 331–332, 391–393). Saunders Publishers: London, **2002**.

75. Tweheyo, M., Lye, K. A., & Weladji, R. B. Chimpanzee diet and habitat selection in the Budongo forest reserve, Uganda. *Forest and Ecological Management*, **2004**, *188*, 267–278.

76. Veiga, V. F. Jr., Rosas, E. C., Carvalho, M. V., Henriques, M. G. M. O., & Pinto, A. C. Chemical composition and anti-inflammatory activity of copaiba oils from *Copaifera cearensis* Huber ex Ducke, *Copaifera reticulata* Ducke and *Copaifera multijuga* Hayne-A comparative study. *Journal of Ethnopharmacology*, **2007**, *112*, 248–254.

77. Venkata, R. B., La, S., Pardha, P. M., Narashimha, R. B., Naga, V. K. A., Sudhakar, M., & Radhakrishnan, T. M. Antibacterial, antioxidant activity and GC-MS analysis of *Eupatorium odoratum*. *Asian Journal of Pharmaceutical Clinical Research*, **2012**, *5*(2), 99–106.

78. Wang, S., Liu, H., Yang, Y., Fan, W., Wilson, L. W., Wang, Q., & Qiu, D. Studies on the production of (E,E)-geranyl linalool in *E. coli*. *Research and Review: Journal of Microbiology and Biotechnology*, **2015**, *4*(3), 1–4.

79. Woode, E., Poku, R. A., & Abotsi, W. K. M. Anticonvulsant effects of a leaf extract of *Ficus exasperate* Vahl (Moraceae) in mice. *International Journal of Pharmacology*, **2011**, *4*, 138–151.

80. World Health Organization (WHO). *Traditional Medicine and Modern Health Care* (Vol. 44, No. 10, p. 67). Progress Report by the Director General, Document A, WHO, Geneva, **1991**.

81. Xiong, L., Peng, C., Zhou, Q., Wan, F., Xie, X., Guo, L., Li, X., He, X., & Dai, O. Chemical composition and antibacterial activity of essential oils from different parts of *Leonurus japonicas* Houtt. *Molecules*, **2013**, *18*, 963–973.

82. Xu, G., Salen, G., Tint, G. S., Batta, A. K., & Shefer, S. Campestanol (24-methyl-5alpha-cholestan-3beta-ol) absorption and distribution in New Zealand White rabbits: Effect of dietary sitostanol. *Metabolism*, **1999**, *48*, 363–368.

83. Xu, X., Ye, H., Wang, W., & Chen, G. An improved method for the quantitation of flavanoids in herbal Leonuri by capillary electrophoresis. *Journal of Agriculture and Food Chemistry*, **2005**, *53*, 5853–5857.

84. Yasukawa, K., Takido, M., Matsumoto, T., Takeuchi, M. J., & Nakagawa, S. Sterol and triterpene derivatives from plants inhibit the effects of tumor promoter and sitosterol and betulinic acid inhibits tumor formation in mouse skin two-stage carcinogenesis. *Journal of Oncology*, **1991**, *41*, 72–76.

CHAPTER 3

THE ROLE OF HERBAL MEDICINES IN FEMALE GENITAL INFECTIONS

DJADOUNI FATIMA

ABSTRACT

Genital infections are severe public health problems worldwide, and infect 100 million women yearly at different ages. The most common infectious diseases of the female are vulvovaginal candidiasis, bacterial vaginosis, trichomoniasis, viral, and parasitic protozoan infections. Many vaginosis treatments are being used by doctors. The treatment protocol varies from the use of synthetic antibiotics to probiotics, the antimicrobial agents, and the antiseptics. The interest in phytopharmaceutical and herbal medicines has been a pattern in recent years, and this approach might be valuable as a reason for developing new control of these infectious diseases. Therefore, this chapter explores available information about vaginal infections and treatment with antibiotics, natural antimicrobials, plant-derived compounds, and health benefits.

3.1 INTRODUCTION

Many human diseases are due to the action of pathogens that develop within a tissue or organ [53]. These germs are of bacterial, viral, mycotic, or protozoan origin, which causes infections or infectious diseases. Among these infections, transmissible genetic infections are public health problems because of the complications, such as: sterility, extrauterine pregnancy, cancer of the cervix, and congenital infections. *Treponema pallidum* (syphilis), *Neisseria gonorrhea*, *Chlamydia trachomatis*, *Trichomonas vaginalis*, *Candida albicans*, scarbies, and herpes simplex are most widespread pathogenic germs [50, 51, 100].

Vaginal infections are a common phenomenon among women of all ages and associated with serious health complications. It may also cause fatal cancerous diseases. Genital infections usually come from outside of the body, usually vagina. Therefore, most of these infections are secondary to diseases transmitted through sexual intercourse. However, other infections sometimes come from the body itself, such as vaginal mycosis [53].

These are generally more or less colored (dirty to brown) and abundant losses, sometimes with itching, foul odors, lower abdominal pain, burning in urination, painful intercourse, and fever or chills can accompany all this. If the infection is not treated, it can continue to the cervix (cervicitis) and then to the uterine cavity (endometritis) and finally to the fallopian tubes (salpingitis). Vaginitis and cervicitis are frequent infections that do not pose problems if properly managed. Rare endometritis is more serious. The most feared, by far, is salpingitis because it is diagnosed too late, poorly treated, or recurrent (repeated salpingitis); and may lead to permanent sterility and peritonitis. The most common types of vaginitis infections are bacterial vaginosis, fungal infections (*Candida*), chlamydia, gonorrhea, trichomoniasis, and viral vaginitis [24, 31, 34, 44, 53, 57, 70, 87]. There are several reasons that lead to vaginal secretions [7, 15, 53, 90].

Importantly, lactobacilli play an important role in maintaining the health condition of the female genital tract while preventing genitourinary infections [6, 53, 70].

There are major causes leading to vaginal discharge with itching, such as: *Tr. vaginalis* [20], *C. albicans*, *N. gonorrhea* [53, 97, 98], candidiasis [45, 73], postmenopausal vaginitis, secondary infection of the wound, scratch or tumor in any part of the lower genital canal and general itching in all parts of the body [46, 97]. There is also vaginal itching without discharge [26, 34, 35]. Other types of vaginal infections are: *Chlamydia trachomatis* [15, 17, 56], *Haemophilus ducreyi* [14], and scabies [35].

This chapter explores available information about vaginal infections and treatment options with antibiotics, natural antimicrobials, plant-derived compounds, and health benefits.

3.2 TREATMENT OF VAGINAL INFECTIONS

Proper management of the infections requires a care file history, physical examination, and laboratory assessment to diagnose the pathogen. Therapy is specific for the causal agent; hence, it is essential that a specific diagnosis is obtained before therapy is initiated. Therefore in past few years, studies of antibacterial vaginosis include: biofilm disruptor candidates, such as antibiotics [11, 32, 35, 53], DNAse, retrocyclins, probiotics, and prebiotics [13, 27, 65, 53, 71], antiseptics, and hygiene [40, 53, 93, 140], natural antimicrobials [4, 33, 53, 73, 84], and plant-derived compounds [10, 11, 16, 53]. Table 3.1 indicates antibiotics therapy for the treatment of vaginal infections.

3.3   HERBAL MEDICINES FOR VAGINAL INFECTIONS

For centuries, medicinal plants or "complementary medicine" have been used for their therapeutic properties and lower side effects to treat various diseases. Plants are the best sources of a wide variety of secondary metabolites, which have antimicrobial properties against several infectious agents [5, 39, 53]. Antibiotics are effective in treating illnesses caused by microbial infections. However, like all drugs, they also have unwanted side effects such as antimicrobial resistance.

Herbal drugs can be used in combination with antibiotics with enhanced activity against bacterial infections [10, 16, 49, 80].

3.3.1   GARLIC

Garlic (*Allium sativum* L.) is a widely used the herb as a food item, spice, and medicine and is considered as a rich resource of ingredients with antimicrobial properties. Clinical indications of garlic include: hemorrhoids, rheumatism, dermatitis, abdominal pain, loss of appetite, and loss of weight [47].

Method to treat vaginitis consists of: a teaspoon of garlic juice added to about four tablespoons of yogurt and then used as a douche of vagina twice a day until the disappearance of inflammation [53, 18].

The second method is to peel one of the garlic cloves, and lobe is placed in the region suffering from vaginal discharge before sleep and get rid of it when waking, taking care not to leave for a long time so as not to cause any kind of negative side effects of the vaginal membrane. Ingesting garlic capsules is an alternative method to cure and prevent a bacterial vaginosis infection [83].

TABLE 3.1 Antibiotics Therapy for Treatment Vaginal Infections

Micro-Organisms	Characteristics	Diseases and Infections	Antibiotics therapy	References
Candida albicans	Yeast, commensal saprophyte, pathogen opportunistic	Candidiasis	Clotrimazol, Miconazol, Premarin pills	[45, 53]
Chlamydia trachomatis	Mandatory intracellular parazitis, pathogen, not stainable in Gram.	Chlamydia urethritis, Lymphogranuloma venereum, pelvic inflammatory disease or salpingitis	Ceftriaxon, Cefoxitin, Cephalosporin, Cefotetan, Clindamycin, Doxycyclin, Gentamicin, Metronidazol, Ofloxacin, Probenecid	[99]
Enterbius vermucularis (pinworm)	Class *Nematoda* (parasite)	Pinworm infection or Enterobiasis	Albendazole, Mebendazole, Pyrantel pamoate	[57]
Gardnerella vaginalis	*Coccobacillus*, gram-positive	Bacterial vaginosis Leucorrhea	Metronidazole	[88]
Haemophilus ducreyi	Bacilli, rod, motionless, gram-negative	Chancroid of Ducrey	Azithromycin, Ceftriaxone Ciprofloxacin Erythromycin	[14, 40]
HSV-1 and HSV-2	Virus, herpes viridae	Lymphadenopathy, cervicitis, proctitis	Acyclovir, Famciclovir, Valacyclovir, Foscarnet	[17, 20, 22, 75]
Human immunodeficiency virus	Virus	Cancer Genital warts Cervical precancer	Imiquimod (Aldara, Zyclara), Salicylic acid, Podofilox (Condylox) Trichloroacetic acid	[17, 20, 22, 76]
Human Papilloma virus	Virus	Condyloma acuminate	Cryotherapy, Conization, Lazer therapy, Creams: Podofilox and Imiquimod	[17, 20, 22, 76]
Mobiluncuscurtisii M. mulieris	Anaerobic, curved, motile gram-negative bacteria	Bacterial vaginosis Irritations	Azithromycin, Clindamycin, Metronidazole	[11, 32, 62]

TABLE 3.1 *(Continued)*

Micro-Organisms	Characteristics	Diseases and Infections	Antibiotics therapy	References
Mycoplasma genitalium M. hominis	Parasitic bacterium, pathogen, variable gram	Pelvic inflammatory disease or salpingitis, cervicitis non-gonococcal urethritis	Macrolides (azithromycin), Tetracyclines, Quinolones (moxifloxacin), Ciprofloxacin, Ofloxacin	[53, 74]
Neisseria gonorrhea	Cocci, gram-negative, aerobic	Gonorrhea, pelvic inflammatory disease or salpingitis	Estrogen is used to treat vaginal atrophy and dryness, vaginal cream (Premarin ointment)	[99]
Sarcoptesscabiei var. *hominins*	Sarcoptes Parasitic arthropod	Skin lesions Skin sores and hematuria	Benzyl benzoate, Crotamiton, Esdepallethrin, Ivermectin, Oral Ivermectina, Malathion, Monosulfiram, Permethrin	[63]
Treponema pallidum	Bacteriumspiral, helical, mobile, gram-negative	Syphilis	Cyclins, Macrolides, Penicillin G	[53]
Trichomonas vaginalis	Protozoan flagellate, parasite of the human being	Trichomoniasis	5-Nitroimidazol, Tinidazol (TNZ), Seconidazol	[46]

3.3.2   CARDAMOM

Cardamom (*Cardamomum elettaria*) is an herbaceous plant. The seeds contain a double amount of oil as the Australian tea tree and terpinene-4-01 compound that has an antibacterial and antifungal effect. It also has a very attractive and wonderful smell [30, 81]. Vaginal creams and preparations containing a large quantity of this compound are quite effective against *C. albicans*. Mix 3 drops of cardamom oil with a large tablespoon of milk and then dips a small piece of clean gauze into a small thread in this oil and uses it as a vaginal plug every night for 6 nights [83].

3.3.3   GOLDENSEAL

The Goldenseal (*Hydrastis canadensis* L.; yellow puccoon or orangeroot) has long being used for the treatment of infections and demonstrated antimicrobial, anti-inflammatory, and anti-carcinogenic properties [24, 42]. Goldenseal is used as vaginal suppositories for *Candida* infections, *Trichomonas*, and *Gardnerella*; or a teaspoon of fresh roots and rhizomes are put in a cup of boiled water that is drunk twice a day until the infection has gone away, or consume 1, 000 mg of goldenseal twice a day [52, 83].

3.3.4   LAVENDER

Lavandula officinalis essential oil is a sedative, carminative, anti-depressive, and anti-inflammatory properties, in addition to its recognized antimicrobial effects [8, 86, 68]. Method of use for vaginal douche: Five drops of lavender oil are diluted in one liter of warm water, or two teaspoons of dried lavender are boiled in one liter of water for three minutes and are left until it becomes warm. This is used to wash twice a day until the inflammation disappears. Or, some drops of oil are mixed with yogurt and applied topically to the vaginal area. Note: The use of vaginal douche is not continued in the case of treatment, because the use of vaginal douche heavily affects the vagina.

3.3.5   YELLOW DOCK

Rumex crispus contains many bioactive compounds (emodin, chrysophanol, physcion, rutin, stilbenes, and rumexoids), and is used for therapeutic medicine for antimicrobial, antioxidant, anti-inflammatory, anti-fertility, and anti-viral activities [55, 89].

- **Method of Use**: One teaspoon of yellow dock root powder is added into a cup of boiled water and is soaked for 10 minutes; the tea is drunk once a day for a week. This preparation is used even as a disinfectant and detoxifier [83].

3.3.6   APPLE CIDER

Apple cider is mostly used in the treatment of different diseases for their anti-infective properties [19, 39]. Ascorbic acid, acetic acid, and polyphenols compounds are a key source of natural antioxidants and antimicrobials agents, since they serve as a natural defense mechanism against microbial infections.

- **Method of Use:** Apple cider vinegar is used as a vaginal bath or shower (douching) to treat bacterial vaginosis. Three cups of apple cider vinegar are added to the hot water bath, and then the woman stays in the bath for at least 20 minutes and cleanses the vaginal canal with this solution once per day until the infection is cleared. For drinking, mix a teaspoon of unfiltered apple cider into a glass of water and drink once or twice a day to prevent infections [83].

3.3.7   MARSH MALLOW

Althaea officinalis is a source of photochemicals with varying biological activities and is used for antimicrobial activity against germs responsible for genital infections. This plant contains a variety of bioflavonoids, vitamins, antioxidants, carvacrol, 2-pentadecanone, dodecanoic acid, n-tetradecanoic acid, n-tetradecanol, n-nonanoic acid, thymol, and methyl hexadecanoate [39, 49].

- **Method of Use**: Leaves are cut and then immersed in an appropriate amount of water and placed on the fire to boil and then cooled and filtered and then is used in the form of a shower vaginal or normal lotion and used two to three times a day [83].

3.3.8 HYSSOP

Hyssopus officinalis has volatile oil, fat, sugar, choline, iodine, tannins, carotene, and xanthophylls. The tops of flower contain ursolic acid and a glucoside diosmin, which produces rhamnose and glucose. Its essential oil has been used as therapeutic agents and possesses properties, including anti-inflammatory, antiviral, antitumor, cytotoxic, antimycosis, and anti-microbial activities [41].

- **Method of Use**: We use flower heads, leaves, the essential oil of the flowering plant and sometimes the roots. It is used as an herbal tea by the addition of 10 g or 1 teaspoon of flowering tops or dried flowers in one liter of boiling water; let infuse ten minutes and drink 2 or 3 cups a day; or is used as bath or a vaginal douche to soothe the irritation of the vagina and eliminate infections; infuse two tablespoons of hyssop into 200 ml of boiling water, then filter and apply to the irritated areas during morning and evening [83].

3.3.9 THYME

Thymus vulgaris is highly recommended due to the therapeutic properties of its essential oils (thymol and carvacrol) that have antimicrobial activity against many pathogens. All parts of plants are used as medicine and in foods [25, 60].

- **Method of Use**: 1–2 g dried thyme is infused in 150 ml hot water 90°C for ten minutes, and one cup is taken each night and rinse the vagina with the filtered and cooled thyme infusion for three times a day. One can also dilute a few drops of liquid extract in water and use it as a vaginal douche for 15 minutes in the evening before sleeping. The second method consists to reduce vaginal yeast infection by rinsing the vagina with a mixture of one teaspoon of sea salt, 5 drops of lavender essential oil and 5 drops of thyme essential oil [94].

3.3.10 MARJORAM

The herb is used as a spice in many types of foods, and its leaves, flowers, and essential oils are used as medicine. It contains many compounds with a wide range of biological activities such as flavonoids, triterpenes, and cafeic acid derivatives [61].

- **Method of Use**: The marjoram herb is used for local functions in a disinfecting, anti-inflammatory, anti-septic way and tightens the mucous membrane in a delicate way. It is used as a vaginal steaming or douching, tea, and suppositories [83].

3.3.11 ROSEMARY

Rosmarinus officinalis L. is known for its wide use as an herb, medicine, and cosmetics. The major bioactive components of rosemary are essential oil and polyphenols, which exhibit high antimicrobial activity against harmful microorganisms [95]. It also possesses several therapeutic effects, including anticancer, anti-inflammatory, and antioxidant properties [19].

- **Method of Use**: A teaspoon of rosemary and a teaspoon of fennel seeds are infused for 15 minutes in hot water, and two cups are drunk daily after eating for 30 days. Rosemary is also prepared as a tea and drunk after lunch and dinner for 30 days. One can make rosemary tea with fresh or dried rosemary [83].

3.3.12 FENNEL

Foeniculum vulgare adequately controls various infectious disorders of bacterial, fungal, viral, mycobacterium, and protozoal beginning [21, 75]. Fennel is used in combination with rosemary to treat vaginosis infections and other plants as *Ribulus terrestris*, *Myrtus commuis*, and *Tamarindus indica* [75, 83].

3.3.13 CHAMOMILE

Matricaria chamomilla L. is an ancient herb, rich in many antimicrobial, antioxidants, antiviral, antidiabetic, antiseptic, and anti-inflammatory bioactive substances with aromatics, therapeutics, cosmetics, and pharmacological properties.

- **Method of Use**: The vagina is washed with chamomile tea three to four times a day until the infection disappears [9, 83].

3.3.14 PARSLEY

Petroselinum crispum plant is widely used for its health benefits and medicinal properties. Parsley seeds are extensively used as a spice and herbal medicine due to high contents of essential oil with antimicrobial and antioxidant effects [19, 37, 49].

- **Method of Use**: 10 g parsley seeds are boiled in one liter of water and then using for vaginal douching three times a day until the infection disappears [83].

3.3.15 WALNUTS

Juglans regia is characterized by its high nutritional value and human health benefits. Essential oils, flavonoids, tocopherols, and polyphenol compounds showed numerous biological activities as nutritional, antioxidant, antibacterial, antifungal, and anticancer, which make this plant indispensable for medicinal and pharmaceutical applications [2, 34, 49].

- **Method of Use:** 110 g of walnut seeds are added to 1 liter of water and boiled for 15 minutes, then filtered and is used as a tea two times daily [83].

3.3.16 EUCALYPTUS

Bioactive compounds in *Eucalyptus camaldulensis* are sterol, triterpenoids, saponins, flavonoids, and phenols. Eucalyptus essential oil has potent medicinal benefits: it is a natural antiseptic with antifungal, antiviral, anticancer, anti-inflammatory and antibacterial properties making it highly effective in treating various diseases and improves health [92, 77].

- **Method of Use**: Add about 3–4 tablespoons of fresh eucalyptus leaves (eucalyptus essential oil can be used instead of fresh eucalyptus) to one liter of water and boiled for ten minutes, then filtered and is used as warm bath once a day and repeated use until the infection disappears [83].

3.3.17 ROSE PLANT

Rose plant is known for its sweet smell and beauty, and generally is used as a decoration in household and public gardens. The petals, stems, leaves, and roots of a rose plant contain many metabolites and nutrients [85]. Extracts from different parts of rose plants show substantial anti-bacterial, anti-fungal, and high antioxidant properties [36].

- **Method of Use**: 1 to 2 cups of fresh, pesticide-free rose petals are added to 3 cups of boiled water and left for 15 minutes then filtered and are drunk 2 to 3 times a day after meals for 30 days [83].

3.3.18 FENUGREEK

Trigonellafoenum-graecum is exceptionally nutritious. The plant contains a variety of bioactive components including flavonoid, alkaloid, polyphenol, galactomannan, diosgenin, and 4-hydroxy isoleucine. In addition, fenugreek possesses gastroprotective, antimicrobial, antioxidant, anticancer, lactation aid, immunomodulatory, and anti-diabetic effects [3, 48, 64, 83].

- **Method of Use:** It is infused by 2 tablespoons of fenugreek in a cup of water overnight and drink the water on an empty stomach. The second technique is to crush 2 tablespoons of fenugreek side to blend into yogurt and eat twice every day to dispense with side effects of bacterial vaginosis. For fenugreek tea, soak 1 tablespoon of fenugreek in some heated water, and filtered to expel all seeds and drink this solution at 2–3 times each day to get the best outcome [83].

3.3.19 TEA TREE

Medical advantages of tea (*Melaleuca alternifolia*) utilization include anti-oxidant properties, anticancer activities, reduction risk of cardiovascular diseases, treat respiratory diseases, corrects skin disorder, anti-diabetic, gives a lift to resistance, treats joint pain, and burns fat [54, 66, 82].

- **Method of Use:** The tea tree oil can clear the infection quickly without causing any unwanted side effects. As an added value, it puts off a minty aroma that will overwhelm any foul smell that may originate from the contamination.
- **Run Steaming Shower:** Include 10–12 drops of tea tree oil, and splash for up to 30 minutes, coat a tampon in coconut oil or a protected ointment before adding a couple of drops to the outside if one lean toward this strategy, and move like clockwork, yet evacuate instantly and don't rehash on the off chance that one may encounter any burning or stinging [83].

3.3.20 ARTEMISIA

Artemisia herba-alba is used as antihelminthic and antimalaria in solution. *Artemisia herba-alba* has been known for its helpful and restorative/biomedical properties. Artemisia species are utilized for the treatment of hepatitis, cancer, inflammation, and infections by fungi, bacteria, and viruses. Besides, a few types of Artemisia are utilized as a part of society's medication [1, 2, 12].

- **Technique to Utilize:** One liter of boiling water is poured onto 5 g dried leaves (substantial tablespoonful, free, or teaspoon if squashed to a powder) of *Artemisia annua*. It is permitted to mix for 10 to 15 minutes, and afterward poured through a strainer. This tea is then smashed in four segments over the span of the day. The time of treatment is about 5 and 7 days. It is additionally utilized for vagina douching and steaming [19, 83].

3.3.21 LEMON VERBENA

Aloysia triphylla has a lemon-like aroma discharged from its leaves and is used as natural tea, which has antispasmodic, antipyretic, calming, stomach related action, antimicrobial, and cancer prevention agent properties. This plant has a long history of people to treat asthma, fits, icy, fever, tooting, colic, loose bowels, heartburn, a sleeping disorder, and uneasiness [4, 28, 31, 72].

- **Technique to Utilize:** Make a hot tea by adding 1/4 cup of leaves in hot water, soak, and appreciate. Consider including mint for a genuine treat; Use it for vagina douching in the case of inflammation [19, 83].

3.3.22 WHITE HOREHOUND

Marrubium vulgare L. is most commonly used as fungicide, bacteriostatics, antioxidant, and antimicrobial agents. The plant is reported to possess hypoglycemic, antibacterial, antidiabetic, gastroprotective activity, and antiodematogenic activities. Bioactivities of *Marrubium vulgare* are attributed to an array of diterpenoids, sterols, marrubenol, flavonoids, and phenylpropanoids. Antioxidant, anticoagulant, antiplatelet, and anti-inflammatory effects have been attributed to the presence of marrubium [38, 43, 58].

- **Method of Use:** White horehound is used for vaginal steaming once a day for 10 minutes before sleeping for 10 consecutive days [19, 83].

3.3.23 OTHERS HERBAL MEDICINAL PLANTS USED FOR TREATMENT OF VAGINAL INFECTIONS

Many plant species are available for the treatment of vaginal infections such as [53, 59, 83]:

- *Bridelia cathartica* subsp *cathartica*;
- *Bryophyllum pinnatum*;
- *Cladostemon kirkii*;
- *Clematis brachiata*;
- *Euphorbia hypericifolia*;
- *Erianthemum dregei*;
- *Euphorbia hypericifolia*;
- *Hypoxis hemerocallidia*;
- *Ipomoea batatas*;

- *Krauseola mosambicina;*
- *Mimusops caffra;*
- *Opuntia stricta;*
- *Origanum vulgare;*
- *Pyrenacantha kaurabassana;*
- *Ranunculus multifidus;*
- *Sarcophyte sanguinea* subsp. *sanguinea;*
- *Senecio serratuloides.*

3.4   SUMMARY

Medicinal plants are notable for their capacities to advance injury mending and avert disease without grave reactions compared to synthetic drug treatments. Thus, herbal therapy may be an alternative strategy for the treatment of diseases. This chapter includes herb plants to avoid the adverse effects of vaginal infections and diseases.

KEYWORDS

- **alkaloid**
- **antibiotic therapy**
- **flavonoids**
- **herbal medicine**
- **polyphenol**
- **tocopherols**
- **vaginal infections**

REFERENCES

1. Abou El-Hamd, H. M., Magdi, A. E. S., Mohamed, E. H., Soleiman, E. H., Abeer, M. E., & Naglaa, S. M. Chemical constituents and biological activities of *Artemisia herba-alba*. *Records of Natural Products*, **2010**, *4*(1), 1–25.
2. Abu Taha, N., & Al-wadaan, M. A. Utility and importance of walnut, *Juglans regia* Linn: A review. *African Journal of Microbiology Research*, **2011**, *5*(32), 5796–5805.

3. Alfonso, D., Vargas-Pereza, I., Aguilar-Cruza, L., Calva-Rodríguezb, R., Treviñoc, S., Venegasd, B., & Contreras-Moraa, R. I. A mixture of chamomile and star anise has anti-motility and antidiarrheal activities in mice. *Revista Brasileira of Farmacognosia*, **2014**, *24*, 419–424.

4. Al-Harbi, N. O. Effect of marjoram extracts treatment on the cytological and biochemical changes induced by cyclophosphamide in mice. *Journal of Medicinal Plants Research*, **2011**, *5*(23), 5479–5485.

5. Allam, A. A., Maodaa, S. N., Abo-Eleneen, R., & Ajarem, J. Protective effect of parsley juice (*Petroselinum crispum*) against cadmium deleterious changes in the developed albino mice newborn (*Mus musculus*) brain. Oxidative medicine and cellular longevity, **2016**, p. 15. E-article ID 2646840. http://dx.doi.org/10.1155/2016/2646840 (Accessed on 30 September 2019).

6. Alsemari, A., Alkhodairy, F., Aldakan, A., Al-Mohanna, M., Bahoush, E., Shinwari, Z., & Alaiya, A. The selective cytotoxic anti-cancer properties and proteomic analysis of *Trigonella Foenum-Graecum*. *BMC Complementary and Alternative Medicine*, **2014**, *14*, 114.

7. Amabye, T. G., Bezabh, A. M., & Mekonen, F. Phytochemical and antimicrobial potentials leaves extract of *Eucalyptus Globulus* oil from maichewtigray Ethiopia. *International Journal of Complementary and Alternative Medicine*, **2016**, *2*(3), 56–60.

8. Amalaradjou, M. A. R., & Venkitanarayanan, K. Natural approaches for controlling urinary tract infections. *InTech*, **2011**, 228–244.

9. Amariei, S., & Mancas, V. Total phenolic content and antioxidant activity of infusions from two by-products: Walnut (*Juglans regia* L.) shell and onion (*Allium cepa* L.). *Journal of Faculty of Food Engineering*, **2015**, *XV*(2), 180–186.

10. Ardekani, M. R. S., Roja, R., Behjat, J., Leila, A., & Mahnaz, K. Relationship between temperaments of medicinal plants and their major chemical compounds. *Journal of Traditional Chinese Medicine*, **2011**, *31*(1), 27–31.

11. Aroutcheva, A., Gariti, D., Simon, M., Shott, S., Faro, J., Simoes, J. A., Gurguis, A., & Faro, S. Defense factors of vaginal Lactobacilli. *American Journal of Obstetrics and Gynecology*, **2001**, *185*(2), 375–379.

12. Banaee, M., Soleimany, V., & Haghi, B. N. Therapeutic effects of marshmallow (*Althaea officinalis* L.) extract on plasma biochemical parameters of common carp infected with *Aeromonas hydrophila*. *Veterinary Research Forum*, **2017**, *8*(2), 145–153.

13. Becker, M., Stephen, J., Moses, S., et al. Etiology and determinants of sexually transmitted infections in Karnataka state, South India. *Sexually Transmitted Diseases*, **2010**, *37*(3), 159–164.

14. Beheshti, Z., Chan, Y. H., Nia, H. S., Hajihosseini, F., Nazari, R., Shaabani, M., & Omran, M. T. S. Influence of apple cider vinegar on blood lipids. *Life Science Journal*, **2012**, *9*(4), 2431–2440.

15. Behmanesh, F., Pasha, H., Sefidgar, A. A., Taghizadeh, M., Moghadamnia, A. A., Rad, H. A., & Shirkhani, L. Antifungal effect of lavender essential oil (*Lavandula angustifolia*) and clotrimazole on *Candida albicans*: An *in vitro* study. *Scientifica*, **2015**, *2*, 5–10.

16. Benetti, C., & Manganelli, F. Clinical experiences in the pharmacological treatment of vaginitis with a chamomile extract vaginal douche. *Minerva Ginecol*, **1985**, *37*, 799–801.

17. Bhardwaj, M., Singh, B. R., Sinha, D. K., Kumar, V., Prasanna, V. O. R., Varan, S. S., Nirupama, K. R., & Pruthvishree, A. S. B. S. Potential of herbal drug and antibiotic combination therapy: A new approach to treat multidrug resistant bacteria. *Pharmaceutica Analytica Acta*, 2016, *7*, 11–16.

18. Bhawana, P., & Sheetal, S. Evaluation of antimicrobial potential of *Eucalyptus camaldulensis* L. *Journal of Pharmaceutical, Chemical and Biological Sciences*, **2014**, *2*(3), 166–171.

19. Bilardi, J., Walker, S., McNair, R., Mooney-Somers, J., Temple-Smith, M., Bellhouse, C., Christopher, F., Marcus, C., & Catriona, B. Women's management of recurrent bacterial vaginosis and experiences of clinical care: A qualitative study. *PLoS One*, 2016, *11*(3), e0151794, pages 10, E-article.

20. Biljana, B. P. Historical review of medicinal plants usage. *Pharmacognosy Review*, 2012, *6*(11), 1–5.

21. Bokaeian, M., Saboori, E., Saeidi, S., Niazi, A. A., Amini-Borojeni, N., & Khaje, H. S. B. Phytochemical analysis, antibacterial activity of *Marrubiumvulgare* L. against *Staphylococcus aureusin vitro*. *Zahedan Journal of Research in Medical Sciences*, **2014**, *16*(10), 60–64.

22. Borges, S., Silva, J., & Teixeira, P. The role of lactobacilli and probiotics in maintaining vaginal health. *Archives of Gynecology and Obstetrics*, **2014**, *289*, 479–489.

23. Budak, N. H., Guzel-Seydim, Z. B., & Aykın, E. Bioactive components of mother vinegar. *Journal of the American College of Nutrition*, **2015**, *34*(1), 80–89.

24. Camila, G. B., Michael, M., Cheng, Y. C., Sally, R., & Oriol, M. Epidemiology of *Haemophilus ducreyi* infections. *Emerging Infectious Diseases*, 2016, *22*(1), 1–8.

25. Carlos, F. B., Patricia, M. L., Luis, O. G., Mario, J. M. M., Pilar, S. L., & Fernando, V. Chlamydial primary care group, prevalence of genital *Chlamydia trachomatis* infection among young men and women in Spain. *BMC Infectious Diseases*, **2013**, *13*, 388–396.

26. Cavanagh, H. M. A., & Wilkinson, J. Lavender essential oil: A review. *Australian Infection Control*, **2005**, *10*(1), 35–37.

27. Chauhan, G., Sharma, M., Varma, A., & Khanrkwal, H. Phytochemical analysis and anti-inflammatory potential of fenugreek, medicinal plants. *International Journal of Phytomedicine and Related Industries*, **2010**, *2*(1), 39–44.

28. Chillemi, S. The complete herbal guide: A natural approach to healing the body: Heal your body naturally and maintain optimal health using alternative medicine, herbals, vitamins, fruits and vegetables. Lulu Enterprise, Kindle edition (amazon.com), **2013**, p. 922.

29. Chouhan, S., Sharma, K., & Guleria, S. Antimicrobial activity of some essential oils: Present status and future perspectives. *Medicines*, 2017, *4*, 58–62.

30. Christine, J., & Lawrence, C. Current concepts for genital herpes simplex virus infection: Diagnostics and pathogenesis of genital tract shedding. *Clinical Microbiology Reviews*, 2016, *29*, 149–161.

31. Cirkovic, I., Jovalekic, M., & Jegorovic, B. *In vitro* antibacterial activity of garlic and synergism between garlic and antibacterial drugs. *Archives of Biological Sciences*, **2012**, *64*(4), 1369–1375.

32. Dadashi, M., Kashani, M. A., Eslami, G., Goudarzi, H., Fallah, F., Owlia, P., Adhami, N., & Mousavi, T. Evaluation of antibacterial effect of green tea extract on bacterial

isolated from patients with vaginal infection. *International Journal of Molecular and Clinical Microbiology*, **2015**, *5*(2), 537–542.

33. Daniel, M., & Breitkopf, M. D. A review of genital chlamydial infections. *Hospital Physician*, **2000**, 27–34.

34. De Wet, H., Nzama, V. N., & Van Vuuren, S. F. Medicinal plants used for the treatment of sexually transmitted infections by lay people in northern Maputaland, South Africa. *South African Journal of Botany*, **2012**, *78*, 12–20.

35. Depuydt, C. E., Leuridan, E., Van Damme, P., Bogers, J., Vereecken, A. J., & Donders, G. G. Epidemiology of *Trichomonas vaginalis* and human papillomavirus infection detected by real-time PCR in flanders. *Gynecology and Obstetrics Investigations*, **2010**, *70*(4), 273–280.

36. Derya, A., Cem, A., & Celalettin, K. The Effect of external apple vinegar application on varicosity symptoms, pain, and social appearance anxiety: A randomized controlled trial. *Evidence-Based Complementary and Alternative Medicine*, **2016**, *1*, 8–12.

37. Djahra, A. B., Bordjiba, O., & Benkherara, S. Extraction, separation et activitéanti-bactérienne des tannins de marrube blanc (*Marrubium vulgare* L.). *Phytothérapie*, **2013**, *11*, 348–352.

38. Douglas, T. F., & Judith, N. W. From epidemiological synergy to public health policy and practice: The contribution of other sexually transmitted diseases to sexual transmission of HIV infection. *Sexually Transmitted Infections*, 1999, *75*, 3–17.

39. Dua, A., Garg, G., & Mahajan, R. Polyphenols, flavonoids and antimicrobial properties of methanolic extract of fennel (*Foeniculum vulgare* Miller). *European Journal of Experimental Biology*, 2013, *3*(4), 203–208.

40. Dunne, E., & Markowitz, L. Genital human Papillomavirus infection. *Clinical Infectious Diseases*, 2006, *43*, 624–629.

41. Egan, J. M., Kaur, A., Raja, H. A., Kellogg, J. J., Oberlies, N. H., & Cech, N. B. Antimicrobial fungal endophytes from the botanical medicine goldenseal (*Hydrastis canadensis*). *Phytochemistry Letters*, **2016**, *17*, 219–225.

42. El-Moursi, A., Mahmoud, T. I., & Balbaa, K. L. Physiological effect of some antioxidant polyphenols on sweet marjoram (*Majorana hortensis*) plants. *Nusantara Bioscience*, 2012, *4*(1), 11–15.

43. Eqbal, M. A. D., & Aminah, A. Medicinal and functional values of thyme (*Thymus vulgaris* L.) herb. *Journal of Applied Biology and Biotechnology*, 2017, *5*(2), 17–22.

44. Fahami, R. Abnormal vaginal discharge. *British Medical Journal*, **2013**, *347*, 4975.

45. Falagas, M. E., Betsi, G. I., & Athanasiou, S. Probiotics for the treatment of women with bacterial vaginosis. *Clinical Microbiology and Infection*, 2007, *13*, 657–664.

46. Fathiazad, F., Mazandarani, M., & Hamedeyazdan, S. Phytochemical analysis and antioxidant activity of *Hyssopus officinalis* L. from Iran. *Advanced Pharmaceutical Bulletin*, 2011, *1*(2), 63–67.

47. Gattuso, S., Van Baren, C. M., Gil, A., Bandoni, F. A. G., & Gattuso, M. Morpho-histological and quantitative parameters in the characterization of lemon verbena (*Aloysiacitriodorapalau*) from Argentina. *Boletín Latinoamericano y del Caribe de Plantas Medicinales y Aromáticas*, **2008**, *7*(4), 190–198.

48. Gilbert, G. G. D., Annie, V., Eugene, B., Alfons, D., Geert, S., & Bernard, S. Definition of a type of abnormal vaginal flora that is distinct from bacterial vaginosis: aerobic vaginitis. *International Journal of Obstetrics and Gynecology*, **2002**, *109*, 34–43.

49. Han, X., & P. Tory, L. Cardamom (*Elettaria cardamomum*) essential oil significantly inhibits vascular cell adhesion molecule and impacts genome-wide gene expression inhuman dermal fibroblasts. *Cogent Medicine*, **2017**, *4*, e1308066.

50. Hanaa, F. M. A., Hossam, S. E. B., & Nasr, F. N. Evaluation of antioxidant and antimicrobial activity of *Aloysiatriphylla*. *Electronic Journal of Environmental, Agricultural and Food Chemistry*, **2011**, *10*(8), 2689–2699.

51. Hanamanthagouda, M. A., Kakkalameli, S. B., Naik, P. M., Nagella, P., Seethara-mareddy, H. R., & Murthy, H. N. Essential oils of *Lavandula bipinnata*and their antimicrobial activities. *Food Chemistry*, **2010**, *118*, 836–839.

52. Hart, P. H. Terpinen-4-ol, the main component of the essential oil of *Melaleuca alternifolia* (tea tree oil), suppresses inflammatory mediator production by activated human monocytes. *Inflammation Research*, **2000**, *49*, 619–626.

53. Hay, P. National guideline for the management of bacterial vaginosis. Clinical effectiveness group British Association for sexual health and HIV (BASHH), London, **2012**, p. 15.

54. Hayek, S. A., Rabin, G., & Salam, A. I. Antimicrobial natural products. *Formatex*, 2013, 910–921.

55. Hayes, D., Angovea, M. J., Tuccia, J., & Dennis, C. Walnuts (*Juglans regia*) Chemical composition and research in human health. *Critical Reviews in Food Science and Nutrition*, 2015, 1–39.

56. Helen, M. Vaginal discharge: Causes, diagnosis, and treatment. *British Medical Journal*, 2004, 328, 1306–1308.

57. Helen, C., & Manoj, M. Vaginal discharge: Recommended management in general practice. Drug review. *Journal of Prescribing and Medicines Management*, 2013, 24(3), 19–32.

58. Hicks, M. I., & Elston, D. M. Scabies. *Dermatologic Therapy*, **2009**, *22*, 279–292.

59. Hirulkar, N. B., & Agrawal, M. Antimicrobial activity of rose petals extract against some pathogenic bacteria. *International Journal of Pharmaceutical and Biological Archives*, **2010**, *1*(5), 478–484.

60. Hosein-Farzaei, M., Abbasabadi, Z., Ardekani, R. S. M., Rahimi, R., & Farzaei, F. Parsley: A review of ethnopharmacology, phytochemistry and biological activities. *Journal of Traditional Chinese Medicine*, 2013, *15*, 33(6), 815–826.

61. Hussein, A. H. S., Zahid, K. A., Sabra, A. S., & Kirill, G. T. Essential oil composition of *Hyssopus officinalis* L. cultivated in Egypt. *International Journal of Plant Science and Ecology*, 2015, *1*(2), 49–53.

62. Jae-Sook, R., & Duk-Young, M. *Trichomonas vaginalis* and trichomoniasis in the Republic of Korea. *Korean Journal of Parasitology*, 2006, *44*(2), 101–116.

63. Jeanelle, B., & Rui, H. L. Apple phytochemicals and their health benefits: A review. *Nutrition Journal*, **2004**, *3*, 5–10.

64. Junio, H. A., Sy-Cordero, A. A., Ettefagh, K. A., Burns, J. T., Micko, K. T., & Cech, N. B. Synergydirected fractionation of botanical medicines: A case study with goldenseal (*Hydrastis canadensis*). *Journal of Natural Products*, **2011**, *74*(7), 1621–1629.

65. Kadri, A., Zarai, Z., Békir, A., Gharsallah, N., Damak, M., & Gdoura, R. Chemical composition and antioxidant activity of *Marrubium vulgare* L. essential oil from Tunisia. *African Journal of Biotechnology*, **2011**, *10*(19), 3908–3914.

66. Kent, M. E., & Romanelli, F. Re-examining syphilis: An update on epidemiology, clinical manifestations, and management. *Annals of Pharmacotherapy*, 2008, *42*(2), 226–236.

67. Khalil, A. F., Elkatry, H. O., & El Mehairy, H. F. Protective effect of peppermint and parsley leaves oils against hepatotoxicity on experimental rats. *Annals of Agricultural Science*, 2015, *60*(2), 353–359.

68. Khan, J. A., & Tiwari, S. A study on antibacterial properties of *Rosa indica* against various pathogens. *Asian Journal of Plant Science and Research*, **2011**, *1*(1), 22–30.

69. Kim, J., & Sudbery, P. *Candida albicans*, a major human fungal pathogen. *Journal of Microbiology*, **2011**, *49*, 171–177.

70. Kissinger, P. *Trichomonas vaginalis*: A review of epidemiologic, clinical and treatment issues. *BMC Infectious Diseases*, 2015, *15*, 307–311.

71. Klebanoff, M. A., Nansel, T. R., Brotman, R. M., Zhang, J., Yu, K. F., Schwebke, J. R., et al. Personal hygienic behaviors and bacterial vaginosis. *Sexually Transmitted Diseases*, **2010**, *37*(2), 94–99.

72. Kopec, A., Piatkowska, E., Leszczynska, T., & Sikora, E. Healthy properties of garlic. *Current of Nutrition and Food Sciences*, **2013**, *9*(1), 59–64.

73. Kor, N. M., Didarshetaban, M. B., & Saeid, P. H. R. Fenugreek (*Trigonellafoenum-graecum* L.) as a valuable medicinal plant. *International journal of Advanced Biological and Biomedical Research*, **2013**, *1*(8), 922–931.

74. Kumar, R., & Trivedi, N. A literature-based study on *Juglans regia*: A review. *Pharmacoaerena: An International Journal of Pharmaceutical Research*, **2016**, *1*(4), 131–136.

75. Lan, P. T., Lundborg, C. S., Phuc, H. D., Sihavong, A., & Unemo, M. Reproductive tract infections including sexually transmitted infections: A population-based study of women of reproductive age in a rural district of Vietnam. *Sexually Transmitted Infections*, **2008**, *84*(2), 126–132.

76. Lewis, D. A., & Ison, C. A. Chancroid. *Sexually Transmitted Infections*, 2006, *82*, 19–20.

77. Leyte-Lugo, M., Britton, E. R., Foil, D. H., Brown, A. R., Todd, D. A., Rivera-Chávez, J., Oberlies, N. H., & Cech, N. B. Secondary metabolites from the leaves of the medicinal plant goldenseal (*Hydrastis canadensis*). *Phytochemistry Letters*, **2017**, *20*, 54–60.

78. Machado, D., Castro, J., Palmeira-de-Oliveira, A., Martinez-de-Oliveira, J., & Cerca, N. Bacterial vaginosis biofilms: Challenges to current therapies and emerging solutions. *Frontiers in Microbiology*, **2016**, *6*, 1528–1536.

79. Mahdieh, M., Mitra, N., & Simin, H. Studies of *in vitro* adventitious root induction and flavonoid profiles in *Rumexcrispus*. *Advances in Life Sciences*, 2015, *5*(3), 53–57.

80. Mahmood, T., Akhtar, N., Ali Khan, B. The morphology, characteristics, and medicinal properties of *Camellia sinensis*' tea. *Journal of Medicinal Plants Research*, **2010**, *4*(19), 2028–2033.

81. Mak, J. C. Potential role of green tea catechins in various disease therapies: Progress and promise. *Clinical and Experimental Pharmacology and Physiology*, **2012**, *39*(3), 265–273.

82. Maksimović, Z., Kovačević, N., Lakušić, B., & Ćebović, T. Antioxidant activity of yellow dock (*Rumexcrispus* L.) fruit extract. *Phytotherapy Research*, **2010**, *25*(1), 101–105.

83. Malhotra, S. K. Fennel and fennel seed. *Indian Council of Agricultural Research* (pp. 275–302). New Delhi, 2012.

84. Manonmani, R., & Abdul, K. V. M. Antibacterial screening on *Foeniculum vulgare* Mill. *International Journal of Pharmacology and Biological Sciences*, 2011, *2*(4), 390–394.

85. Marija, B., Zdravkovica, N., Ivanovica, J., Janjica, J., Djordjevica, J., Starcevica, M., & Baltica, M. Z. Antimicrobial activity of thyme (*Tymus vulgaris*) and oregano (*Origanum vulgare*) essential oils against some food-borne microorganisms. *Procedia Food Science*, 2015, *5*, 18–21.

86. Meghwal, M., & Goswami, T. K. A review on the functional properties, nutritional content, medicinal utilization and potential application of fenugreek. *Journal of Food Processing and Technology*, **2012**, *3*, 9–15.

87. Mehani, M., Segni, L., Terzi, V., Morcia, C., Ghizzoni, R., Goudgil, B., & Benchikh, S. Antifungal activity of *Artemisia herba-alba* on various *Fusarium* (*Activité antifongique de l'armoise blanche sur différents Fusarium*). *Phytothérapie*, **2016**, E-article online, doi: 10.1007/s10298–016–1071–2.

88. Miller, W. C., & Zenilman, J. M. Epidemiology of chlamydial infection, gonorrhea, and trichomoniasis in the United States. *Infectious Diseases Clinics of North America*, **2005**, *19*(2), 281–296.

89. Mitchell, H. ABC of sexually transmitted infections vaginal discharge: Causes, diagnosis, and treatment. *British Medical Journal*, **2004**, *328*, 29–36.

90. Murtaza, M., Illzam, E. M., Muniandy, R. K., Sharifah, A. M., Nang, M. K., & Ramesh, B. Herpes simplex virus infections, pathophysiology and management. *IOSR Journal of Dental and Medical Sciences*, 2016, *15*(7), 85–91.

91. Nanova, Z., Slavova, Y., Nenkova, D., & Ivanova, I. Microclonal propagation of hyssop (*Hyssopus officinalis* L.). *Bulgarian Journal of Agricultural Science*, 2007, *13*, 213–219.

92. Nikolic, M., Glamoclija, J., Ferreira, I. C., Calhelha, R. C., Fernandes, Â., Markovic, T., Markovic, D., Giweli, A., & Sokovic, M. Chemical composition, antimicrobial, antioxidant and antitumor activity of *Thymus serpyllum* L., *Thymus algeriensis* Boiss and Reut and *Thymus vulgaris* L. essential oils. *Indian Crop Production*, 2014, *52*, 183–190.

93. Nowak, K., & Ognowski, J. Marjoram oil, its characteristics and application. *CHEMIK*, 2010, *64*(7/8), 539–548.

94. Nyirjesy, P., McIntosh, M. J., Steinmetz, J. I., Schumacher, R. J., & Joffrion, J. L. The effect of intravaginal clindamycin and metronidazole therapy on vaginal *Mobiluncus* morphotypes in patients with bacterial vaginosis. *Sexually Transmitted Diseases*, 2007, *34*(4), 197–202.

95. Otero, L., Varela, J. A., Espinosa, E., Sánchez, C., Junquera, M. L., Del Valle, A., & Vázquez, F. *Sarcoptesscabiei* in a sexually transmitted infections unit: A 15-year study. *Sexually Transmitted Diseases*, **2004**, *31*(12), 761–765.

96. Oueslati, H. A., & Ghédira, K. *Notes ethnobotanique et phytopharmacologique sur Trigonellafoenumgraecum* (Ethnobotanical and phytopharmacological notes

on *Trigonellafoenumgraecum*). *Phytothérapie*, **2015**, *13*, 234–238. https://doi. org/10.1007/s10298-015-0934-2 (Accessed on 30 September 2019).

97. Pandey, A., Kumar, K., & Kumari, D. A comparative analysis of antibacterial properties of different varieties of *Rosa indica* leaves and petals against various pathogens. *International Journal of Pharmaceutical Research and Development*, **2012**, *3*(12), 39–47.

98. Parma, M., Stella, V. V., Bertini, M., & Candiani, M. Probiotics in the prevention of recurrences of bacterial vaginosis. *Alternative Therapies in Health and Medicine*, **2014**, *20*, 52–57.

99. Parmar, N., Rawat, M., & Kumar, J. V. *Camellia Sinensis* (Green Tea): A review. *Global Journal of Pharmacology*, **2012**, *6*(2), 52–59.

100. Pazyar, N., Yaghoobi, R., Bagherani, N., & Kazerouni, A. A review of applications of tea tree oil in dermatology. *International Journal of Dermatology*, 2013, *52*, 7–11.

101. Peel, T. N., Bhatti, D., De Boer, J., Stratov, I., & Spelman, D. Chronic cutaneous ulcers secondary to *Haemophilus ducreyi* infection. *Medical Journal of Australia*, **2010**, *192*, 348–350.

102. Pendharkar, S., Brandsborg, E., Hammarström, L., Marcotte, H., & Larsson, P. G. Vaginal colonization by probiotic lactobacilli and clinical outcome in women conventionally treated for bacterial vaginosis and yeast infection. *BMC Infectious Diseases*, **2015**, *15*, 255–260.

103. Pengelly, A., Bennett, K., Spelman, K., & Tims, M. *Appalachian Plant Monographs: Goldenseal (Hydrastis canadensis L.)* (p. 42). Appalachian Center for Ethnobotanical Studies, **2012**, online. https://www.researchgate.net/publication/236659731_An_ Appalachian_Plant_Monograph_Goldenseal_Hydrastis_canadensis_L (Accessed on 30 September 2019).

104. Potroz, M. G., & Cho, N. J. Natural products for the treatment of *Trachoma* and *Chlamydia trachomatis*. *Molecules*, **2015**, *20*, 4180–4203.

105. 105.Prusinowska, R., & Śmigielski, K. A review: Composition, biological properties and therapeutic effects of lavender (*Lavandula angustifolia* L.). *Herba Polinica*, **2014**, *60*, 2–10.

106. Rašković, A., Milanović, I., Pavlović, N., Ćebović, T., Vukmirović, S., & Mikov, M. Antioxidant activity of rosemary (*Rosmarinus officinalis* L.) essential oil and its hepatoprotective potential. *BMC Complementary and Alternative Medicine*, 2014, *14*, 225–230.

107. Ratho, S. D., Krupp, K., Klausner, J. D., Arun, A., Reingold, A. L., & Madhivanan, P. Bacterial vaginosis and risk for *Trichomonas vaginalis* infection: A longitudinal analysis. *Sexually Transmitted Diseases*, **2011**, *38*(9), 882–886.

108. Ravel, J., Gajer, P., Abdo, Z., Schneider, G. M., Koenig, S. S. K., McCulle, S. L., et al. Vaginal microbiomeof reproductive age women. *Proceedings of the National Academy of Sciences*, **2011**, *108*, 4680–4687.

109. Reid, G., & Bruce, A. W. Urogenital infections in women: Can probiotics help? *Postgraduate Medical Journal*, **2003**, *79*, 428–432.

110. Rice, A., El Werdany, M., Hadoura, E., & Mahmood, T. Vaginal discharge. *Obstetrics, Gynaecology and Reproductive Medicine*, 2016, *26*(11), 317–323.

111. Rojas, L. B., Velasco, J., Díaz, T., Gil, R. O., Carmona, J., & Usubillaga, A. Chemical composition and antibacterial effects of the essential oil of *Aloysiatriphylla* against

genito-urinary pathogens Bol. *Latinoamericano y del Caribe de Plantas Medicinales y Aromaticas*, **2010**, *9*(1), 56–62.

112. Ross, J. D. C., & Jensen, J. S. *Mycoplasma genitalium* as a sexually transmitted infection: Implications for screening, testing, and treatment. *Sexually Transmitted Infections*, 2006, *82*(4), 269–271.

113. Sardi, J. C. O., Scorzoni, L., Bernardi, T., Fusco-Almeida, A. M., & Mendes, G. M. J. S. *Candida* species: Current epidemiology, pathogenicity, biofilm formation, natural antifungal products and new therapeutic options. *Journal of Medical Microbiology*, 2013, *62*, 10–24.

114. Satterwhite, C. L., Torrone, E., Meites, E., et al. Sexually transmitted infections among US women and men: Prevalence and incidence estimates, 2008. *Sexually Transmitted Diseases*, 2013, *40*, 187–193.

115. Sauerbrei, A. Herpes genitalis: Diagnosis, treatment and prevention. *Geburtshilfe Frauenheilkd*, 2016, *76*(12), 1310–1317.

116. Savan, E. K., & Küçükbay, F. Z. Essential oil composition of *Elettaria cardamomum* L. Maton. *Journal of Applied Biological Sciences*, **2013**, *7*, 42–45.

117. Schiller, J. T., Lowy, D. R., & Markowitz, L. E. Human papillomavirus vaccines. In: Plotkin, S. A., Orenstein, W. A., & Offit, P. A., (eds.), *Vaccines* (6th edn., pp. 235–256). China: Saunders, 2012.

118. Seydim, A. C., Guzel-Seydim, Z. B., Doguc, D. K., M. Savas, C., & Budak, H. N. Effects of grape wine and apple cider vinegar on oxidative and antioxidative status in high cholesterol-fed rats. *Functional Foods in Health and Disease*, **2016**, *6*(9), 569–577.

119. Shagal, M. H., Kubmarawa, D., Tadzabia, K., & Dennis, K. I. Evaluation of phytochemical and antimicrobial potentials of roots, stem-bark and leaves extracts of *Eucalyptus camaldulensis*. *African Journal of Pure and Applied Chemistry*, **2012**, *6*(5), 74–77.

120. Shahanara, B., Sushmita, R., & Md. Abdullah, Y. Anaerobic bacteria: Infection and management. *IOSR Journal of Dental and Medical Sciences*, **2015**, *14*(12), 69–72.

121. Shahram, S., & Omid, A. German and Roman chamomile. *Journal of Applied Pharmaceutical Science*, 2011, *1*(10), 1–5.

122. Shamkant, B. B., Vainav, V. P., & Atmaram, H. B. *Foeniculum vulgare* mill: A review of its botany, phytochemistry, pharmacology, contemporary application, and toxicology. *Biomed. Research International*, 2014, *1*, 32–36.

123. Sharafati, F. C., & Sharafati, R. C. *In vitro* antibacterial and antioxidant properties of *Elettaria cardamomum* Maton extract and its effects, incorporated with chitosan, on storage time of lamb meat. *Veterinarski Arhiv*, **2017**, *87*(3), 301–315.

124. Sharangi, A. B. Medicinal and therapeutic potentialities of tea (*Camellia sinensis* L.): A review. *Food Research International*, **2009**, *42*, 529–535.

125. Sharifi-Rad, M., Mnayer, D., Tabanelli, G., Stojanovic-Radic, Z. Z., Sharifi-Rad, M., Yousaf, Z., Vallone, L., Setzer, W. N., & Iriti, M. Plants of the genus Allium as antibacterial agents: From tradition to pharmacy. *Cellular and Molecular Biology*, 2016, *62*, 57–68.

126. Shayoub, M. E., Dawoud, A. H., Abdelmageed, M. A., Ehassan, A. M., & Ehassan, A. M. Phytochemical analysis of leaves extract of *Eucalyptus camaldulensis* Dehnh. *Omdurman Journal of Pharmaceutical Science*, **2015**, *2*(1), 64–71.

127. Silva, B. M., Palmeira-de-Oliveira, A., Palmeira-de-Oliveira, R., Martinez-de-Oliveira, J., & Duarte, A. P. Tea (*Camellia sinensis* L.): A putative antimicrobial agent in sexually transmitted infections. In: Méndez-Vilas, A., (ed.), *Microbial Pathogens and Strategies for Combating Them: Science, Technology and Education, Formatex* (pp. 1343–1351). **2013**.

128. Śmigielski, K., Prusinowska, R., Krosowiak, K., & Sikora, M. Comparison of qualitative and quantitative chemical composition of hydrolate and essential oils of lavender (*Lavandula angustifolia*). *Journal of Essential Oil Research*, **2013**, *25*(4), 291–299.

129. Sohayeb, M., Hameed, E. S. S. A., Bazaid, S. A., & Maghrabi, I. Antibacterial and antifungal activity of *Rosa damascene* MILL. Essential oil, different extracts of rose petals. *Global Journal of Pharmacology*, **2014**, *8*(1), 1–7.

130. Soleimany, V., Banaee, M., Mohiseni, M., Nematdoost, H. B., & Mousavi, D. L. Evaluation of pre-clinical safety and toxicology of *Althaea officinalis* extracts as naturopathic medicine for common carp (*Cyprinus carpio*). *Iranian Journal of Fisheries Sciences*, 2016, *15*(2), 613–629.

131. Solomon, H. The therapeutic potential of rosemary (*Rosmarinus officinalis*) diterpenes for Alzheimer's disease. *Evidence-Based Complementary and Alternative Medicine*, 2016, *1*, 14–18.

132. Sonali, N., & Pratyush, K. D. Medicinal efficacy of rose plant: A mini review. *PharmaTutor*, **2015**, *3*(10), 23–26.

133. Spence, D., & Melville, C. Vaginal discharge. *British Medical Journal*, **2007**, *335* (7630), 1147–1151.

134. Srivastava, J. K., Shankar, E., & Gupta, S. Chamomile: An herbal medicine of the past with bright future. *Molecular Medicine Reports*, 2010, *3*, 895–901.

135. Swidsinski, A., Loening-Baucke, V., Swidsinski, S., & Verstraelen, H. Polymicrobial *Gardnerella* biofilm resists repeated intravaginal antiseptic treatment in a subset of women with bacterial vaginosis: A preliminary report. *Archives of Gynecology and Obstetrics*, **2015**, *291*, 605–609.

136. Syed Mudassir, J., Umer, F., Ajai, P. G., & Surrinder, K. L. Phytochemical evaluation of major bioactive compounds in different cytotypes of five species of *Rumex* L. *Industrial Crops and Products*, **2017**, *109*, 897–904.

137. Tânia Maria, M. V., Juraci, A. C., Raúl, A. M. S., & Elisabeth, B. S. Pathological vaginal discharge among pregnant women: Pattern of occurrence and association in a population-based survey. *Obstetrics and Gynecology International*, 2013, *1*, 7–11.

138. Thaylise, V. P., Agueda, P. De Castagna, V., Carina, K., Érico, M. M. F., Bernardo, B., Berta, M. H., José, V. O., Ariana, S. P., & Mariane, M. Chemical composition and antibacterial activity of *Aloysiatriphylla* (L'Hérit) extracts obtained by pressurized CO_2 extraction. *Brazilian Archives of Biology and Technology*, **2013**, *56*(2), 283–292.

139. Vázquez-Conzález, D., Persguia-Ortiz, A. M., Hundeiker, M., & Bonifaz, A. Opportunistic yeast in infections: Candidiasis, cryptococcosis, trichosporonosis and geotrichosis. *Journal der Deutschen Dermatologischen Gesellschaft*, 2013, *11*, 381–395.

140. Vecchio, M. G., Loganes, C., & Minto, C. Beneficial and healthy properties of *Eucalyptus* Plants: A great potential use. *The Open Agriculture Journal*, **2016**, *10*(1: M3), 52–57.

141. Verstraelen, H., Verhelst, R., Roelens, K., & Temmerman, M. Antiseptics and disinfectants for the treatment of bacterial vaginosis: A systematic review. *BMC Infections Diseases,* **2012**, *12*, 148–156.

142. Viuda-Martos, M., Mohamady, M. A., Fernández-López, J., Abd ElRazik, K. A., Omer, E. A., Pérez-Alvarez, J. A., & Sendra, E. *In vitro* antioxidant and antibacterial activities of essentials oils obtained from Egyptian aromatic plants. *Food Control,* **2011**, *22*, 1715–1722.

143. Wang, W., Li, N., Luo, M., Zu, Y., & Efferth, T. Antibacterial activity and anticancer activity of *Rosmarinus officinalis* L. essential oil compared to that of its main components. *Molecules,* 2012, *17*, 2704–2713.

144. Wang, Z. L., Fu, L., Xiong, Z., Qin, Q., Yu, T., Wu, Y., Hua, Y., & Zhang, Y. Diagnosis and microecological characteristics of aerobic vaginitis in outpatients based on preformed enzymes. *Taiwanese Journal of Obstetrics and Gynecology,* **2016**, *55*, 40–44.

145. Wangnapi, R. A., Soso, S., Unger, H. W., Sawera, C., Ome, M., Umbers, A. J., et al. Prevalence and risk factors for *Chlamydia trachomatis, Neisseria gonorrheae* and *Trichomonas vaginalis* infection in pregnant women in Papua New Guinea. *Sexually Transmitted Infections,* **2015**, *91*(3), 194–200.

146. Wani, S. A., & Kumar, P. Fenugreek: A review on its nutraceutical properties and utilization in various food products. *Journal of the Saudi Society of Agricultural Sciences* (p. 7). **2016**, E-article. http://dx.doi.org/10.1016/j.jssas.2016.01.007 (Accessed on 30 September 2019).

147. Williams, J. A., Ofner, S., Batteiger, B. E., Fortenberry, J. D., & Van Der Pol, B. Duration of polymerase chain reaction-detectable DNA after treatment of *Chlamydia trachomatis, Neisseria gonorrheae,* and *Trichomonas vaginalis* infections in women. *Sexually Transmitted Diseases,* **2014**, *41*(3), 215–219.

148. Wilson, J. Managing recurrent bacterial vaginosis. *Sexually Transmitted Infections,* 2004, *80*, 8–11.

149. Workowski, K. A., & Berman, S. Prevention: Sexually transmitted diseases treatment guidelines. *Morbidity and Mortality Weekly Report* (pp. 1–110). Centers for Disease Control and Prevention (CDC), U. S. Department of Health and Human Services, Atlanta, GA, **2010**, *59*(RR-12). https://www.cdc.gov/std/treatment/2010/std-treatment-2010-rr5912.pdf.

150. World Health Organization. *Global Incidence and Prevalence of Selected Curable Sexually Transmitted Infections* (p. 20). World Health Organization (WHO), Dept. of Reproductive Health and Research, Rome, **2008**. ISBN 978924150383–9.

151. Yuan, L., Meng, L., Ma, W., Xiao, Z., Zhu, X., Feng, J. F., Yu, H., & Xiao, R. Impact of apple and grape juice consumption on the antioxidant status in healthy subjects. *International Journal of Food Sciences and Nutrition,* **2011**, *62*(8), 844–850.

PART II
Functional Activities of Selected Plants

FROM DESIGNER FOOD FORMULATION TO OXIDATIVE STRESS MITIGATION: HEALTH-BOOSTING CONSTITUENTS OF CABBAGE

FAIZA ASHFAQ, MASOOD SADIQ BUTT, AHMAD BILAL, KANZA AZIZ AWAN, and HAFIZ ANSAR RASUL SULERIA

ABSTRACT

Currently, the oxidant-antioxidant status of the junk food seekers is facing an imbalance in such a way that free radicals are dominating, ultimately exhausting the endogenous stores of free radical trapping agents. This condition is further aggravated by the sedentary life-style, thus compromising various body organs leading to non-alcoholic fatty liver disease, heart complications, kidney dysfunctions, etc. The prevailing epidemiological shift has caused switching to functional or designer foods that not only enhance endogenous antioxidant status but also fight against ailments of a contemporary lifestyle. In this regard, cabbage owing to its rich nutritive value and phytochemistry based on phenolic acids and flavonoids along with hypocaloric nature and economic accessibility is positively linked to heart, liver, and kidney health. Furthermore, such vegetables possess high nutritive value in terms of minerals and antioxidant vitamins. Resultantly, sulfur-containing ingredients in cabbage are negatively associated with oncogenic events. However, cabbage juice is considered as a remedy against gastric ulcer. Besides, the health-boosting ingredients are related to anti-inflammation and anti-oxidative stress activities.

4.1 INTRODUCTION

Designer or functional foods refer to normal or processed foods that confer additional health benefits beyond complementary nutrition. These foods are enriched with a plethora of health-boosting moieties via the nutrification process as part of a regular diet. In China, 3000 varieties of designer foods are currently available as therapeutic foods. Recent research on nutraceuticals has captured the attention of the scientific community in designer foods like broccoli containing isothiocyanates, eggs enriched with Ω-3 fatty acids, grains fortified with minerals and vitamins, drinking yogurt containing probiotics, foods enriched with phytosterols or essential amino acids, etc. From a health standpoint, designer foods address adjunctive therapeutic benefits under various physiological and pathological conditions. Beyond nutritional enhancement, these foods can improve health in addition to a reduction in disease severity [79, 131].

Large population survey in China has inversely associated the sufficient intake of vegetables and legumes with diseases owing to the presence of biologically active ingredients including dietary fiber, phytoceutics, and minerals that aid in preventing uncontrolled free radical mechanisms [12, 63, 67]. Amongst various vegetables, brassica vegetables (such as broccoli, cabbage, cauliflower, and Brussels sprouts) contribute positively towards human nutrition and health perspectives [14, 109, 149].

Cabbage (*Brassica oleracea* L.) has abundant proportions of minerals, vitamins, fiber, and nutraceuticals [4, 12]. The most familiar varieties of cabbage include red/purple, green, Chinese, and Savoy [76]. The antioxidant capacity of the cabbage head is strongly influenced by genome, geographical locations and environment [3, 43, 53].

It has been reported that red cabbage possesses maximum amounts of anthocyanins; compared to that in black carrot, blackcurrant, grape skin, and elderberry [16, 44]. Anthocyanins are potent antioxidants and their effective dose in the regular diet may reduce various chronic ailments [43]. Besides anthocyanins, cabbage also possesses flavonoids, ascorbic acid, lutein, zeaxanthin, isothiocyanates, hydroxycinnamic residues, β-carotenes, etc. [5, 12, 76].

Commonly, cruciferous vegetables are consumed as salad [5]. Previous reports have confirmed positive health attributes of cabbage as salads, juiced or steamed. The ascorbic acid in cabbage juice and polysaccharides in cabbage leaves has been negatively associated with the management of heart diseases, hypertension, and immunity disorders [124]. Furthermore, the cardioprotective effect of brassica plants is related to ascorbate, α-tocopherol, and carotenoids [149]. In Arabic folkloric practices, cabbage juice is considered effective to protect against hyperlipidemia, obesity, stomach ulcer, liver cirrhosis, obstructive jaundice, hepatitis, renal stress, and tachycardia. The cabbage juice possesses agents involved in detoxification mechanisms, therefore ameliorative against hepatic stress [5]. Recently, the phenolic compounds of cabbage have gained immense attention to limit LDL oxidation, i.e., a major determinant of atherosclerotic events and possess higher reducing power as compared to vitamin C [67].

In restaurants or at the domestic level, vegetables are normally processed considering the taste regardless of nutritional value and antioxidant potential. In this context, scientists have developed broccoli based – bars and apple juice-enriched with red cabbage extract [19, 128]. The lower eating preference for veganism despite health and cost-effective relevance has attracted the attention of food researchers to focus on the concept of designer vegetable supplemented meat products to counterbalance the negative aspects associated with overconsumption of conventional meat-based products. Generally, consumers presumed unprocessed foods relatively healthier than processed. The positive aspects of processed foods are linked with shelf-life extension or activation of some health-boosting compounds [19]. Furthermore, the scientific fraternity seems curious in searching for novel natural antioxidants to replace synthetic constituents [81, 91].

Advancement in food technology has overcome the challenge of food wastage on the one hand but increased the access to empty caloric products on the other side. The rising reliance on fast foods hydrogenated edible oil and sugar-rich foods have begun to dominate the globe, ultimately shifting the dietary pattern towards metabolic disorders [123, 147]. In an elaborate manner, consumption of junk food may lead to the augmented generation of reactive oxygen species/metabolites (ROS/ROM) [28]. One of these radicals is superoxide anion that scavenges nitric oxide (NO) leading to atherosclerosis and cardiac stress [53]. Furthermore, the hyper-caloric diet suppresses superoxide dismutase (SOD) and catalase (CAT): sensitive enzymatic antioxidants involved in the detoxification of free radicals (superoxide anion and hydrogen peroxide) and lipid peroxides to non-toxic metabolites. Moreover, it decreases non-enzymatic glutathione (GSH), reactive non-protein thiol; ultimately up-regulating lipid peroxidation associated oxidative stress [23, 67].

High lipid load in hepatocytes and cardiomyocytes is the prime factor for the induction of hepatotoxicity and cardiotoxicity, respectively [67, 139]. Globally, cardiovascular incidents are considered as the primary cause of mortality [76]. On the other side, liver is a primary organ that is involved in detoxification of toxic exogenous and endogenous substances into less toxic components for excretion. Due to its involvement in the detoxification mechanisms, it is at the highest risk of being affected by free radicals [102]. The multiple problems associated with the liver are on the rise, and non-alcoholic fatty liver diseases have covered up to 3–30% of the globe. Besides, heart ailments have been faced by 27–48% of the Asian Pacific and 40–58% of the western populations. Furthermore, the death rate associated with chronic kidney ailments has reached up to 1.5% of the total, worldwide [77, 130, 167]. Hence, the scientific fraternity is showing keen interest for the preparation of hepato-, cardio-, and renal-protective agents from plant extract with minimal side-effects due to their similar mode of action as that of synthetic medicines [146, 166].

In this context, the application of polyphenols in the preparation of designer foods has become the center of focus to improve the quality of life by neutralizing the deleterious effects of dietary fat [5, 67]. These healthy dietary interventions provide sufficient amounts of exogenous antioxidants that not only strengthen the endogenous antioxidants but also improve the overall health and natural defense against numerous ailments including diabetes, obesity, cardiovascular disorders, insulin resistance, etc. [14, 53].

The objectives of this chapter are:

- Consider consumer attitude in adopting functional/designer foods in their routine menu.
- Compare phytochemistry of green and red cabbage.
- Evaluate the health perspectives of cabbage as designer food with special emphasis to its inclusion in meat-based products.
- Bioefficacy assessment of cabbage against high cholesterol diet mediated oxidative stress.

4.2 CONSUMER PERCEPTION FOR FUNCTIONAL/DESIGNER FOODS

The increasing demand for functional/designer foods is due to the escalating health care cost and desire of older people for optimum life quality.

The increase in consumer awareness and focus of health care professionals towards fruits and vegetables for obtaining bioactive ingredients and antioxidant minerals has raised the trend of functional ingredients in the market. Moreover, the scientific discoveries are attracting the interest of food consumers for the attainment of balanced calorific content in addition to health and nutritional values hence booming the concept of functional/designer foods [24, 40]. The market for these foods is growing, and as innovative products are launching continuously, the competition is becoming more intense. In this regard, incorporation of healthy ingredients in non-healthy versions is considered more justified rather than the enrichment of those products, i.e., normally perceived healthy per se. The positive response is achieved when enrichment is done with such a functional ingredient, i.e., inherent to the product. Furthermore, consumer willingness to try or like the new product depends on basic ingredient and carrier compatibility for that moiety [8].

4.2.1 HEALTH-CONSCIOUS CONSUMERS

There are two groups of consumers who are taking health seriously:

1. **Health Active:** Those believing in good health when they grow older and concerned about family nutrition. As a response, they consume plenty of fruits and vegetables; take medication only if necessary, and exercise twice a week.
2. **Health Aware:** Over age 18, persons are considering health as the former group but do not exercise regularly. Additionally, these groups underscore their desire for nutraceuticals or active ingredients from natural plants despite extracted moieties or laboratory sourced synthetic substitutes [33].

4.2.2 BUYERS OF FUNCTIONAL/DESIGNER FOODS

In the selection of functional foods, women, and middle-aged or older people (45 to 55 years) tend to be more health-oriented as compared to young persons. Multivariate analyses have emphasized that women are more concerned to health due to their heightened responsibility being the dominant role in food purchasing and cooking. Likewise, the reason behind why healthy food is attracting middle-aged or older people is due to their appearing disease symptoms or high purchasing potential like in the US. Moreover, claims that emphasize on improved performance are of the highest interest among men, especially breakfast cereals; however, claims on labels like constipation prevention and low caloric content are more attracted by the female. On the other hand, youngsters (18 to 34 years) prefer high energy claims. Moreover, consumer's food choice is changing; they are more likely to accept food, i.e., well known to them than experimenting with something new. In this context, ingredients with well-acknowledged health perspectives such as vitamin C, n-3, Fe, and Ca are gaining preference over practically unknown moieties such as selenium, xylitol, probiotics, oligosaccharides, and phytoestrogens. Few consumers are willing to accept functional food, i.e., inferior in taste this is because the public always evaluates a functional food first as food and its functional attributes could not outweigh its sensory perspectives [22, 24, 25].

4.2.3   FUNCTIONAL/DESIGNER FOODS: CONSUMER SURVEY

In a survey among Irish consumers, 90% of the individuals of the age group 35 to 44 responded that they are well aware of the concept of functional foods. Additionally, it was surveyed that while purchasing a functional food, around 47% of the consumers look at health claims, whereas 89% of the individuals look for an expiry date and 72% for price followed by a list of ingredients. Price and cooking instructions were more attracted by men or the age group 18 to 24 years while 70% female and older were more interested in nutritional information, ingredient list, and manufacture name and origin. Furthermore, 63% of the consumers were interested in purchasing functional drinks over pills, and those interested in pills were mostly male. However, 69% of the consumers were willing to pay a high cost for functional/designer foods, whereas many others believe to consume fruits and vegetables over functional/designer foods [25].

4.2.4   HEALTH RELATED LEGISLATIVE CLAIMS AND CONSUMER RESPONSE

For the legislative purposes, there are two physiologically oriented food claims; (1) enhanced function: focus on health promotion, for instance, cognitive performance (2) reduced disease risk: focus on prevention of specific disease or associated complication, e.g., cardiac ailments. Amongst these claims, enhanced functional claims evoke a positive signal in the memory thus lucrative to consumers. Likewise, for the food industry, the enhanced functional health claims have a more persuasive impact due to their promotional focus. On the other hand, disease risk reduction claims activate negative perception to the masses. However, consumers consider specifically designed foods more appealing to confront an illness if it already exists. In fact, these claims provide an opportunity to the customers, either to fight the disease they are already victimized or to prevent its aggravating complications [135, 150, 157].

Nowadays, consumers are reducing fat, sodium, and cholesterol in their diets, along with the adoption of regular exercise for optimal fitness. They are discouraging the concept of extreme diet practices and replacing it with a balanced and healthy lifestyle. They are more inclined to positive health messages like calcium to build strong bones in contrast to negative notions like calcium prevent osteoporosis. Previous surveys have shown that the majority of the USA adults are unaware of functional/designer foods, whereas 95% of them believe in diet-health linkages such as oat bran to suppress heart diseases and colorectal cancer, milk to down-regulate osteoporosis and cranberry juice to attenuate urinary tract infections. Additionally, 75% of the females and 72% of the males believe in the negative association between fat and heart ailments [25].

4.3 FUNCTIONAL/DESIGNER FOODS

4.3.1 HISTORICAL PERSPECTIVE

Now, consumers know that eating right could improve their mental and physical well-being automatically promoting life quality and longevity and disease-fighting tendency. The functional/designer foods are recognized as second after fat-reduced foods. The concept of functional foods originated back in Chinese traditional medicine (1000 BC), and then the term was initially coined by Japanese in 1985. Afterward, the concept was introduced by Paul La Chance, who developed the Tang beverage with Ca-fortification in the US (the 1960s). However, the first functional foods in the US and Europe were the fortified breakfast cereals and drinks [25].

4.3.2 CONCEPT AND DEFINITIONS

Despite their contribution to positive health outcomes, the concept of functional/designer foods has not yet been emphasized by the food industry. As a response, the practical application of these foods is somewhat limited in developing countries where nutritional security needs to be fixed first by health-related authorities [131]. Functional/designer food is a food like conventional food, i.e., consumable on a daily basis and ensures health beyond the basic provision of nutrients. In Japan (1991), the health ministry has named functional foods as Foods for Specific Health Uses (FOSHU). It is not a medicine, but normal food, i.e., modified to serve normal physiology or beneficial against modern worries or technology-related health impacts. Generally, consumers tend to incorporate healthy foods in their routines but at the same time find it difficult to change their eating patterns. However, they could be convinced by providing detailed information regarding ingredients and their health response [10, 22, 27, 40, 83, 150]. Designer foods are processed, or cooked foods prepared by using scientific intelligence with special emphasis of processing on antioxidant activity. Furthermore, these foods contain a balance of compositional constituents, nutrients, vitamins, and minerals and polyphenols needed for healthy survival [30, 114]. According to the European legislation, functional food is a concept based on three aspects: (1) optimal wellness or health gains and disease prevention for a better life; (2) technological process; and (3) nutritional function [24, 25].

Besides, functional food is an ambiguous phrase that has undergone revision for several times. There is no legislative or unitary accepted definition for this term that could draw a borderline between traditional and functional/designer foods. ADA has recommended an eating plan, i.e., moderation, wholesome, and healthier; polyphenols rich diet based on varied nomenclature including nutraceuticals, functional or designer foods, hyper-nutritious or superfoods, pharma-foods, medical or longevity foods. Though, there is no official term that exists in US food regulations except medical foods, i.e., regulated by FDA [33, 40, 156].

In most of the economies, there are no specific rules for functional/ designer foods nevertheless, under Food and Drugs Act and Regulations, food is defined as any article represented for use as food, drink, chewing gum or food mixed with ingredients for some purpose [10, 24, 131]. The designer foods are called as modified conventional foods that ensure health benefits in contrast to un-modified versions [92]. Unique features of functional food include: optimum nutrition, a blend of nutrients and non-nutrients, natural routine foods enriched with bioactive molecules or naturally extracted components, and removal of toxicants, aiming to maximize physical and mental health, improve bioavailability and biokinetics and reduce health care cost [24, 58, 68, 107, 134, 135, 145].

4.3.3 NUTRACEUTICALS

Many persons are now turning towards nutraceuticals owing to their role in controlling various ailments. The term nutraceutical was coined by Stephen Defelice in 1989, derived from two words: nutrition and pharmaceutics. It is a component, extracted from food or part of food that ensures health enhancement and disease protection. Nutraceutical is defined as any pharmacologically active substance or single natural nutrient of food, not a drug that promotes health and prevents disease by modulating metabolic functions of the body. Nutraceuticals encompass isolated components, dietary supplements, and processed designer foods like beverages, soup, etc. Later, the Dietary Supplement Health and Education Act (1994) included amino acids, minerals, antioxidant vitamins, and plant extracts under the umbrella of nutraceuticals. Initially, nutraceuticals are potential nutraceuticals, claiming to have health benefits and transformed to established nutraceuticals, if supported by sufficient clinical data to demonstrate the claimed benefit [11, 30, 58, 114, 134].

4.3.4 MARKETING OF NUTRACEUTICALS

Nutraceuticals are the fast-growing segment of the food industry. Nutrition Business Journal (NBJ) has identified US$80 billion markets of nutraceuticals in 1995, escalating 5% per annum [10, 58, 114]. Globally, the market of functional foods has reached US$ 33 billion, whereas contribution by Europe was only 1% of the whole. In 2000, the market value of FOSHU reached US$2 billion, with approvals of 174 FOSHU labels. Most of the functional foods in Germany in 2001 were soft drinks, confectionery, dairy, bakery items, and baby foods [24, 97]. In 2003, the market value of functional foods was around US$11.7 billion in Japan, compared to US$10.5 billion in the US [22]. As functional foods are not well defined, so the global market of these products varied from US$ 25 to 250 billion, whereas the US market for such products worth US$43 billion with annual growth rate 5–10% and Europe contributed US$1.37 billion in 1997 [25]. Globally, the food regulatory system varies widely, 600 FOSHU products were allowed in Japan (birthplace) in 2006 while only limited products are accepted in Canada [10, 24, 58, 68, 150].

4.3.5   OPPORTUNITIES AND CHALLENGES FOR MARKET SUCCESS

Scientifically developed functional foods have become feasible, but unable to meet market acceptance or success. The functional/designer food concept has persuaded people to make healthy food selection and convinced the food industries to attain larger profit margins by reformulating the conventional foods by incorporating clinically proven effective dose of functional ingredients. In this regard, the functional beverage is considered as the most popular and easily formulated functional food besides a complex food matrix. Most innovative products failed before reaching to market because product development is a difficult and risky process, and it cannot be advertised to the public without approving its health claims. In order to achieve niche in market place, the road to success need some basic steps such as: to identify the common diseases in the targeted area, followed by legislative framework to approve a product such as nutrition information on labels, checking functional ingredients for their mode of action, efficacy, and safety on biological markers, educating consumers regarding diet-disease linkages and focusing on consumer perception regarding taste and trust on health claims [10, 24, 33, 39, 40, 54, 55, 150, 156].

4.3.6   FUNCTIONAL FOODS: CLASSIFICATION AND EXAMPLES

The functional foods are classified as essential macronutrients (resistant starch and omega-3 fatty acids), essential micronutrients (minerals and antioxidant vitamins), and non-essential nutrients (oligosaccharides and phytochemicals). These are involved in growth and development, maintenance of body weight, defend against free radicals, resist diseases, and maintain lipoprotein profile, intestinal physiology, and psychological functions [10, 83]. Some examples of functional foods include: fortified foods containing iodine, vitamin C, E, B$_9$, Zn, Fe, Ca, n–3 fatty acid, sterols, soluble fiber, oligosaccharides, probiotics (LAB: lactic acid bacteria), soy-proteins, and peptides to medical foods (enteral administration under the supervision of a medical doctor to curb specific diseases). The functional meat products prepared by reformulating fatty acid profile, the addition of antioxidants, dietary fibers and probiotics, etc. open up a new horizon for the meat industry [10, 24, 58, 150].

Mainly functional foods have been launched in dairy, confectionery, beverage, bakery, and baby food sectors. Normally, the fruit juices are fortified using vitamin C, E, B$_9$, Zn, and Ca, infant foods are enriched with pro-/pre-biotics, food alteration is done by removing deleterious components, incorporation of dietary fiber as fat replacer in meat products (intended as weight loss foods), naturally enhanced foods such as eggs containing omega-3 fatty acids and hypoallergenic foods such as lactose or gluten-free products.

In Japanese and European markets, there is a dominant influence of probiotics, especially LAB and bifidobacteria, launching around 370 products in 2005. The functional/designer foods enriched with soy could overcome the protein deficiencies faced by vegetarians, lactose intolerant could fulfill nutritional gaps by incorporating Ca-fortified juices in their diet, and those who dislike seafood could switch to genetically enriched omega-3 eggs. Traditional technology involves formulation/blending procedures to develop functional foods, whereas technologically designed foods are based on: (1) microencapsulation to extend the shelf life of functional ingredients, and (2) concept of personalized functional food development involving the principles of nutrigenomics; the interaction between foods and human genome [24, 25].

4.3.7 SUCCESSFULLY MARKETED FUNCTIONAL/DESIGNER FOODS

The market for functional food products is divided into four groups: enhance performance, self-treatment, lessening disease risk, and prevention of existing disease. The new generation functional foods have arrived in the market with the name of cosmetic foods containing collagen and vitamin C. In Japan, 70% of the functional food market is comprised of drinks while the dominant functional food in the US is sports-drink carrying vitamins, minerals, amino acids, nucleic acids, and electrolytes. In this context, energy-boosting functional beverages are called ACE vitamin drinks, such as: Red Bull and Lucozade. In Denmark, the active food sector for functional food development is dairy especially probiotic to cholesterol-lowering yogurt. Moreover, European functional food market basically consists of bakery products: bread, biscuits, pasta enriched with vitamins such as vitamins C and E, minerals like Ca, cholesterol-lowering soluble fiber, e.g., Kellogg's products that are now upgraded by adding bran, called 'All Bran Plus' or omega-3 fatty acids [25].

4.4   COMPARISON BETWEEN VEGETABLE AND MEAT-BASED FOODS

4.4.1   VEGETABLE-BASED FOODS

Some decades back, man has included processed foods based on high fat and simple sugars and reduced the use of naturally existing fruits and vegetables. This trend has brought negative outcomes in terms of metabolic syndromes [42]. Some believe that it does not make much difference what we eat until we eat in moderation, hence we are advised to eat less at a time and think about serving size. Vegetables are considered as natural, wholesome, and healthy in terms of dietary fiber, phytochemicals, and minerals; hence, it is recommended to consume @ ½ kg/day [63, 107]. The WHO estimates that worldwide consumption of fruits and vegetables is only 20 to 50% of the recommended daily minimum amount of 400 g per person [133]. Most often, isolated nutraceuticals derived from fruits and vegetables were assessed against disease but ingesting isolated moieties will not have a similar response as that of eating a whole plant matrix [42].

Vegetables are conferred with the status of functional foods, and their habitual consumption is capable of delivering health beneficial moieties to restore the physiological needs of the body. *In vitro* studies have strongly suggested that foods enriched with polyphenols (flavonoids especially anthocyanins and catechins) and antioxidant vitamins (ascorbic acid, tocopherol, and β-carotene) possess strong free radical scavenging potency thus protecting against various ailments [72, 105]. In Pakistan, polyphenols in commonly consumed vegetables have been investigated, and results indicated maximum polyphenols in cabbage followed by cauliflower, spinach, yellow turnip, white turnip, peas, and carrot [152].

A minimum of five servings of green and yellow vegetables are recommended [144]. The vegetables in relevance to fruits are not that pleasant in taste or texture; thus not favorably consumed on a frequent basis, but their entrance into the human diet has proved relatively healthy. The nutraceutical worth has been acknowledged among vegetables (garlic, spinach, onion, cruciferous vegetables, green peppers, tomatoes, and root vegetables), berries in fruits, walnuts in nuts and spicy aromatic plants [21]. Based on 206 human epidemiological analyzes and 22 animal trials, the American Dietetic Association have demonstrated that vegetables including cruciferous, allium, and tomatoes have the potential to fight against oncogenesis of the gastrointestinal tract [30]. Cruciferous vegetables like cabbage contain isothiocyanates and sulforaphane that are responsible to induce phase II enzymes, involved in the detoxification of reactive DNA metabolites, thus prevent cancer initiation or propagation [102].

A different genus of cabbage (green, red, savory, and Chinese) belongs to the same family and contributes positively towards human nutrition and health perspectives. Brassica vegetables are processed or cooked using different methods or may be eaten as a raw ingredient in different salads [120, 149]. These vegetables contribute glucosinolates (GLS) in our diet besides the provision of dietary fiber, and vitamin C. Americans were reported to consume 3 billion pounds of cabbage in 2001. Around 88% of the consumers relied on fresh cabbage, and remaining have adapted taste for sauerkraut [37]. It is documented that major antioxidants in cruciferous vegetables are hydrophilic in nature while hydrophobic antioxidants are responsible for 20% of total antiradical capacity [86]. From a health point of view, cabbage is considered for its antioxidant, anticancer, antiplatelet, antihyperthyroidism, antihyperglycemic, and antihyperlipidemic effects. It can attenuate bronchoconstriction and inflammation. In Arab folkloric practices, cabbage juice is recommended against various disorders like hyperlipidemia, obesity, stomach ulcer, liver dysfunctions, nephritis, and tachycardia [5].

Recently, numerous medicinal reasons such as headaches, diarrhea, gout, and peptic ulcer are negatively associated with red/purple cabbage being rich in pigmented compounds. According to the findings of previous researchers, suppressive effects of red cabbage were measured against lipid peroxidation and plasma protein carbonylation induced by peroxynitrite [79, 116].

Besides health, natural antioxidants in vegetables have been reported to counterbalance microbes, especially in meat-based products. In this context, the use of cabbage powder or extract serves as bio-preservatives in food applications [91].

4.4.2 MEAT-BASED FOODS

On the other hand, consumers often recognize meat products as unhealthy, unlike dairy products due to high-fat content or cancer stimulating the role of red meat. Alongside, the addition of sodium chloride during meat processing is responsible to promote hypertension. These views disregard the fact that functional meat plays an important role in health maintenance. Though, there are some hurdles in designing and marketing functional meat products. Therefore, there is an urgent need to make the consumer well-aware of the limited scientific information that meat itself contains valuable healthy minerals like Fe, vitamins; vitamin $B_{12}$and B_9, bioactive compounds (carnosine, anserine, L-carnitine, and conjugated linolenic acid), bioactive peptides, antihypertensive peptides and meat protein with higher biological value and bioavailability. There are number of routes in the development of functional meat such as reduction of cholesterol, fatty acids, calories, sodium, and nitrites, the addition of vegetables/fish oil and fiber and inclusion of functional ingredients (vegetable or soy proteins, antioxidants, and probiotics)[10].

4.5 PHYTOCHEMICALS AND ANTIOXIDANT VITAMINS: ANTIOXIDANT ASSAYS

4.5.1 EXTRACTION OF PHYTOCHEMICALS

The plants possess a mixture of hydrophilic and lipophilic constituents either in free or bound form, thus binary or multiple solvent systems are normally employed rather than single solvent for solubilization of varied components. However, for a specific component, a single solvent may work effectively [135]. To facilitate the extraction of bioactive compounds, there is a need to increase the contact area of the sample with that of solvent involving numerous steps (pre-washing, drying or freeze-drying, and size reduction). Furthermore, the selection of solvent largely depends on the nature or type of biologically active molecule. The hydrophilic compounds are extracted through polar solvents, particularly methanol, ethanol, or ethyl-acetate while the lipophilic compounds need non-polar solvents such as dichloromethane or hexane [38, 154]. The modern methods including microwave-assisted extraction, pressurized-liquid extraction, supercritical-fluid extraction, etc. are gaining importance in the nutraceutical world due to certain advantages such as extraction proficiency and optimization, less use of organic solvent, less possibility of sample degradation, no supplementary clean-up and concentration stages before characterization and safer for clinical trials in the attainment of particular targets [62, 154].

For the laboratory analyzes of plant extracts, solvent extraction is the most frequently employed extraction procedure. The pH of the solvent is an important factor for extraction purposes. Generally, acidic conditions (using weak acid or strong acid of lower concentration) are preferred for higher stability and easy extractability of phenolic compounds via organic solvents. On the other hand, highly acidic conditions may cause hydrolysis of glycosides or acylglycosides altering the native polyphenols picture [154].

Considering anthocyanins, the glycosylated anthocyanins have better extractability via polar solvents [50]. Earlier, Lapornik et al., [84] explored that anthocyanins are naturally polar compounds, and their mass transfer is related to solvent polarity. They also found that methanol and ethanol are equally effective and suitable solvents for the extraction of anthocyanins. Earlier researchers found that methanol is better for extraction of phenolics and anthocyanins than ethanol while three to four times lower extraction efficiency was achieved through water. The higher concentration of methanol is related to its smaller size, relative to ethanol, thereby penetrating deep into plant matrix where ethanol cannot reach. However, ethanol residues are considered safe for food-based applications [26, 115].

4.5.2 EXTRACTION OF CABBAGE NUTRACEUTICALS

Hydrophilic antioxidants are major active moieties in cruciferous vegetables with a total contribution of nearly 99% [119]. Accordingly, polar solvents are considered as a preferable choice for mass transfer of hydrophilic compounds [140]. These compounds include pigmented molecules (anthocyanins), phenolic acids, vitamin C and GLS, whereas hydrophobic substances include carotenoids and vitamin E, or some lipophilic flavonoids; therefore, non-polar solvents are employed for their extraction [43, 53, 116, 119, 140].

4.5.3   QUANTIFICATION OF PHYTOCHEMICALS

For quantification, spectrophotometric, and chromatographic techniques are commonly employed to achieve valuable information about phenolic acids and flavonoids [154]. The crude plant extracts contain various bioactive compounds with different polarities that could be detected using different techniques including chromatographic (TLC, column chromatography, Sephadex chromatography, and HPLC system) and non-chromatographic (polyphenol photometric assay and Fourier-transform infrared spectroscopy (FTIR)) for the determination of structural features and bioactivity. Currently, HPLC analysis is gaining popularity due to varying migration rates of active molecules in the column and high resolution. Furthermore, the extent of separation is based on the choice of mobile and stationary phases. Generally, the isocratic system is employed for mass transfer phytochemicals while gradient elution is desirable if more than one moieties are under study, varying widely in their retention time under specific conditions. Each molecule possesses a distinctive peak under specific settings. Depending on active compound, the chromatographer chooses conditions to identify a substance of interest via HPLC system [140].

On the contrary, polyphenols assessment via spectrophotometric method is an easy approach to quantify total active molecules in plants extracts. However, HPLC is sophisticated equipment and is employed to characterize individual moieties with high resolution even though it is hectic in terms of sample preparation and system checks [140, 153].

4.5.4   DIETARY PHYTOCEUTICALS

Antioxidants are considered as the first line of defense against oxidative stress in response to free radicals including vitamins C and E, carotenoids, phenolic acids, polyphenols, and flavonoids [47, 91]. Dietary phytoceuticals are secondary metabolites of plants with >8,000 structural variants and are divided into four classes: phenolic acids, flavonoids, stilbenes, and lignans. The phenolic acids include hydroxybenzoic acids and hydroxycinnamic acid, constituting 1/3rd of total polyphenols. The flavonoid encompasses >4,000 different kinds of components. Amongst which anthocyanins are the most abundant, glycosylated derivative of anthocyanidins exist in colorful plants and anthoxanthins (colorless compounds including flavanols like catechins, flavonols (myricetin, fisetin, quercetin, and kaempferol), flavones, isoflavones, flavanones, and their glycosides) comprise of $2/3^{rd}$ of total polyphenols. The health benefits of polyphenols are due to their safe, long term administration and defensive role against various degenerative diseases associated with redox-mediated tissue damage.

The polyphenols have antioxidant activity due to the presence of aromatic rings containing OH moieties or serve as metal scavengers hence suppress lipoproteins from oxidation or improve protective HDL levels or down-regulate lipid peroxidation ultimately protecting plasma vitamin E. Alongside, these moieties may inhibit the activities of hepatic HMG-CoA reductase (3-hydroxy-3-methyl-glutaryl-coenzyme A reductase), ACAT (Acyl-CoA:cholesterol acyltransferase)and stearoyl-CoA desaturase (key enzymes involved in lipid synthesis) or increase fecal cholesterol excretion or up-regulate LDL receptor and genetic expressions of endogenous antioxidant enzymes (SOD, CAT, and GPx (glutathione peroxidase)). Moreover, they have the potential to activate intracellular defensive mechanisms by modulating signal transduction pathways and glutathione synthesis [20, 31, 35, 57, 66, 75, 132, 164, 180].

4.5.5   QUANTIFICATION OF CABBAGE PHYTOCHEMICALS

Researchers found that total polyphenols in white, green or common cabbage vary from 12.58 to 153.00 mg/100 g F.W., whereas 34.41 to 322.00 mg/100 g F.W. for red cabbage. Furthermore, the total flavonoids in white and red cabbage vary from 13.13 to 18.95 and 9.01 to 141.21 mg/100 g F.W., respectively. The higher antioxidant activity of red cabbage is attributed to anthocyanin content. Besides, these differences are due to different localities like, USA, England, Poland, Germany, and India, variation in extracting solvents (methanol, acetone, etc.) and different cultivars and lines of cabbages [15, 36, 43, 72, 86–88, 119–122, 125, 148, 149, 161, 171, 173].

4.5.6   ANTHOCYANINS

There are about 600 naturally occurring anthocyanins that vary based on the number and position of methoxyl and hydroxyl groups on anthocyanidin skeleton, and positions and number at which sugar molecules are attached [29, 116]. Anthocyanin is the most ubiquitous flavonoids, or pigmented compounds, responsible for red, blue, and purple look; hence is considered as the natural choice of coloring drinks, jams, and jellies and confectionery products. Their esterification of sinapic-, ferulic-, and *p*-coumaric acids enhances antioxidant activity; thus, anthocyanins rich phenolic mixture has more nutritional value.

Anthocyanins are water-soluble glycosylated and non-acetylated compounds that are present abundantly in red cabbage, grapes, berries, and apples. However, anthocyanins in red cabbage are basically cyanidin as aglycone (glycosylated with glucose) and/or sophorose that are acylated with diverse aromatic and aliphatic acids [40, 79]. Naturally, anthocyanins are the largest group of hydrophilic vacuolar pigments. The name derived from two Greek words: anthos (flower) and kyanos (dark blue). The anthocyanins in red cabbage are negatively related to hepatotoxicity, hyperglycemia, neurotoxicity, and hypercholesterolemia [116, 139]. Furthermore, pure or crude extract of anthocyanins demonstrated health benefits against obesity, cardiovascular diseases, visual, and brain disorders, ulcers, and cancer [46].

4.5.7   ANTHOCYANINS IN RED CABBAGE

Previous researchers have reviewed anthocyanins in different cultivars and lines of red cabbage; and they found the range from 14 to 495 mg/100 g F.W. [43, 61, 79, 86, 95, 118–120, 143, 162, 168, 172, 174, 176, 178].

4.5.8   ANTIOXIDANT VITAMINS

Ascorbic acid (also called as vitamin C) is a free radical scavenger, reduce lipid peroxides, and function parallel to glutathione hence possesses therapeutic role against CVD, Parkinson's, Huntington's and Alzheimer's disparities. Vitamin E prevents LDL oxidation by fighting against free radical-induced oxidative stress, especially in the hydrophobic environment of cells and its membrane [42, 106]. It serves as an inhibitor of HMG-CoA reductase [32]. Carotenoids is a large family involving >700 molecules of different structure; and are hydrophobic pigments, majorly existing in vegetables. They include *β*-carotene (pro-vitamin A activity and pro-oxidant, i.e., itself act as an oxidant in higher/toxic dose), lutein, lycopene, zeaxanthin, and cryptoxanthin. They are potent free radical scavengers and negatively linked with diabetes, obesity, low sperm motility, and hearing loss and function as immune-boosters and anti-cancer agents [42].

4.5.9   ANTIOXIDANT VITAMINS IN CABBAGE

It is documented that almost ½ cup of raw and cooked red cabbage can provide correspondingly 20–25 mg of vitamin C, protecting from lipid peroxidation by-products and oncogenic events [14, 112]. Moreover, dehydroascorbic acid is considered as the dominant form of vitamin C in cabbage, i.e., almost 4 times to that of ascorbic acid [120]. Red cabbage possesses more vitamin C than oranges and fulfills 61% of vitamin K requirement and 20% of the RDA of vitamin A along with healthy amounts of vitamin B_5, B_6 and B_1[43].

The amounts of vitamin C in green and red cabbage were quantified in various lines, and it ranged from 22.72 to 51.65 and 36.57 to 129.90 mg/100 g F.W., respectively [116]. Furthermore, vitamin C in white and red cabbage was 9.65 and 24.38 mg/100 g F.W [148]. Podsedek et al., [119] measured the ascorbic acid as 18.00 to 35.64 in white cabbage and 62.00 to 72.52 mg/100 g F.W. in red cabbage.

The red and green cabbage also contains natural antioxidants, such as [15, 34, 37, 60, 82, 116, 119, 120, 126, 148, 149]:

Red Cabbage, mg/100 g F.W.	Green Cabbage, mg/100 g F.W.
ɤ-tocopherol0.00 to 0.006	apigenin as 0.8 to 6.1
α-carotene 0.00 to 0.002	kaempferol 0 to 23.9 ± 0.7
α-tocopherol 0.03 to 0.69	lutein + zeaxanthin 0.03 to 0.15
β-carotene 0.01 to 0.13	quercetin 0.9 to 5.1
lutein + zeaxanthin 0.08 to 0.45	
vitamin C 5.7 to 695.6	

These natural antioxidants not only maintain vegetable quality but also play a nutritive role in our daily diet.

Gaafar et al., [53] and Al-Dosari [5] documented the presence of rutin, ferulic, benzoic, acacetin, myricetin, coumarin, luteolin, genistein, pyrogallol, gallic acid, catechins, p-coumaric acid, chlorogenic acid, vanillic acid, caffeic acid, and protocatechuic acid. They also mentioned that flavonols (quercetin and kaempferol) are influenced by various extraneous determinants including variation in type and growth pattern of plant, seasonal variability, degree of maturity and food processing conditions [37]. Conclusively, the variation in phenolic content, anthocyanins, and vitamin C was attributed to differences in cultivar, geographical conditions, agricultural practices, maturity stage, seasons, storage conditions and analytical methods [37, 120].

4.5.10   ANTIOXIDANT ASSAYS

Numerous tests have acknowledged the assessment of antioxidant potential [99]. Commonly, radical trapping capacity is measured by ABTS and DPPH (free radical producing) reagents because polyphenols can react with these free radicals, ensuring stability [86]. Additionally, FRAP assay is a reducing assay by which reductional potential (ability to donate electron or hydrogen) is estimated, also known as total antioxidant capacity [7]. Earlier research on fifty popular fruits and vegetables and beverages showed that antioxidant assessment of pigmented and hydrophilic phenolics is better represented via ABTS assay than that of DPPH reagent [52].

4.5.11   ANTIOXIDANT ASSAYS WITH SPECIAL EMPHASIS TO CABBAGE

Previously, DPPH scavenging ability of different cultivars of red cabbage were measured as 26.6 to 70.9% compared to 2 to 8% for white cabbage [86]. The DPPH scavenging ability of freshly harvested and stored white cabbage was 2.76 to 10.33 and 2.49 to 15.14%, respectively [87]. Xu et al., [173] determined the DPPH quenching ability of fresh-cut red cabbage as 95%. Moreover, the DPPH scavenging ability of white and red cabbage were16.095 and 22.533 for conventional and 66.370 and 73.316% for organic, respectively [138].

The TEAC of white cabbage was measured as 1.73 μM Trolox/g F.W. [15]. Furthermore, the TEAC of green and red cabbage were 4.92 and 13.77 μM Trolox/g F.W., respectively [125]. Podsedek et al., [119] measured the ABTS values for white and red cabbage as 1.34 to 1.81 and 9.81 to 12.64 μM TEAC/g F.W., respectively.

Earlier, the FRAP values for green and red cabbage were estimated at 6.94 and 18.70 μM Fe^{+2}/g F.W., respectively [125]. Recently, the FRAP value of green and red cabbage were measured as 17.00 and 80.87 μM TE/g F.W., respectively [136]. The FRAP values of white and red cabbage were found as 1.93–2.25 and 0.20–0.30 mM/100 g F.W. [169].

These antioxidant assays serve as an effective tool to validate various health benefits; hence, these give an estimation of an effective dose of antioxidants and their incorporation in designer foods.

4.6 IMPACT OF PROCESSING ON ANTIOXIDANT CAPACITY OF VEGETABLES

Food is a mixture of nutrients, phytochemicals, and fiber; however, their health-promoting ability depends on content, activity, and bioavailability of these ingredients that are altered based on processing history. This aspect is of great consideration, as in Pakistan only small proportions of vegetables are consumed in their raw state, whereas most of them are often processed. Processing and preservation procedures are responsible for the improvement or depletion of health-promoting antioxidant vitamins and polyphenols. However, blanching treatment retains the original antioxidant profile in most of the cases. Other operations, including peeling and slicing contribute to enzymatic oxidation leading to a modification of inherent antioxidants or decrement of their antioxidant activity.

Furthermore, prolonged storage promotes enzymatic and chemical oxidation of phenolic moieties. High antioxidant capacity is attributed to a higher ability to donate a hydrogen atom from antioxidant (aromatic hydroxyl group) to an unpaired electron system of free radicals [105]. The concept that antioxidant activity of fruits and vegetables are lost during processing is due to the removal of skin [145].

There is a general trend of consuming semi-cooked or boiled vegetables in western countries, whereas, in Asian countries like Pakistan, the vegetables are normally subjected to various cooking procedures prior to consumption including shallow or deep-frying, steaming, roasting, and stewing. These treatments affect the efficacy of active constituents in vegetables either positively or sometimes negatively. Cauliflower, cabbage, spinach, carrot, yellow, and white turnip are often consumed in Pakistani cuisine. These vegetables contain rich nutritive but low caloric value along with the presence of a myriad of antioxidants. An earlier study revealed that cabbage exhibits varied phenolic compounds compared to other vegetables; thus, there is a need to explore its nutraceutical worth and designer food applications [152].

The cooking methods such as boiling, steaming, and microwaving are generally practiced in western society while stir-frying is considered as a common dietary habit among Chinese. It is assumed that domestic cooking methods resulted in significant alterations in chemical composition and bioavailability of active compounds in cruciferous vegetables [173].

In general, processing or cooking methods impact the stability of polyphenols of brassica vegetables. Some of the cooking methods do not cause any significant alteration in antioxidant profile, whereas it may deplete or form novel antioxidant metabolites that may have high or low bioactivity or bioavailability [155, 162]. Therefore, food manufacturers should focus on the antioxidant potential of designer foods to ensure health promotion [19].

Roy et al., [137] and Wachtel-Galor et al., [163] found high extractability of broccoli nutraceuticals during steaming, resulting in more antioxidant capacity in contrast to raw equivalents. These findings were contradictory to the earlier notion demonstrating the reduction of polyphenols in response to thermal treatments [120]. In addition, processing converts GLS to isothiocyanates, thus suppressing urinary mutagens derived from meat, tobacco carcinogens and gastric ulcers. Besides, cabbage also comprises of an anti-inflammatory amino acid (glutamine) [43].

4.7 CABBAGE-BASED DESIGNER PRODUCTS

Incorporation of cabbage in meat-based products ensures the provision of a myriad of natural phytoceuticals including β-carotene, ascorbic acid, and α-tocopherol hence overcoming the injudicious incorporation of synthetic antioxidants to prolong meat quality against oxidative degradation. Furthermore, cabbage may overcome the calcium deficiency in meat products. Additionally, the use of fiber from cruciferous vegetables resulted in fat-reduced meat products. On the other hand, cabbage may mask the meaty flavor. Besides, cabbage powder or extracts are considered as bio-preservatives [81, 91]. Mostly in dieting programs, cabbage is given a major share of the dietary plan being low in calories [5]. The calorie content in brassica vegetables was approximately 24–34 kcal/100 g, according to Heimler et al., [59]. Recently, the calorific values of control and 6% cabbage powder-based meat patties were reported to be 194.65 and 187.18 kcal/100 g, respectively [91].

The inclusion of cruciferous vegetables enhances DPPH and ABTS radical scavenging activity due to the presence of natural antioxidants such as carotenoids, tocopherols, ascorbic acid, and flavonoids [81]. On a commercial scale, anthocyanins from red cruciferous vegetables have gained immense use in designer foods, including jams, juices, preserves, and ice-cream to curtail escalating health disparities [29]. In this context, Barakat, and Rohn [19] prepared broccoli-based bars with higher organoleptic acceptability via frying; however, maximum retention of flavonoids was noticed in baked and steamed bars. In another study, Radziejewska-Kubzdela et al., [128] incorporated anthocyanin-rich red cabbage extract in apple juice and found increment in antioxidant activity due to an increase in polyphenols up to 3.1 to 4.9 times.

Banerjee et al., [18] developed broccoli powder enriched nuggets and found a linear increasing trend in radical scavenging potential and reducing power with increment in broccoli dosage.

Llorach et al., [89] prepared soup modified with cauliflower by-products and found higher antiradical activity attributed to polyphenols. Malav et al., [91] evaluated the general appearance of cabbage enriched mutton patties and found that a decrease in appearance during storage was attributed to the breakdown of pigmented compounds, depletion of nitrites and occurrence of browning reactions. Verma et al., [159] studied the sensory response of green cabbage @ 15 and 25% incorporated meatballs. They assessed gradual decrement in appearance, flavor, and juiciness over prolonged storage.

4.8 EPIDEMIOLOGICAL TRANSITION

The Asian world has shifted from nutritional deficiencies to chronic ailments such as CVD, diabetes, and cancer, i.e., termed as "epidemiological transition" in response to urbanization. The marked dependence on hypercaloric diets and decrement in energy expenditure has raised the probabilities of non-communicable diseases even below the age of 50 years. Concomitantly, the global influences in terms of television viewing or increased access to junk foods are positively associated with BMI (body mass index). The higher energy proportion from fat sources has led to escalated rates of hyperlipidemia, hyperglycemia, obesity, insulin resistance, hypertension, and strokes [51, 65, 171].

Among modern societies, high reliance on calorie-dense foods is responsible for obesogenic environment, generating ROS and weakening of antioxidant defense system; hence 3000 kcal need to be cut down to 2200 kcal [13, 69, 110, 165]. According to WHO reports (2007), overweight people were estimated at around 1.7 billion, involving 155 million children [42]. This situation leads to hyperlipidemia and related disparities. This postulation is found to be consistent with animal models where the accumulation of lipids occurs in the liver, heart, and kidney. Moreover, the available therapies to alleviate hyperlipidemia are responsible to aggravate liver toxicity; and more than 900 drugs have been implicated to affect liver.

Throughout the world, 5% of all hospitalized and 50% of acute liver failures are the culprits of drug-induced liver damage still all these disorders are without appropriate therapies [113]. Thus, scientists are looking for hypolipidemic therapies that could increase hepatic LDL receptors expression [80]. The cross-sectional and longitudinal studies have reported that consumption of junk foods is linked to insulin resistance and obesity that further exacerbates to non-alcoholic fatty liver disease, i.e., prevalent worldwide in 10–24% [113, 127, 158].

In normal metabolic processes, Reactive Oxygen/Nitrogen Species/ Metabolites (ROSs/RNSs/ROMs) are continuously being produced, i.e., neutralized by detoxification system of antioxidants and antioxidative enzymes. In response to the over-consumption of junk food, this system is insufficient to combat free radicals. This results in the generation of ROS/ROM (superoxide anion, hydrogen peroxide, and hydroxyl ion) via numerous pathomechanisms inducing sequential events of oxidative injury especially in the fragile organs of the body [28, 139]. Basically, the over-accumulation of free radicals and their covalent bonding with bio-molecules induce peroxidative damage of lipophilic (rich in PUFA) membrane structure leading to the synthesis of lipid peroxides that are involved in multiple pathologies.

Furthermore, the pro-inflammatory end products of lipid peroxidation (MDA (malondialdehyde) and 4-hydroxynonenal (aldehyde by-products)) enhances LDL deposition around the vasculature and are responsible for hepato-pathological features. Furthermore, prolonged consumption of hypercaloric diet depletes antioxidant reserves or compromises innate antioxidant defense system of oxygen-dependent organisms that fight against pro-oxidants (active oxygen molecules) by converting them to lipid peroxides and finally to non-toxic alcohol. The first line of defense against oxidative damage is SOD: catalyzes superoxide anions, O^{2-} (generated by one-electron reduction in oxygen molecule by transition metal ion) to hydrogen peroxide (H_2O_2) that get converted to (OH^-) radicals, promptly reacting with cellular components (lipids, proteins, and nucleic acids) leading to lipid peroxidation and cell death. However, presence of CAT (hemeprotein) and GSH are considered as second line of defense and its related enzymes (GPx and GST) detoxify (OH^-) radicals to non-toxic metabolites like H_2O [2, 30, 35, 41, 64, 66, 67, 73, 78, 98, 104, 129, 132, 151, 160, 164, 179, 180, 182].

Initially, the intracellular enzymatic antioxidants become activated as cells' well-being becomes a threat due to the over-generation of ROS being sensitive to redox state. All the enzymes (SOD, CAT, GP_x, GR (glutathione reductase) and GST (glutathione-S-transferase)) work in synergy with non-enzymatic antioxidants (endogenous GSH and exogenous polyphenols or antioxidant vitamins; ascorbate, tocopherols, carotenes, and retinol in diet) to neutralize redox state of cell by converting free radicals to less toxic/reactive compounds or maintaining the free radicals within the tolerable limits. The supportive team of free radical scavengers (CAT, SOD, GST, GPx) and PON1 (paraoxonase 1 – an antioxidant enzyme linked with HDL, calcium-dependent esterase that detoxifies lipid peroxides and widely distributed in numerous tissues) mutually fight against markers of oxidative stress; and products of lipid and protein oxidation include thiobarbituric acid reactive substances, hydroperoxides, and carbonyl proteins.

The hypertriglyceridemia inactivates antioxidant enzymes by cross-linking with MDA or rapid consumption/exhaustion of enzymes in fighting against free radicals further aggravating lipid peroxidation leading to cell injury that in turn causes the release of cytokines like TNF-α, Tumor necrosis factor-α, IL-6, *Interleukin-6* and CRP, C-reactive protein (a pro-inflammatory marker of cardiac stress). Additionally, cellular proteins are targeted by free radicals resulting in the formation of protein carbonyl (PCO: an indicator of damaged proteins) content in the liver, heart, and kidney. Under stress, PON-1 gets inactivated because of impaired synthesis/secretion of HDL due to alteration in lecithin cholesterol acyltransferase (LCAT) activity or S-glutathionylation, redox mechanism forming mixed disulfide linkage between protein thiol and oxidized glutathione [12, 42, 74, 108, 111, 174].

Hyperlipidemia produces ROS that results in the formation of cardio-vascular diseases as well as NAFLDs (non-alcoholic fatty liver disease) [85, 129, 151, 179]. Previously, high cholesterol diet @ 2% to male white New Zealand rabbits depicted damage to tissues of the liver (fatty degener-ation, inflammation, and necrosis) and heart (vacuolar degeneration, disor-ganized myofibrils, and necrosis). Previous studies reported that sustained hyperlipidemia is responsible for elevated cholesterol, especially LDL, redox-mediated oxidative stress in the liver, heart, and kidney and lower activities of endogenous radical scavengers.

Due to the modern lifestyle, people are unable to manage their cholesterol levels; therefore, medication is considered as an option. However, these hypolipidemic drugs may contribute to liver damage, myopathy, and drug-drug interactions. In this context, natural antioxidants are gaining much more attention being potentially safe, economical, and effective radical scavengers to attenuate lipid peroxidation (process involved in the conversion of unsaturated fatty acids to free radicals by subtracting hydrogen) and apoptosis (programmed cell death process involved in the elimination of defective and harmful cells). This revitalized the concept of herbal remedy to prevent hyperlipidemia in individuals with cholesterol at borderline levels and to improve the antioxidant status of the body. The antioxidant-rich diet particularly targets symptoms of liver, heart, and kidney hence termed as hepato-, cardio-and nephroprotective diets [1, 6, 64, 98, 117, 127, 151, 175].

The experimental evidences have suggested that small changes in atti-tudes could bring appreciable response like watching television one hour less per day could result in risk reduction of heart attacks up to 2.5 and 4% in men and women, respectively [65]. Moreover, fast food establishments are adversely affecting health quality by offering options that tend to have lower nutritional value (minerals, vitamins, and dietary fiber and higher in fat, sodium, and sugars). In the USA (2007), 37.4% of the food eaten was purchased from fast food establishments [51, 100, 101, 141]. The unhealthy meal in poor countries is basically attracted by its low price, but there is a need to realize the fact that a cheaper pro-inflammatory diet is not cost-saving; rather it adds burden in terms of medical costs. Based on an obesity survey in America (2010), it was observed that obese men and women besides poor work productivity, pay an extra medical bill up to $1,152 and $3,613, respectively [103].

4.8.1 TRANSLATED ANIMAL MODEL FOR METABOLIC SYNDROMES (RABBITS AND RATS)

Earlier studies on laboratory animals have shown that cession of cholesterol feeding or the introduction of antioxidant-rich foods resulted in a reverse of oxidative stress, normalizing insulin release, and lowering cholesterol accretion in hepatocytes. Therefore, animals are considered as ideal models to induce chronic ailments in response to diet to validate the biological efficacy of antioxidant moieties [13].

According to the previous review, the rat is considered as a resistant animal to cholesterol feeding and moderate hyperlipidemia demand for extra cholic acid with double frequency and time frame. On the other hand, the provision of 1% cholesterol to rabbits could increase plasma cholesterol, phospholipid, and triglyceride levels and liver cholesterol to higher extents. It is found that aortic ACAT (microsomal enzyme catalyzes the formation of esterified cholesterol) activity of rabbits is much higher than rats. However, hepatic ACAT activity is similar in both animal models. Furthermore, cholesterol feeding increases plasma lipid peroxidation in rabbits, whereas it may increase, decrease, or remain constant in rats. On the other hand, mild hepatic lipid peroxidation is detected in both animals.

Also, high cholesterol intake decreases rabbit liver GSH concentration and GSH-related enzyme activities (GPx and GST with a minor change in hepatic SOD activities), while no changes in GSH and SOD were observed in rats rather decrement in GPx and GST activities. Moreover, antioxidant enzyme activities in the rat liver detected an increase in response to high cholesterol diet due to an adaptive increase in rats to protect the liver but the obvious decrease was observed in the case of rabbits. Similarly, high levels of aortic and plasma MDA levels for a longer time have an additive effect in the inducement of atherogenesis in rabbits (@0.2–1.3% cholesterol in 8 to 14-week period) in contrast to rats [9, 17, 64, 93, 104, 132, 160].

Furthermore, rabbits and rats have different metabolic responses to cholesterol-rich diets due to the difference in lipid peroxidation and antioxidant enzyme activities. The basal SOD and CAT activities are 50 and 100% higher in rats than rabbits, respectively; thus, rabbits are regarded as less resistant animals. In rabbits, cholesterol causes severe hyperlipidemia and increases in oxysterols (7a-, 7β-hydroxycholesterol, a-epoxy, β-epoxycholesterol, cholestanetriol, 7-keto and 27-hydroxycholesterol) in contrast to rats where increment was observed in $7a$- and 7β-hydroxycholesterol only. In rabbits, serum thromboxane (TXA2) level was increased by cholesterol supplementation; however, the inverse was observed in rats. The decrease in triglyceride in the rabbit liver is attributed to its accumulation in VLDL in high amounts.

The plasma lipid peroxidation was more in rabbits because anti-oxidant protection by vitamin E is insufficient to capture free radicals. Thus, lower vitamin E: cholesterol ratio in rabbits strongly predicts poor antioxidant protection resulting in higher lipid peroxidation probabilities. Furthermore, cholesterol in hypercholesterolemic rabbits is around 30% polyunsaturated fatty acids, thus an easy target for free radicals. According to research studies, two types of effects were noted: (1) SOD activity in cholesterol-fed rabbits remained the same; however, there was a decrease in GSH-Px that deals with hydrogen peroxide causing an increment in CAT activity as an adoption. If enhancement of CAT is not sufficient, then lipid peroxidation would dominate; (2) Significant decrease in SOD activity of rabbits was in response to higher levels of superoxide anions thus more lipid peroxidation, and both GSH-Px and CAT become inactive. On the other hand, fewer changes were observed in hypercholesterolemic rats due to the protective effect of antioxidant defense system against lipid peroxidation and formation of oxysterols [49, 57, 64, 85, 90, 127].

Cholesterol fed rabbit models have various advantages, such as: cholesterol-based diet is not a special induction treatment rather it reflects junk food consumption pattern of general public; fatty liver disorder occur in association with β-oxidation in mitochondria without obesity or insulin resistance hence clarifies the mechanism of hyperlipidemia induced NAFLD; and slender fibrosis (increase in LPO and number of HSCs) resembles strongly with that observed in NAFLD patients. In NAFLD patients, mixed macro-vesicular fats (benign; a mild decrease of mitochondrial β-oxidation) and micro-vesicular fats (severely impaired mitochondrial β-oxidation) were found. On the other hand, micro-vesicular lipids are more prominent in rabbits representing the severe condition of fat deposition. Moreover, the cholesterol-fed rabbit model may also be used to study hyperlipidemia induced CVD [69].

The rabbits are used as translated animal models for hyperlipidemia induced atherosclerosis because their lipoprotein profile is like humans. Some other researchers depicted that feeding cholesterol to rabbits for 12 weeks led to metabolic syndromes like obesity, hypertension, and fibrosis [71, 175]. Rabbits are the most sensitive species to dietary cholesterol induction therapy being herbivores, and it is the only animal in which hyperlipidemia could be induced within a few days [70, 165].

A study conducted on New Zealand rabbits fed on 1% high cholesterol diet for 4 weeks that depicted:

Increment in:	Decrement in:
Total cholesterol (from 74 ± 8.2 to 688.75 ± 20.02 mg/dL); LDL (from 199.5 ± 23.6 to 621.00 ± 20.9 mg/dL); triacylglycerols (from 35.75 ± 2.25 to 168.00 ± 16.04 mg/dL); and TBARS (from 7.39 ± 0.35 to 14.43 ± 0.95 nmol/mL)	HDL (from 266.50 ± 23.67 to 36.25 ± 3.52 mg/dL); and GSH (from 0.58 ± 0.002 to 0.26 ± 0.01 μmol/mL).

Furthermore, kidney functioning test showed significant alteration in response to high cholesterol diet especially increase in (a) creatinine from 0.7 ± 0.16 to 1.73 ± 0.23 mg/dL and (2) BUN from 25.25 ± 3.47 to 59.25 ± 0.48 mg/dL [104].

Marinou et al., [93] elucidated that 1% of the cholesterol in rabbits resulted in altered biochemical and liver biomarkers. The body weight was increased from 3.17 ± 0.21 to 3.49 ± 0.19 kg, total cholesterol was up-regulated from 67.88 ± 5.66 to 1347.00 ± 149.33 mg/dL, and HDL was decreased from 24.62 ± 2.38 to 46.37 ± 4.20 mg/dL, decrement in triacylglycerol from 97.63 ± 12.93 to 770.13 ± 68.48 mg/dL. However, the increments in MDA levels were from 1.14 ± 0.14 to 3.47 ± 0.47 nmol/L; similarly increase in ALT, AST, and γGT was from 7.75 ± 2.12, 11.75 ± 2.87 and 14.25 ± 2.71 to 37.25 ± 5.42, 47.75 ± 6.69 and 215.13 ± 67.34 IU/L, respectively.

Al-Naqeep et al., [6] fed 1% cholesterol to New Zealand male white rabbits that induced hyperlipidemia and related atherosclerosis. Yu et al., [175] studied the effect of hypercholesterolemic diet-induced hyperlipidemia in rabbits for the 12-week duration and results depicted increment in body weight from 2.51 ± 0.06 to 2.95 ± 0.07 kg and cardiac stress marker (CRP) from 16.76 ± 3.53 to 81.51 ± 27.90 mg/dL. Furthermore, a study was conducted on New Zealand white rabbits fed with 1% cholesterol for 3 months and it presented early fibrotic lesions in hepatocytes and development of foam cells in arteries due to cholesterol deposition that was further exacerbated in response to systemic inflammation leading to early lesions in liver and aorta of rabbits [70, 78].

4.8.2   HEALTH PROTECTIVE MOIETIES OF CABBAGE

Currently, the application of polyphenols and designer foods has become the center of focus to improve quality of life by neutralizing deleterious effects of dietary fat responsible to exert oxidative stress in hepatic, cardiac, and renal tissues [5, 67]. In this context, functional ingredients present in cabbage may control the negative biological impact of free radicals by enhancing enzymatic and non-enzymatic antioxidant defense systems hence maintaining normal cell structure and functions [67].

Numerous studies have depicted that cruciferous vegetables possess a complex combination of bioactive ingredients, minerals, and antioxidant vitamins that could scavenge ROS, ultimately improving the levels of GSH, SOD, and CAT. Besides other ingredients, the GLS-myrosinase system exists inherently in brassica vegetables, and myrosinase hydrolyzes GLS to isothiocyanates. The myrosinase is either natively present within the compartments of plant cells or in the mammalian intestine. It is still unclear that these S-containing compounds are either responsible for lowering cholesterol levels or the synergistic effect of several antioxidants in cabbage to improve the overall antioxidant status. Recent investigation on broccoli has associated isothiocyanates and sulforaphane with cholesterol-lowering activity [96].

Red cabbage contains anthocyanins as the main active ingredient predominantly cyanidin-3-diglucoside-5-glucoside derivatives. These are water-soluble pigments that contribute in health promotion. However, other natural antioxidants (such as ascorbic acid, β-carotene, α-tocopherol, and lutein) could further improve its disease modulatory role. Previous data directly linked these antioxidants with hypocholesterolemic, hypoglycemic, hepatoprotective, cardioprotective, nephroprotective, and neuroprotective activities. An earlier study based on anthocyanins rich diet to oxidatively stressed rats showed a significant decrement in lipid profile by eliminating cholesterol and triglyceride through feces or by inhibiting intestinal absorption. The anthocyanins were assumed to reduce cholesterol and triglyceride absorption in the intestine. This postulation is further supported by reviewers who reported that herbal extracts are primarily responsible for the catabolism of cholesterol to bile acids, thus eliminating subsequently through feces [139].

It has been reported that cabbage extract could reverse oxidative damage leading to the restoration of GSH. The biological action of phenolics is based on free radical quenching and metal chelating ability [67].

Cruciferous vegetables are commonly consumed dietary vegetables owing to their consumer preference, easy in accessibility, and cost-effectiveness. It is considered for its hypocholesterolemic, hypoglycemic, and anticancer activities.

The principle antioxidants are ascorbic acid, anthocyanins, and isothiocyanates. Animal and human interventional studies suggested its chemoprotective aptitude due to GLS and their hydrolytic product (isothiocyanate), responsible in inhibiting phase I enzymes hence preventing the activation of carcinogens, while inducing phase II enzymes that are involved in detoxification of xenobiotic [53]. Also, cabbage extracted moieties are believed to protect plasma protein and lipid profiles from hydrogen peroxide-induced oxidative stress [79].

4.8.3   EFFICACY OF CABBAGE EXTRACTS AGAINST OXIDATIVE STRESS BIOMARKERS

A recent study has presented convincing evidences in favor of cabbage extract @ 300 to 500 mg/kg B.W. against drug-induced hepatotoxicity. Alongside this, an inclining trend was observed in endogenous antioxidants (SOD, CAT, and GPx by 46.48, 58.89, 27.06, and 26.16%) [23]. Earlier, another study depicted dose-dependent hepatoprotective role of indole-3-carbinal (present in cruciferous vegetables) on cell viability against oxidative damage and subsequent leakage of oxidative stress biomarkers (ALT–alanine transaminase, AST–aspartate transaminase, ALP–alkaline phosphatase, GST, and LDH–lactate dehydrogenase) from liver into the bloodstream [56].

Morsy et al., [102] explored the hepatoprotective properties of red cabbage and broccoli extract @ 10% against hepatic cancer induced by N-Nitrosodiethyamine (NDEA) and carbon tetrachloride (CCl_4). During 30 days of experimentation phase, the common cabbage extract @ 240 mg/kg B.W./day showed an inclining trend in hepatic SOD and CAT up to 28 and 24%, respectively; however, hepatic lipid peroxidation demonstrated significant decline by 36%. The positive impact of cabbage extract on GSH is attributed to chlorogenic-, Gallic-, protocatechuic-, caffeic-, and vanillic acid [67].

In a rat model trial, co-ingestion of anthocyanins derived from red cabbage plus high fat diet reported decreased serum biochemical parameters including cholesterol 57%, triglyceride 23%, LDL 70%, VLDL

27%, atherogenic index 72%, ALT 32% and ALP 35%, whereas HDL was raised up to 32%. The red cabbage extract was found to suppress hepatic lipid peroxidation by 44% along with the significant rise in the activity of enzymatic and non-enzymatic antioxidants; SOD 44%, CAT 47% and reduced glutathione as compared positive control animals. However, alterations in the histological architecture of oxidatively stressed liver tissues were counteracted by co-administration of red cabbage extract. Furthermore, serum creatine kinase (CK), creatine kinase-MB (CK-MB), LDH, and AST showed a reduction of up to 40, 42, 32, and 31%, respectively. The decrement in MDA of heart tissues was up to 40%, whereas momentous increment was observed in the endogenous antioxidants of heart tissues (SOD 46% and CAT 47%) as compared to the positive control group. Moreover, histopathology of damaged cardiac tissues was significantly reversed with anthocyanin-rich diet [139].

Al-Dosari[5] found a decrement in cholesterol, triacylglycerol, LDL, and VLDL up to 10 and 33, 16 and 28, 12 and 42, 17 and 29% at different doses (250 and 500 mg/kg/day of lyophilized red cabbage juice), respectively. On the other hand, HDL showed an inclining trend up to 16 and 26% in the corresponding doses. The decrement in cellular leakage of the corresponding liver enzymes ALT, ALP, ɤ-GT (γ-glutamyl transferase) and bilirubin levels were 5.3 and 12.3, 11 and 14.9, 14 and 36 and 9 and 23%. The decrement in hepatic MDA was 57 to 71%. Briefly, red cabbage has the potential to modulate redox-sensitive dyslipidemia and associated hepatic injury in a dose-dependent manner. The suppression in cardiac indicators (CK, LDH, and AST) was 15 and 33, 22 and 33 and 2.3 and 16% at two different doses, respectively. Apart from this, the MDA level in heart tissues was decreased to 55 and 66%, whereas creatinine and urea were reduced to 25 and 50 and 12 and 18%. The ethanolic extract of common cabbage @ 500 mg/kg B.W. along with high-fat diet decreased serum cholesterol, triglyceride, LDL, and MDA by 23.23, 4.54, 3.81 and 31.7% during 12-week trial; though, anti-atherogenic index (AAI) showed an increase of 49% compared with hyperlipidemic rats ultimately protecting against life-threatening modalities such as myocardial infarction and atherosclerosis [166].

In an animal study, red cabbage powder @ 10% and red cabbage extract @ 100 mg/kg B.W. was tested against paracetamol-induced hepatotoxicity (Figure 4.1). The serum cholesterol, LDL, and VLDL were reduced up to 38 and 43, 53 and 61 and 29 and 29%; whereas ALT, ALP, AST, and ɤ-GT

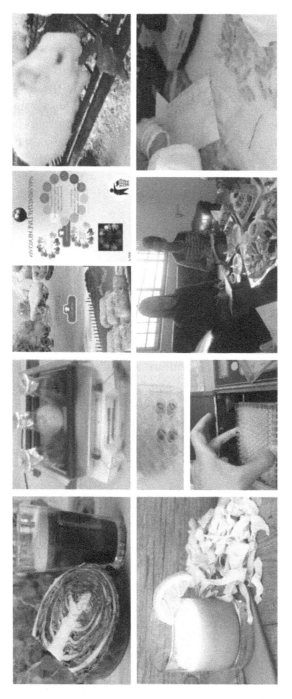

FIGURE 4.1 Comparison between green and red cabbage via *in vitro*, sensory, and *in vivo* studies.

were down-regulated by 20.9 and 24.3, 25 and 32.5, 16.9 and 31.3 and 41 and 44% via red cabbage powder @ 10% and red cabbage extract @ 100 mg/kg B.W., respectively. Furthermore, the increment in hepatic SOD and decrement in MDA was up to 35.5 and 54.5 and 47 and 59.5% on feeding red cabbage powder and extract, respectively [45].

The general recommended golden standard for lifestyle modifications against obesity and related NAFLD and CVD include: caloric restriction by 50 to 70% (delaying metabolism responsible for redox-mediated lipid peroxidation), inclusion of antioxidants and soluble fiber-enriched functional foods or dietary supplements along with habit of regular physical exercise [2, 9, 20, 30, 48, 94, 110, 127, 142, 151, 170, 181, 182].

4.9 SUMMARY

Cabbage promises splendid nutritional and nutraceutical importance, hence their incorporation in already existing conventional edibles could replenish them, supporting the concept of designer foods. Additionally, cabbage inclusion in meat-based products not only offer an opportunity to the food manufacturers to improve the conventional edibles but also encourage the inclusion of nutritious and cost-effective vegetables in our routine menu by overcoming the associated organoleptic disliking. Further, this chapter may go a long way in guiding vegetable growers to prefer red cabbage production over green equivalent to promote health and prevent disease incidences with special emphasis on anthocyanins.

KEYWORDS

- **designer foods**
- **flavonoids**
- **green cabbage**
- **oxidative stress**
- **phenolic acids**
- **red cabbage**

REFERENCES

1. Abdel-Rahman, G. H. Taurine attenuates hepatic and cardiac damage and apoptosis in rabbits fed a high-fat diet. *OnLine Journal of Biological Sciences*, **2014**, *14*(1), 12–20.
2. Adaramoye, O. A., & Akanni, O. O. Effects of methanol extract of breadfruit (*Artocarpusaltilis*) on atherogenic indices and redox status of cellular system of hypercholesterolemic male rats. *Advances in Pharmacological Sciences*, **2014**, E-article, http://dx.doi.org/10.1155/2014/605425:1–11 (Accessed on 30 September 2019).
3. Aires, A. Brassica composition and food processing. In: *Processing and Impact on Active components in Food* (1st edn., p. 10). Academic Press, New York, **2015**, E-article. https://doi.org/10.1016/B978–0–12–404699–3.00003–2:17–25 (Accessed on 30 September 2019).
4. Akbar, M. F., Haq, M. A., Parveen, F., Yasmin, N., & Khan, M. F. U. Comparative management of cabbage aphid (*Myzuspersicae* (Sulzer) through bio-and synthetic-Insecticides. *The Pakistan Entomologist*, **2010**, *32*(1), 12–17.
5. Al-Dosari, M. S. Red cabbage (*Brassica oleracea* L.) mediates redox-sensitive amelioration of dyslipidemia and hepatic injury induced by exogenous cholesterol administration. *The American Journal of Chinese Medicine*, **2014**, *42*(01), 189–206.
6. Al-Naqeep, G., Al-Zubairi, A. S., Ismail, M., Amom, Z. H., & Esa, N. M. Antiatherogenic potential of *Nigella sativa* seeds and oil in diet-induced hypercholesterolemia in rabbits. *Evidence-Based Complementary and Alternative Medicine* (p. 8). **2011**, E-article online. http://dx.doi.org/10.1093/ecam/neq071:1–8 (Accessed on 30 September 2019).
7. Antolovich, M., Prenzler, P. D., Patsalides, E., McDonald, S., & Robards, K. Methods for testing antioxidant activity. *Analyst*, **2002**, *127*(1), 183–198.
8. Ares, G., & Gámbaro, A. Influence of gender, age and motives underlying food choice on perceived healthiness and willingness to try functional foods. *Appetite*, **2007**, *49*(1), 148–158.
9. Arhan, M., Öztürk, H. S., Turhan, N., Aytac, B., Güven, M. C., Olcay, E., & Durak, İ. Hepatic oxidant/antioxidant status in cholesterol-fed rabbits: Effects of garlic extract. *Hepatology Research*, **2009**, *39*(1), 70–77.
10. Arihara, K. Strategies for designing novel functional meat products. *Meat Science*, **2006**, *74*(1), 219–229.
11. Aruoma, O. I. The impact of food regulation on the food supply chain. *Toxicology*, **2006**, *221*(1), 119–127.
12. Assad, T., Khan, R. A., & Feroz, Z. Evaluation of hypoglycemic and hypolipidemic activity of methanol extract of *Brassica oleracea*. *Chinese Journal of Natural Medicines*, **2014**, *12*(9), 648–653.
13. Auberval, N., Dal, S., Bietiger, W., Pinget, M., Jeandidier, N., Maillard-Pedracini, E., Schini-Kerth, V., & Sigrist, S. Metabolic and oxidative stress markers in Wistar rats after 2 months on a high-fat diet. *Diabetology and Metabolic Syndrome*, **2014**, *6*(1), 1–9.
14. Bacchetti, T., Tullii, D., Masciangelo, S., Gesuita, R., Skrami, E., Brugè, F., Silvestri, S., Orlando, P., Tiano, L., & Ferretti, G. Effect of black and red cabbage on plasma carotenoid levels, lipid profile and oxidized low density lipoprotein. *Journal of Functional Foods*, **2014**, *8*, 128–137.

15. Bahorun, T., Luximon-Ramma, A., Crozier, A., & Aruoma, O. I. Total phenol, flavonoid, proanthocyanidin and vitamin C levels and antioxidant activities of Mauritian vegetables. *Journal of the Science of Food and Agriculture*, **2004**, *84*(12), 1553–1561.

16. Bakowska-Barczak, A. Acylated anthocyanins as stable, natural food colorants-a review. *Polish Journal of Food and Nutrition Sciences*, **2005**, *14/55*(2), 107–116.

17. Balkan, J., Doğru-Abbasoğlu, S., Aykaç-Toker, G., & Uysal, M. The effect of a high cholesterol diet on lipids and oxidative stress in plasma, liver and aorta of rabbits and rats. *Nutrition Research*, **2004**, *24*(3), 229–234.

18. Banerjee, R., Verma, A. K., Das, A. K., Rajkumar, V., Shewalkar, A. A., & Narkhede, H. P. Antioxidant effects of broccoli powder extract in goat meat nuggets. *Meat Science*, **2012**, *91*(2), 179–184.

19. Barakat, H., & Rohn, S. Effect of different cooking methods on bioactive compounds in vegetarian, broccoli-based bars. *Journal of Functional Foods*, **2014**, *11*, 407–416.

20. Barakat, L. A., & Mahmoud, R. H. The antiatherogenic, renal protective and immuno-modulatory effects of purslane, pumpkin and flax seeds on hypercholesterolemic rats. *North American Journal of Medical Sciences*, **2011**, *3*(9), 411–417.

21. Beceanu, D. Nutritive, nutraceutical, medicinal and energetic value of fruits and vegetables. *Cercetări Agronomiceîn Moldova*, **2008**, *41*(2), 65–81.

22. Bech-Larsen, T., & Scholderer, J. Functional foods in Europe: Consumer research, market experiences and regulatory aspects. *Trends in Food Science and Technology*, **2007**, *18*(4), 231–234.

23. Bhavani, R., Kotteeswaran, R., & Rajeshkumar, S. Hepatoprotective effect of *Brassica oleracea* vegetable and its leaves in Paracetamol induced liver damage in albino rats. *International Journal of ChemTech Research*, **2014**, *6*(7), 3705–3712.

24. Bigliardi, B., & Galati, F. Innovation trends in the food industry: The case of functional foods. *Trends in Food Science and Technology*, **2013**, *31*(2), 118–129.

25. Bogue, J., & Ryan, M. *Market-Oriented New Product Development: Functional Foods and the Irish Consumer* (pp. 1–35). Bulletin, University College, Department of Food Economics, Cork, **2000**.

26. Bridgers, E. N., Chinn, M. S., & Truong, V. D. Extraction of anthocyanins from industrial purple-fleshed sweet potatoes and enzymatic hydrolysis of residues for fermentable sugars. *Industrial Crops and Products*, **2010**, *32*(3), 613–620.

27. Bui, D. T. Consumer acceptance of functional foods in Ho Chi Minh city. *Eurasian Journal of Business and Economics*, **2015**, *8*(16), 19–34.

28. Burton, G. J., & Jauniaux, E. Oxidative stress. *Best Practice and Research Clinical Obstetrics and Gynecology*, **2011**, *25*(3), 287–299.

29. Cavalcanti, R. N., Santos, D. T., & Meireles, M. A. A. Non-thermal stabilization mechanisms of anthocyanins in model and food systems-an overview. *Food Research International*, **2011**, *44*(2), 499–509.

30. Cencic, A., & Chingwaru, W. The role of functional foods, nutraceuticals, and food supplements in intestinal health. *Nutrients*, **2010**, *2*(6), 611–625.

31. Chen, C. C., Hsu, J. D., Wang, S. F., Chiang, H. C., Yang, M. Y., Kao, E. S., Ho, Y. C., & Wang, C. J. Hibiscus sabdariffa extract inhibits the development of atherosclerosis in cholesterol-fed rabbits. *Journal of Agricultural and Food Chemistry*, **2003**, *51*(18), 5472–5477.

32. Chen, L., Haught, W. H., Yang, B., Saldeen, T. G., Parathasarathy, S., & Mehta, J. L. Preservation of endogenous antioxidant activity and inhibition of lipid peroxidation as common mechanisms of antiatherosclerotic effects of vitamin E, lovastatin and amlodipine. *Journal of the American College of Cardiology*, **1997**, *30*(2), 569–575.

33. Childs, N. M. Functional foods and the food industry: Consumer, economic and product development issues. *Journal of Nutraceuticals, Functional and Medical Foods*, **1997**, *1*(2), 25–43.

34. Ching, L. S., & Mohamed, S. Alpha-tocopherol content in 62 edible tropical plants. *Journal of Agricultural and Food Chemistry*, **2001**, *49*(6), 3101–3105.

35. Choi, C. S., Chung, H. K., Choi, M. K., & Kang, M. H. Effects of grape pomace on the antioxidant defense system in diet-induced hypercholesterolemic rabbits. *Nutrition Research and Practice*, **2010**, *4*(2), 114–120.

36. Chu, Y. F., Sun, J. I. E., Wu, X., & Liu, R. H. Antioxidant and antiproliferative activities of common vegetables. *Journal of Agricultural and Food Chemistry*, **2002**, *50*(23), 6910–6916.

37. Chun, O. K., Smith, N., Sakagawa, A., & Lee, C. Y. Antioxidant properties of raw and processed cabbages. *International Journal of Food Sciences and Nutrition*, **2004**, *55*(3), 191–199.

38. Cos, P., Vlietinck, A. J., Berghe, D. V., & Maes, L. Anti-infective potential of natural products: How to develop a stronger *in vitro* 'proof-of-concept.' *Journal of Ethnopharmacology*, **2006**, *106*(3), 290–302.

39. Costa, A. I., & Jongen, W. M. F. New insights into consumer-led food product development. *Trends in Food Science and Technology*, **2006**, *17*(8), 457–465.

40. Day, L., Seymour, R. B., Pitts, K. F., Konczak, I., & Lundin, L. Incorporation of functional ingredients into foods. *Trends in Food Science and Technology*, **2009**, *20*(9), 388–395.

41. Dimitrova-Shumkovska, J., Veenman, L., Ristoski, T., Leschiner, S., & Gavish, M. Chronic high fat, high cholesterol supplementation decreases 18 kDa Translocator Protein binding capacity in association with increased oxidative stress in rat liver and aorta. *Food and Chemical Toxicology*, **2010**, *48*(3), 910–921.

42. Domínguez-Avila, J. A., Alvarez-Parrilla, E., Laura, A., Martínez-Martínez, A., González-Aguilar, G. A., Gómez-García, C., & Robles-Sánchez, M. Effect of fruit and vegetable intake on oxidative stress and dyslipidemia markers in human and animal models. In: *Phytochemicals-Bioactivities and Impact on Health* (pp. 227–252). Online InTech, **2011**.

43. Draghici, G., Alexandra, L. M., Aurica-Breica, B., Nica, D., Alda, S., Liana, A., Gogoasa, I., GergenI., & Despina-Maria, B. Red cabbage, millennium's functional food. *Journal of Horticultural Science and Biotechnology*, **2013**, *17*(4), 52–55.

44. Dyrby, M., Westergaard, N., & Stapelfeldt, H. Light and heat sensitivity of red cabbage extract in soft drink model systems. *Food Chemistry*, **2001**, *72*(4), 431–437.

45. El-Mowafy, E. M. Treatment effect of red cabbage and cysteine against paracetamol induced hepatotoxicity in experimental rats. *Journal of Applied Sciences Research*, 2012, *8*(12), 5852–5859.

46. Espín, J. C., García-Conesa, M. T., & Tomás-Barberán, F. A. Nutraceuticals: Facts and fiction. *Phytochemistry*, **2007**, *68*(22–24), 2986–3008.

47. Farida, S. S. Y., & Sari, D. P. P. Antioxidant activity of white and red cabbage (*Brassica oleracea* L. var capitata L.) using DPPH. In: *2nd International Conference on Pharmacy and Advanced Pharmaceutical Sciences in Yogyakarta* (p. 3). **2011**. http://dosen.univpancasila.ac.id/dosenfile/208721101213885579190lJanuary2014. pdf (Accessed on 30 September 2019).

48. Farrell, G. C., & Larter, C. Z. Nonalcoholic fatty liver disease: From steatosis to cirrhosis. *Hepatology*, **2006**, *43*(1), 99–112.

49. Feillet-Coudray, C., Sutra, T., Fouret, G., Ramos, J., Wrutniak-Cabello, C., Cabello, G., Cristol, J., & Coudray, C. Oxidative stress in rats fed a high-fat high-sucrose diet and preventive effect of polyphenols: Involvement of mitochondrial and NAD (P) H oxidase systems. *Free Radical Biology and Medicine*, **2009**, *46*(5), 624–632.

50. Ferreiro-González, M., Carrera, C., Ruiz-Rodríguez, A., Barbero, G. F., Ayuso, J., Palma, M., & Barroso, C. G. A new solid phase extraction for the determination of anthocyanins in grapes. *Molecules*, **2014**, *19*(12), 21398–21410.

51. Fleischhacker, S. E., Evenson, K. R., Rodriguez, D. A., & Ammerman, A. S. A systematic review of fast food access studies. *Obesity Reviews*, **2011**, *12*(5), 460–471.

52. Floegel, A., Kim, D. O., Chung, S. J., Koo, S. I., & Chun, O. K. Comparison of ABTS/DPPH assays to measure antioxidant capacity in popular antioxidant-rich US foods. *Journal of Food Composition and Analysis*, **2011**, *24*(7), 1043–1048.

53. Gaafar, A. A., Hanan, F. A., Zeinab, A. S., & Naima, Z. M. Hypoglycemic effects of white cabbage and red cabbage (*Brassica oleracea*) in STZ induced type 2 diabetes in rats. *World Journal of Pharmaceutical Research*, **2014**, *3*(4), 1583–1609.

54. German, B., Schiffrin, E. J., Reniero, R., Mollet, B., Pfeifer, A., & Neeser, J. R. The development of functional foods: Lessons from the gut. *Trends in Biotechnology*, **1999**, *17*(12), 492–499.

55. Grunert, K. G., Bredahl, L., & Brunsø, K. Consumer perception of meat quality and implications for product development in the meat sector-a review. *Meat Science*, **2004**, *66*(2), 259–272.

56. Guo, Y., Wu, X. Q., Zhang, C., Liao, Z. X., Wu, Y., Xia, Z. Y., & Wang, H. Effect of indole-3-carbinol on ethanol-induced liver injury and acetaldehyde-stimulated hepatic stellate cells activation using precision-cut rat liver slices. *Clinical and Experimental Pharmacology and Physiology*, **2010**, *37*(12), 1107–1113.

57. Han, X., Shen, T., & Lou, H. Dietary polyphenols and their biological significance. *International Journal of Molecular Sciences*, **2007**, *8*(9), 950–988.

58. Hardy, G. Nutraceuticals and functional foods: Introduction and meaning. *Nutrition*, **2000**, *16*(7/8), 688–692.

59. Heimler, D., Vignolini, P., Dini, M. G., Vincieri, F. F., & Romani, A. Antiradical activity and polyphenol composition of local Brassicaceae edible varieties. *Food Chemistry*, **2006**, *99*(3), 464–469.

60. Hertog, M. G., Hollman, P. C., & Katan, M. B. Content of potentially anticarcinogenic flavonoids of 28 vegetables and 9 fruits commonly consumed in the Netherlands. *Journal of Agricultural and Food Chemistry*, **1992**, *40*(12), 2379–2383.

61. Hodges, D. M., DeLong, J. M., Forney, C. F., & Prange, R. K. Improving the thiobarbituric acid-reactive-substances assay for estimating lipid peroxidation in plant tissues containing anthocyanin and other interfering compounds. *Planta*, **1999**, *207*(4), 604–611.

62. Huie, C. W. Review of modern sample-preparation techniques for the extraction and analysis of medicinal plants. *Analytical and Bioanalytical Chemistry*, **2002**, *373*(1/2), 23–30.

63. Hussein, E. A. Potential therapeutic effects of dried cabbage and eggplant on hypercholestromic rat. *Food Chemistry*, **2012**, *96*, 572–579.

64. Hussein, S. A., Abdel-Mageid, A. D., & Abu-Ghazalla, A. M. Biochemical study on the effect of alpha-lipoic acid on lipid metabolism of rats fed high fat diet. *Benha Veterinary Medical Journal*, **2015**, *28*(1), 109–119.

65. Jakes, R. W., Day, N. E., Khaw, K. T., Luben, R., Oakes, S., Welch, A., Bingham, S., & Wareham, N. J. Television viewing and low participation in vigorous recreation are independently associated with obesity and markers of cardiovascular disease risk: EPIC-Norfolk population-based study. *European Journal of Clinical Nutrition*, **2003**, *57*(9), 1089–1096.

66. Jeon, S. M., Bok, S. H., Jang, M. K., Lee, M. K., Nam, K. T., Park, Y. B., Rhee, S. J., & Choi, M. S. Antioxidative activity of naringin and lovastatin in high cholesterol-fed rabbits. *Life Sciences*, **2001**, *69*(24), 2855–2866.

67. Ji, C., Li, C., Gong, W., Niu, H., & Huang, W. Hypolipidemic action of hydroxycinnamic acids from cabbage (*Brassica oleracea* L. var. capitata) on hypercholesterolaemic rat in relation to its antioxidant activity. *Journal of Food and Nutrition Research*, **2015**, *3*(5), 317–324.

68. Jones, P. J., & Jew, S. Functional food development: concept to reality. *Trends in Food Science and Technology*, **2007**, *18*(7), 387–390.

69. Kainuma, M., Fujimoto, M., Sekiya, N., Tsuneyama, K., Cheng, C., Takano, Y., Terasawa, K., & Shimada, Y. Cholesterol-fed rabbit as a unique model of nonalcoholic, nonobese, non-insulin-resistant fatty liver disease with characteristic fibrosis. *Journal of Gastroenterology*, **2006**, *41*(10), 971–980.

70. Karbiner, M. S., Sierra, L., Minahk, C., Fonio, M. C., De Bruno, M. P., & Jerez, S. The role of oxidative stress in alterations of hematological parameters and inflammatory markers induced by early hypercholesterolemia. *Life Sciences*, **2013**, *93*(15), 503–508.

71. Karimi, I. Animal models as tools for translational research: focus on atherosclerosis, metabolic syndrome and type-II diabetes mellitus. *Lipoproteins-Role in Health and Diseases* (p. 10). **2012**, online. http://dx.doi.org/10.5772/47769 (Accessed on 30 September 2019).

72. Kaur, C., & Kapoor, H. C. Anti-oxidant activity and total phenolic content of some Asian vegetables. *International Journal of Food Science and Technology*, **2002**, *37*(2), 153–161.

73. Kertész, A., Bombicz, M., Priksz, D., Balla, J., Balla, G., Gesztelyi, R., Varga, B., Haines, D. D., Tosaki, A., & Juhasz, B. Adverse impact of diet-induced hypercholesterolemia on cardiovascular tissue homeostasis in a rabbit model: Time-dependent changes in cardiac parameters. *International Journal of Molecular Sciences*, **2013**, *14*(9), 19086–19108.

74. Keyamura, Y., Nagano, C., Kohashi, M., Niimi, M., Nozako, M., Koyama, T., Yasufuku, R., Imaizumi, A., Itabe, H., & Yoshikawa, T. Add-on effect of probucol in atherosclerotic, cholesterol-fed rabbits treated with atorvastatin. *PLoS One*, **2014**, *9*(5), 1–10.

75. Khademi, F., Danesh, B., Delazar, A., Nejad, D. M., Ghorbani, M., & Rad, J. S. Effects of quince leaf extract on biochemical markers and coronary histopathological changes in rabbits. *ARYA Atherosclerosis*, **2013**, *9*(4), 223–231.

76. Khan, R. A., Feroz, Z., Jamil, M., & Ahmed, M. Hypolipidemic and antithrombotic evaluation of *Myrtuscommunis* L. in cholesterol-fed rabbits. *African Journal of Pharmacy and Pharmacology*, **2014**, *8*(8), 235–239.

77. Kim, D., & Kim, W. R. Nonobese fatty liver disease. *Clinical Gastroenterology and Hepatology*, **2017**, *15*, 474–485.

78. Kim, E. J., Kim, B. H., Seo, H. S., Lee, Y. J., Kim, H. H., Son, H. H., & Choi, M. H. Cholesterol-induced non-alcoholic fatty liver disease and atherosclerosis aggravated by systemic inflammation. *PloS One*, **2014**, *9*(6), 1–11.

79. Kolodziejczyk, J., Saluk-Juszczak, J., Posmyk, M. M., Janas, K. M., & Wachowicz, B. Red cabbage anthocyanins may protect blood plasma proteins and lipids. *Central European Journal of Biology*, **2011**, *6*(4), 565–574.

80. Kong, W., Wei, J., Abidi, P., Lin, M., Inaba, S., Li, C., Wang, Y., Wang, Z., Si, S., & Wang, S. Berberine is a novel cholesterol-lowering drug working through a unique mechanism distinct from statins. *Nature Medicine*, **2004**, *10*(12), 1344–1351.

81. Kumar, V., Biswas, A. K., Sahoo, J., Chatli, M. K., & Sivakumar, S. Quality and storability of chicken nuggets formulated with green banana and soybean hulls flours. *Journal of Food Science and Technology*, **2013**, *50*(6), 1058–1068.

82. Kurilich, A. C., Tsau, G. J., Brown, A., Howard, L., Klein, B. P., Jeffery, E. H., Kushad, M., Wallig, M. A., & Juvik, J. A. Carotene, tocopherol, and ascorbate contents in subspecies of *Brassica oleracea*. *Journal of Agricultural and Food Chemistry*, **1999**, *47*(4), 1576–1581.

83. Kwak, N. S., & Jukes, D. J. Functional foods. Part 1: The development of a regulatory concept. *Food Control*, **2001**, *12*(2), 99–107.

84. Lapornik, B., Prošek, M., & Wondra, A. G. Comparison of extracts prepared from plant by-products using different solvents and extraction time. *Journal of Food Engineering*, **2005**, *71*(2), 214–222.

85. Lee, L. S., Cho, C. W., Hong, H. D., Lee, Y. C., Choi, U. K., & Kim, Y. C. Hypolipidemic and antioxidant properties of phenolic compound-rich extracts from white ginseng (Panax ginseng) in cholesterol-fed rabbits. *Molecules*, **2013**, *18*(10), 12548–12560.

86. Leja, M., Kamińska, I., & Kołton, A. Phenolic compounds as the major antioxidants in red cabbage. *Folia Horticulturae*, **2010**, *22*(1), 19–24.

87. Leja, M., Mareczek, A., Adamus, A., Strzetelski, P., & Combik, M. Some antioxidative properties of selected white cabbage DH lines. *Folia Horticulturae*, **2006**, *18*(1), 31–40.

88. Leja, M., Wyżgolik, G., & Mareczek, A. Phenolic compounds of red cabbage as related to different forms of nutritive nitrogen. *Sodininkystėir Daržininkystė*, **2005**, *24*(3), 421–428.

89. Llorach, R., Tomás-Barberán, F. A., & Ferreres, F. Functionalization of commercial chicken soup with enriched polyphenol extract from vegetable by-products. *European Food Research and Technology*, **2005**, *220*(1), 31–36.

90. Mahfouz, M. M., & Kummerow, F. A. Cholesterol-rich diets have different effects on lipid peroxidation, cholesterol oxides, and antioxidant enzymes in rats and rabbits. *The Journal of Nutritional Biochemistry*, **2000**, *11*(5), 293–302.

91. Malav, O. P., Sharma, B. D., Kumar, R. R., Talukder, S., & Ahmed, S. R. Antioxidant potential and quality characteristics of functional mutton patties incorporated with cabbage powder. *Nutrition and Food Science*, **2015**, *45*(4), 542–563.

92. Manjula, K., & Suneetha, C. Designer foods-their role in preventing lifestyle disorders. *International Journal of Science and Nature*, **2011**, *4*, 878–882.

93. Marinou, K. A., Georgopoulou, K., Agrogiannis, G., Karatzas, T., Iliopoulos, D., Papalois, A., Chatziioannou, A., Magiatis, P., Halabalaki, M., Tsantila, N., & Skaltsounis, L. A. Differential effect of Pistaciavera extracts on experimental atherosclerosis in the rabbit animal model: An experimental study. *Lipids in Health and Disease*, **2010**, *9*(1), 1–9.

94. Masoro, E. J. Caloric restriction and aging: an update. *Experimental Gerontology*, **2000**, *35*(3), 299–305.

95. Mazza, G. *Anthocyanins in Fruits, Vegetables, and Grains* (p. 328). CRC Press, Boca Raton–FL, **1993**. https://www.nal.usda.gov/ (Accessed on 30 September 2019).

96. Melega, S., Canistro, D., De Nicola, G. R., Lazzeri, L., Sapone, A., & Paolini, M. Protective effect of Tuscan black cabbage sprout extract against serum lipid increase and perturbations of liver antioxidant and detoxifying enzymes in rats fed a high-fat diet. *British Journal of Nutrition*, **2013**, *110*(6), 988–997.

97. Menrad, K. Market and marketing of functional food in Europe. *Journal of Food Engineering*, **2003**, *56*(2/3), 181–188.

98. Mohamed, O. S., Said, M. M., Ali, Z. Y., Atia, H. A., & Mostafa, H. S. Improving effect of dietary oat bran supplementation on oxidative stress induced by hyperlipidemic diet. *Evaluation*, **2011**, *53*, 1–10.

99. Molyneux, P. The use of the stable free radical DPPH for estimating antioxidant activity. *Songklanakarin Journal of Science and Technology*, **2004**, *26*(2), 211–219.

100. Moore, L. V., Diez Roux, A. V., Nettleton, J. A., Jacobs, D. R., & Franco, M. Fast-food consumption, diet quality, and neighborhood exposure to fast food: The multi-ethnic study of atherosclerosis. *American Journal of Epidemiology*, **2009**, *170*(1), 29–36.

101. Morland, K., Roux, A. V. D., & Wing, S. Supermarkets, other food stores, and obesity: The atherosclerosis risk in communities' study. *American Journal of Preventive Medicine*, **2006**, *30*(4), 333–339.

102. Morsy, A. F., Ibrahima, H. S., & Shalabyb, M. A. Protective effect of broccoli and red cabbage against hepatocellular carcinoma induced by N-nitrosodiethylamine in rats. *American Journal of Science*, **2010**, *6*(12), 1136–1144.

103. Myles, I. A. Fast food fever: Reviewing the impacts of the western diet on immunity. *Nutrition Journal*, **2014**, *13*(1), 1–17.

104. Nader, M. A., El-Agamy, D. S., & Suddek, G. M. Protective effects of propolis and thymoquinone on development of atherosclerosis in cholesterol-fed rabbits. *Archives of Pharmacal Research*, **2010**, *33*(4), 637–643.

105. Nicoli, M. C., Anese, M., & Parpinel, M. Influence of processing on the antioxidant properties of fruit and vegetables. *Trends in Food Science and Technology*, **1999**, *10*(3), 94–100.

106. Niki, E. Do free radicals play causal role in atherosclerosis? Low density lipoprotein oxidation and vitamin E revisited. *Journal of Clinical Biochemistry and Nutrition*, **2010**, *48*(1), 3–7.

107. Niva, M. All foods affect health': Understandings of functional foods and healthy eating among health-oriented Finns. *Appetite*, **2007**, *48*(3), 384–393.

108. Noeman, S. A., Hamooda, H. E., & Baalash, A. A. Biochemical study of oxidative stress markers in the liver, kidney and heart of high fat diet induced obesity in rats. *Diabetology and Metabolic Syndrome*, **2011**, *3*(1), 1–8.

109. Oerlemans, K., Barrett, D. M., Suades, C. B., Verkerk, R., & Dekker, M. Thermal degradation of glucosinolates in red cabbage. *Food Chemistry*, **2006**, *95*, 19–29.

110. Ogawa, T., Fujii, H., Yoshizato, K., & Kawada, N. A human-type nonalcoholic steatohepatitis model with advanced fibrosis in rabbits. *The American Journal of Pathology*, **2010**, *177*(1), 153–165.

111. Olorunnisola, O. S., Bradley, G., & Afolayan, A. J. Protective effect of *T. Violacea* rhizome extract against hypercholesterolemia-induced oxidative stress in Wistar rats. *Molecules*, **2012**, *17*(5), 6033–6045.

112. Padayatty, S. J., Katz, A., Wang, Y., Eck, P., Kwon, O., Lee, J. H., Chen, S., Corpe, C., Dutta A., Dutta S. K., & Levine, M. Vitamin C as an antioxidant: Evaluation of its role in disease prevention. *Journal of the American College of Nutrition*, **2003**, *22*(1), 18–35.

113. Pan, S. Y., Yang, R., Dong, H., Yu, Z. L., & Ko, K. M. Bifendate treatment attenuates hepatic steatosis in cholesterol/bile salt-and high-fat diet-induced hypercholesterolemia in mice. *European Journal of Pharmacology*, **2006**, *552*(1–3), 170–175.

114. Pandey, M., Verma, R. K., & Saraf, S. A. Nutraceuticals: new era of medicine and health. *Asian Journal of Pharmaceutical and Clinical Research*, **2010**, *3*(1), 11–15.

115. Pankaj, C., & Sharma, C. Ultrasonic study of binary solutions of methanol, ethanol and phenol in sulpholane. *Physics and Chemistry of Liquids*, **1991**, *22*(4), 205–211.

116. Park, S., Arasu, M. V., Lee, M. K., Chun, J. H., Seo, J. M., Lee, S. W., Al-Dhabi, N. A., & Kim, S. J. Quantification of glucosinolates, anthocyanins, free amino acids, and vitamin C in inbred lines of cabbage (*Brassica oleracea* L.). *Food Chemistry*, **2014**, *145*, 77–85.

117. Paul, S. M., Mytelka, D. S., Dunwiddie, C. T., Persinger, C. C., Munos, B. H., Lindborg, S. R., & Schacht, A. L. How to improve RandD productivity: The pharmaceutical industry's grand challenge. *Nature Reviews Drug Discovery*, **2010**, *9*(3), 203–214.

118. Piccaglia, R., Marotti, M., & Baldoni, G. Factors influencing anthocyanin content in red cabbage (*Brassica oleracea* var capitata L.). *Journal of the Science of Food and Agriculture*, **2002**, *82*(13), 1504–1509.

119. Podsedek, A., Sosnowska, D., Redzynia, M., & Anders, B. Antioxidant capacity and content of *Brassica oleracea* dietary antioxidants. *International Journal of Food Science and Technology*, **2006**, *41*, 49–58.

120. Podsędek, A. Natural antioxidants and antioxidant capacity of Brassica vegetables: A review. *LWT-Food Science and Technology*, **2007**, *40*(1), 1–11.

121. Podsędek, A., Redzynia, M., Klewicka, E., & Koziołkiewicz, M. Matrix effects on the stability and antioxidant activity of red cabbage anthocyanins under simulated gastrointestinal digestion. *Biomed. Research International*, **2014.**

122. Podsędek, A., Sosnowska, D., Redzynia, M., & Koziołkiewicz, M. Effect of domestic cooking on the red cabbage hydrophilic antioxidants. *International Journal of Food Science and Technology*, **2008**, *43*(10), 1770–1777.

123. Popkin, B. M., Adair, L. S., & Ng, S. W. Global nutrition transition and the pandemic of obesity in developing countries. *Nutrition Reviews*, **2012**, *70*(1), 3–21.
124. Priya, S. L. S. Cabbage-A wonderful and awesome remedy for various ailments. *Research in Pharmaceutical Sciences*, **2012**, *1*, 28–34.
125. Proteggente, A. R., Pannala, A. S., Paganga, G., Buren, L. V., Wagner, E., Wiseman, S., Put, F. V. D., Dacombe, C., & Rice-Evans, C. A. The antioxidant activity of regularly consumed fruit and vegetables reflects their phenolic and vitamin C composition. *Free Radical Research*, **2002**, *36*(2), 217–233.
126. Puupponen-Pimiä, R., Häkkinen, S. T., Aarni, M., Suortti, T., Lampi, A. M., Eurola, M., Piironen, V., Nuutila, A. M., & Oksman-Caldentey, K. M. Blanching and long-term freezing affect various bioactive compounds of vegetables in different ways. *Journal of the Science of Food and Agriculture*, **2003**, *83*(14), 1389–1402.
127. Qin, Y., & Tian, Y. P. Preventive effects of chronic exogenous growth hormone levels on diet-induced hepatic steatosis in rats. *Lipids in Health and Disease*, **2010**, *9*(1), 1–12.
128. Radziejewska-Kubzdela, E., & Biegańska-Marecik, R. A comparison of the composition and antioxidant capacity of novel beverages with an addition of red cabbage in the frozen, purée and freeze-dried forms. *LWT-Food Science and Technology*, **2015**, *62*(1), 821–829.
129. Ragab, S. M., Omar, H. E. D. M., Elghaffar, S. K. A., & El-Metwally, T. H. Hypolipidemic and antioxidant effects of phytochemical compounds against hepatic steatosis induced by high fat high sucrose diet in rats. *Archives of Biomedical Sciences*, **2014**, *2*(1), 1–10.
130. Rajadurai, J., Tse, H. F., Wang, C. H., Yang, N. I., Zhou, J., & Sim, D. Understanding the epidemiology of heart failure to improve management practices: An Asia-Pacific perspective. *Journal of Cardiac Failure*, **2017**, *23*(4), 327–339.
131. Rajasekaran, A., & Kalaivani, M. Designer foods and their benefits: A review. *Journal of Food Science and Technology*, **2013**, *50*(1), 1–16.
132. Rezq, A. Beneficial health effects of fennel seeds (Shamar) on male rats feeding high fat-diet. *The Medical Journal of Cairo University*, **2012**, *80*(2), 101–113.
133. Rickman, J. C., Barrett, D. M., & Bruhn, C. M. Nutritional comparison of fresh, frozen and canned fruits and vegetables. Part 1. Vitamins C and B and phenolic compounds. *Journal of the Science of Food and Agriculture*, **2007**, *87*(6), 930–944.
134. Roberfroid, M. B. Defining functional foods. In: *Functional Foods* (pp. 9–15). Woodhead Publishing Ltd, Abington Hall, Abington, Cambridge, England, **2000**.
135. Roberfroid, M. B. Global view on functional foods: European perspectives. *British Journal of Nutrition*, **2002**, *88*, 133–138.
136. Rokayya, S., Li, C. J., Zhao, Y., Li, Y., & Sun, C. H. Cabbage (*Brassica oleracea* L. var. capitata) phytochemicals with antioxidant and anti-inflammatory potential. *Asian Pacific Journal of Cancer Prevention*, **2013**, *14*(11), 6657–6662.
137. Roy, M. K., Juneja, L. R., Isobe, S., & Tsushida, T. Steam processed broccoli (*Brassica oleracea*) has higher antioxidant activity in chemical and cellular assay systems. *Food Chemistry*, **2009**, *114*(1), 263–269.
138. Sakunasing, P., & Kangsadalampai, K. Different antimutagenicity against urethane between conventionally and organically grown cruciferous vegetables (*Brassica* spp.). *Thai Journal of Toxicology*, **2008**, *23*(2), 126–134.

139. Sankhari, J. M., Thounaojam, M. C., Jadeja, R. N., Devkar, R. V., & Ramachandran, A. V. Anthocyanin-rich red cabbage (*Brassica oleracea* L.) extract attenuates cardiac and hepatic oxidative stress in rats fed an atherogenic diet. *Journal of the Science of Food and Agriculture*, **2012**, *92*(8), 1688–1693.

140. Sasidharan, S., Chen, Y., Saravanan, D., Sundram, K. M., & Latha, L. Y. Extraction, isolation and characterization of bioactive compounds from plants' extracts. *African Journal of Traditional, Complementary and Alternative Medicines*, **2011**, *8*(1), 1–10.

141. Satia, J. A., Galanko, J. A., & Siega-Riz, A. M. Eating at fast-food restaurants is associated with dietary intake, demographic, psychosocial and behavioral factors among African Americans in North Carolina. *Public Health Nutrition*, **2004**, *7*(8), 1089–1096.

142. Savransky, V., Bevans, S., Nanayakkara, A., Li, J., Smith, P. L., Torbenson, M. S., & Polotsky, V. Y. Chronic intermittent hypoxia causes hepatitis in a mouse model of diet-induced fatty liver. *American Journal of Physiology-Gastrointestinal and Liver Physiology*, **2007**, *293*(4), 871–877.

143. Scalzo, R. L., Genna, A., Branca, F., Chedin, M., & Chassaigne, H. Anthocyanin composition of cauliflower (*Brassica oleracea* L. var. botrytis) and cabbage (*B. oleracea*L. var. capitata) and its stability in relation to thermal treatments. *Food Chemistry*, **2008**, *107*(1), 136–144.

144. Schieber, A., Stintzing, F. C., & Carle, R. By-products of plant food processing as a source of functional compounds—recent developments. *Trends in Food Science and Technology*, **2001**, *12*(11), 401–413.

145. Shahidi, F. Nutraceuticals and functional foods: Whole versus processed foods. *Trends in Food Science and Technology*, **2009**, *20*(9), 376–387.

146. Shen, A., Zhang, B., Ping, J., Xie, W., Donfack, P., Baek, S. J., Zhou, X., Wang, H., Materny A., & Hu, J. *In vivo* study on the protection of indole-3-carbinol (I3C) against the mouse acute alcoholic liver injury by micro-Raman spectroscopy. *Journal of Raman Spectroscopy*, **2009**, *40*(5), 550–555.

147. Sibai, A. M., Nasreddine, L., Mokdad, A. H., Adra, N., Tabet, M., & Hwalla, N. Nutrition transition and cardiovascular disease risk factors in Middle East and North Africa countries: Reviewing the evidence. *Annals of Nutrition and Metabolism*, **2010**, *57*(3/4), 193–203.

148. Singh, J., Upadhyay, A. K., Bahadur, A., Singh, B., Singh, K. P., & Rai, M. Antioxidant phytochemicals in cabbage (*Brassica oleracea* L. var. capitata). *Scientia Horticulturae*, **2006**, *108*(3), 233–237.

149. Singh, J., Upadhyay, A. K., Prasad, K., Bahadur, A., & Rai, M. Variability of carotenes, vitamin C, E and phenolics in Brassica vegetables. *Journal of Food Composition and Analysis*, **2007**, *20*(2), 106–112.

150. Siro, I., Kapolna, E., Kápolna, B., & Lugasi, A. Functional food. Product development, marketing and consumer acceptance-a review. *Appetite*, **2008**, *51*(3), 456–467.

151. Suanarunsawat, T., Devakul, N. A. W., Songsak, T., Thirawarapan, S., & Poung-shompoo, S. Lipid-lowering and antioxidative activities of aqueous extracts of *Ocimum sanctum* L. leaves in rats fed with a high-cholesterol diet. *Oxidative Medicine and Cellular Longevity* (p. 9). **2011**, Online. doi:10.1155/2011/962025.

152. Sultana, B., & Anwar, F. Flavonols (kaempeferol, quercetin, myricetin) contents of selected fruits, vegetables and medicinal plants. *Food Chemistry*, **2008**, *108*(3), 879–884.

153. Szôllôsi, R., & Varga, I. S. Total antioxidant power in some species of Labiatae (Adaptation of FRAP method). *Acta Biologica Szegediensis*, **2002**, *46*(3/4), 125–127.
154. Tsao, R. Chemistry and biochemistry of dietary polyphenols. *Nutrients*, **2010**, *2*(12), 1231–1246.
155. Turkmen, N., Sari, F., & Velioglu, Y. S. The effect of cooking methods on total phenolics and antioxidant activity of selected green vegetables. *Food Chemistry*, **2005**, *93*(4), 713–718.
156. Urala, N., & Lähteenmäki, L. Reasons behind consumers' functional food choices. *Nutrition and Food Science*, **2003**, *33*(4), 148–158.
157. Van Kleef, E., & Van Trijp, H. Functional foods: Health claim-food product compatibility and the impact of health claim framing on consumer valuation. *Appetite*, **2005**, *44*(3), 299–308.
158. Van Rooyen, D. M., Larter, C. Z., Haigh, W. G., Yeh, M. M., Ioannou, G., Kuver, R., Lee, S. P., Teoh, N. C., & Farrell, G. C. Hepatic free cholesterol accumulates in obese, diabetic mice and causes nonalcoholic steatohepatitis. *Gastroenterology*, **2011**, *141*(4), 1393–1403.
159. Verma, A. K., Pathak, V., Singh, V. P., & Umaraw, P. Storage study of chicken meatballs incorporated with green cabbage (*Brassica olerecea*) at refrigeration temperature (4±1 C) under aerobic packaging. *Journal of Applied Animal Research*, **2016**, *44*(1), 409–414.
160. Vijayakumar, R. S., Surya, D., & Nalini, N. Antioxidant efficacy of black pepper (*Piper nigrum* L.) and piperine in rats with high fat diet induced oxidative stress. *Redox Report*, **2004**, *9*(2), 105–110.
161. Vinson, J. A., Hao, Y., Su, X., & Zubik, L. Phenol antioxidant quantity and quality in foods: Vegetables. *Journal of Agricultural and Food Chemistry*, **1998**, *46*(9), 3630–3634.
162. Volden, J., Bengtsson, G. B., & Wicklund, T. Glucosinolates, L-ascorbic acid, total phenols, anthocyanins, antioxidant capacities and color in cauliflower (*Brassica oleracea* L. ssp. botrytis): effects of long-term freezer storage. *Food Chemistry*, **2009**, *112*(4), 967–976.
163. Wachtel-Galor, S., Wong, K. W., & Benzie, I. F. The effect of cooking on *Brassica* vegetables. *Food Chemistry*, **2008**, *110*(3), 706–710.
164. Wang, X., Hasegawa, J., Kitamura, Y., Wang, Z., Matsuda, A., Shinoda, W., Miura, N., & Kimura, K. Effects of hesperidin on the progression of hypercholesterolemia and fatty liver induced by high-cholesterol diet in rats. *Journal of Pharmacological Sciences*, **2011**, *117*(3), 129–138.
165. Waqar, A. B., Koike, T., Yu, Y., Inoue, T., Aoki, T., Liu, E., & Fan, J. High-fat diet without excess calories induces metabolic disorders and enhances atherosclerosis in rabbits. *Atherosclerosis*, **2010**, *213*(1), 148–155.
166. Waqar, M. A., & Mahmood, Y. Anti-platelet, anti-hypercholesterolemic and antioxidant effects of ethanolic extracts of *Brassica oleracea* in high fat diet provided rats. *World Applied Sciences Journal*, **2010**, *8*(1), 107–112.
167. Webster, A. C., Nagler, E. V., Morton, R. L., & Masson, P. Chronic kidney disease. *The Lancet*, **2017**, *389*(10075), 1238–1252.
168. Wiczkowski, W., Topolska, J., & Honke, J. Anthocyanins profile and antioxidant capacity of red cabbages are influenced by genotype and vegetation period. *Journal of Functional Foods*, **2014**, *7*, 201–211.

169. Wold, A. B., Rosenfeld, H., Lea, P., & Baugerød, H. The effect of CA and conventional storage on antioxidant activity and vitamin C in red and white cabbage (*Brassica oleracea* var. capitata L.) and Savoy (*Brassica oleracea* var. sabauda L.). *European Journal of Horticultural Science.*2006, *71*, 212–216.

170. Wouters, K., Van Gorp, P. J., Bieghs, V., & Gijbels, M. J. Dietary cholesterol, rather than liver steatosis, leads to hepatic inflammation in hyperlipidemic mouse models of nonalcoholic steatohepatitis. *Hepatology*, **2008**, *48*(2), 474–486.

171. Wu, X., Beecher, G. R., Holden, J. M., Haytowitz, D. B., Gebhardt, S. E., & Prior, R. L. Lipophilic and hydrophilic antioxidant capacities of common foods in the United States. *Journal of Agricultural and Food Chemistry*, **2004**, *52*(12), 4026–4037.

172. Wu, X., Beecher, G. R., Holden, J. M., Haytowitz, D. B., Gebhardt, S. E., & Prior, R. L. Concentrations of anthocyanins in common foods in the United States and estimation of normal consumption. *Journal of Agricultural and Food Chemistry*, **2006**, *54*(11), 4069–4075.

173. Xu, F., Zheng, Y., Yang, Z., Cao, S., Shao, X., & Wang, H. Domestic cooking methods affect the nutritional quality of red cabbage. *Food Chemistry*, **2014**, *161*, 162–167.

174. Xu, J., Yang, W., Deng, Q., Huang, Q., & Huang, F. Flaxseed oil and α-lipoic acid combination reduces atherosclerosis risk factors in rats fed a high-fat diet. *Lipids in Health and Disease*, **2012**, *11*(1), 1–7.

175. Yu, Q., Li, Y., Waqar, A. B., Wang, Y., Huang, B., Chen, Y., Zhao, S., Yang, P., Fan, J., & Liu, E. Temporal and quantitative analysis of atherosclerotic lesions in diet-induced hypercholesterolemic rabbits. *BioMed Research International*, **2012**, p. 7, doi: 10.1155/2012/506159.

176. Yuan, Y., Chiu, L. W., & Li, L. Transcriptional regulation of anthocyanin biosynthesis in red cabbage. *Planta*, **2009**, *230*(6), 1141.

177. Yusuf, S., Reddy, S., & Ôunpuu, S. Global burden of cardiovascular diseases, Part II: Variations in cardiovascular disease by specific ethnic groups and geographic regions and prevention strategies. *Circulation*, **2001**, *104*(23), 2855–2864.

178. Zaidel, D. N. A., Sahat, N. S., Jusoh, Y. M. M., & Muhamad, I. I. Encapsulation of anthocyanin from Roselle and red cabbage for stabilization of water-in-oil emulsion. *Agriculture and Agricultural Science Procedia*, **2014**, *2*, 82–89.

179. Zhong, W., Huan, X. D., Cao, Q., & Yang, J. Cardioprotective effect of epigallocatechin-3-gallate against myocardial infarction in hypercholesterolemic rats. *Experimental and Therapeutic Medicine*, **2015**, *9*(2), 405–410.

180. Zhu, L., Luo, X., & Jin, Z. Effect of resveratrol on serum and liver lipid profile and antioxidant activity in hyperlipidemia rats. *Asian Australasian Journal of Animal Sciences*, **2008**, *21*(6), 890–899.

181. Zivkovic, A. M., German, J. B., & Sanyal, A. J. Comparative review of diets for the metabolic syndrome: Implications for nonalcoholic fatty liver disease. *The American Journal of Clinical Nutrition*, **2007**, *86*(2), 285–300.

182. Zou, Y., Li, J., Lu, C., Wang, J., Ge, J., Huang, Y., Zhang, L., & Wang, Y. High-fat emulsion-induced rat model of nonalcoholic steatohepatitis. *Life Sciences*, **2006**, *79*(11), 1100–1107.

CHAPTER 5

FUNCTIONAL PROPERTIES OF MILK YAM (*IPOMOEA DIGITATA* L.)

K. M. VIDYA, N. S. SONIA, and P. C. JESSYKUTTY

ABSTRACT

Ipomoea digitata L. (Milk yam or *Ksheervidari*) has mature tubers that are grayish-brown in color, bitter in taste, having no characteristic odor and starch-filled parenchyma. In India, many Ayurvedic industries are using yam tubers in several formulations like: *Vidaryadikvatha Churna, Chavanaprasha, Vidaryadi Ghirta, Marma Gutika, Manmathabhra Rasa,* and *Pugakhanda* (Aparah). The tuber pharmacognosy has been studied for the presence of phytoconstituents, such as: alkaloids, tannins, steroids, gums, glycosides, carbohydrates, and saponins. It is considered as a tonic, hypolipidemic, hypoglycemic, galactogogic, antioxidant, antimicrobial, aphrodisiac, demulcent, alterative, cholagogic and also can improve digestion. Tubers are also used to treat debility, spermatorrhoea, fever, bronchitis, scorpion stings, and menorrhagia. The raw tubers are used to treat blood dysentery, astringent, anti-hypertensive, anti-diabetic, etc. Phytochemicals (taraxerol, taraxerol acetate, umbelliferone, β-sitosterol, and scopoletin, and 7-O-β-D-glycopyranosylscopoletin (Scopolin), umbelliferone) have been isolated from the tubers. Structure of two compounds (scopoletin and β-sitosterol) was elucidated using H-NMR and C-NMR spectroscopic analysis. FT-IR analysis of its ethanolic extract confirmed the presence of alkyne, alkane, and C-F bonding.

5.1 INTRODUCTION

Medicinal plants are one of the important components of biodiversity. India is one of the 12 mega bio-diversity nations in the world. This extensive flora has been considered as God's gift from time immemorial by Indian traditional systems of medicine for curing several diseases. Besides using for therapeutic purposes, they are prescribed for preventing diseases and overall human health [15, 63, 111, 112, 118]. Tapping the potentials of a large array of underutilized medicinal plants might benefit with several nutraceutical products, which can provide health benefits in terms of prevention and/or therapy.

The synonyms of *Ipomoea digitata* L. are *I. mauritiana* Jacq., *I. paniculata*, *I. paniculata* var. *paniculata*, *I. eriosperma*, *I. rubrocincta*, *I. rubrocincta* var. *brachyloba*, *Quamoclit digitata*, *Batatas paniculata*, *Convolvulus paniculatus* [26, 29, 34, 48, 103, 116, 118]. Milk yam (*Ipomoea digitata* L.: Figure 5.1) is an underutilized medicinal plant having immense therapeutic and nutraceutical potential [14, 24, 26]. The common names are: milk yam, Vidhari Kand, alligator yam, giant potato, finger leaf morning glory, palmate morning glory, etc. It is an extensive perennial climber found all over India, commonly in Eastern Bihar, West Bengal, Uttar Pradesh, and the west coast from Konkan to Kerala [63, 111]. It has been traditionally used as a medicine in India and parts of Southeast Asia as a general tonic, to treat diseases of the spleen and liver and to prevent fat accumulation in the body. The presence of fixed oil, carbohydrate, tannins, phenolic compounds, alkaloids, saponins, sterols, and flavonoids has been reported in the tubers of the plant. The herb is also known as a lacto-stimulant and libido enhancer. The tubers of the herb contain beta-sitosterol, which is an antioxidant. Ergonovine, an alkaloid found in the herb, is used to stop menstrual bleeding [34, 41, 44, 50, 98, 101]. The root has alterative, aphrodisiac, tonic, stimulant properties and used in male infertility and inflammations. Aqueous infusions of the roots are used for treating epileptic seizures and as an antioxidant in Ayurvedic medicine. Powdered tuber with honey is used for high blood pressure and heart diseases. It is a potential nutraceutical agent not completely explored and is a very useful and priority species facing extinction threats [104, 107, 108, 110].

This chapter explores the therapeutic and nutraceutical potential of milk yam.

5.2 DISTRIBUTION OF MILK YAM

Even though the origin is unknown, milk yam is pantropic, naturalized in moist tropical regions of India and many other parts of the world [64, 65].

It is generally seen in moist areas, riverbanks, marshy areas, vine thickets, monsoon forests, beach forests, gallery forest, semi-evergreen forests, coastal areas mostly in open areas [66, 68, 76, 90]. It is also reported in wastelands, plantations of *Tectona* sp., and *Gmelina* sp. in Africa [82].

It has been reported in West Africa, Gambia, in the riparian forests of Benin, Northern Australia, Taiwan, and Hawaii [33]; and is widely distributed in the hills of South-East Asia, and South America (Caribbean, Dominican Republic, and Haiti). In Australia, it occurs in Northern Territory, Cape York Peninsula, and Northern Queensland and towards South to the coastal central Queensland [7]. It is also found in Paraguay, Uruguay, Argentina, and Mauritius [12].

FIGURE 5.1   Milk yam (*I. digitata* L.).

According to *the Ethnobotany Database of Bangladesh* [59], it is distributed in forests and hill tracts of Chittagong and Sylhet. Also found in Brazil, Cambodia, Ivory Coast, Japan, Pakistan, Philippines, Senegal, and Sri Lanka [22]. In Kerala, it is widely distributed in all the districts like Kasaragod, Kannur, Kozhikode, Malappuram, Thrissur, Kottayam, Pathanamthitta, and Thiruvananthapuram, particularly in low lands and midlands [71]. Even though leaves stem and flowers of milk yam are pharmacologically beneficial, mature tubers of more than two years of age are ideal for the medicinal purpose [76].

5.3 TUBER MORPHOLOGY

Mature plant has a large ovoid and elongated tuberous roots, up to 60 cm long, 30 cm thick, even weighing up to 35 kg (average being 5 to 10 kg) [2]. The tubers exhibit annual rings when cut and exude sticky, milky latex [87]. Milk yam tubers have a dark brown outer surface with a pale white starchy mass inside. It is 16–24 cm long and has 8–10 cm breadth, cylindrical to sub-cylindrical or ellipsoid, oblong or globose in outline, with a tapering base. It has a few thick lateral filial roots [43, 88, 121].

Khan et al., [46] compared the mature and immature yam tubers and reported to possess an outer brownish surface and inner white portion, agreeable odor, mucilaginous and slightly sweet taste with milky latex for tubers at both stages of maturity. Mature tubers are obovoid to sub-cylindrical in shape whereas the immature ones are cylindrical in shape with the tapering end. Dried cut pieces of *I. digitata* L. tubers have a light brown outer surface, transverse warts, and ridges on the epidermis. The cut surface is creamy and fleshy with a smooth texture, sweet taste, and has no smell [106]. The structural variation between the mature and immature tubers was expressed by cutting it transversely by Khan et al., [46] and reported the presence of a large number of growth rings in mature and only one ring in the immature tubers [106].

Longitudinal and transverse sections of milk yam tubers were studied for its laticiferous system and cambial activity by Karthik and Padma [43]. Usually, the laticiferous system contains secondary metabolites like alkaloids, phytosterols, etc. Kuntal Das et al., [51] reported that abundant simple to compound starch grains are densely arranged in *I. digitata* L. tubers. It did not have calcium oxalate prismatic crystals, and it has pitted vessels.

5.4 DOMESTICATION AND CULTIVATION

Milk yam grows mostly in moist areas, monsoon forests and coastal tracts. A survey carried out for germplasm collection of milk yam accessions from different natural growing tracts of Kerala revealed that the plant could flourish under varying soil and climatic conditions ranging from lateritic to sandy loam soils and high altitude hilly regions to coastal tracts [116]. In Kerala, mature tubers are procured from forests [71]. Milk yam is commercially cultivated by marginal landholders of villages in the Natore district of West Bangladesh [92]. It is an important component of trade in the Asian continent [82].

Seeds and vine cuttings, having 1–2 nodes, are usually used for propagation. Both will form a caudex. *In vitro* multiplication, callogenesis, and indirect shoot regeneration in *Ipomoea digitata* L. was studied by Islam and Bari [36], who reported that MS media containing 1.0 mg/l BAP and 0.5 mg/l IAA could have 95% shoot regeneration and highest number of shoots per culture (6 numbers). For internodal explants, profuse callus induction (85%) was observed to be effective in MS media containing 1.0 mg/l 2, 4-D and 1.0 mg/l BAP. Islam et al., [37] developed a protocol for *in-vitro* callus induction, by culturing young stems in Murashige-Skoog's (MS) medium supplemented with 6-benzyl amino purine (BAP), 2-4-dichlorophenoxyacetic acid (2, 4-D), indole-3-butyric acid (IBA), etc. MS medium fortified with BAP (2.0 mg/ml) produced the best callus having a whitish green color, granular, and hard in nature. MS medium supplemented with 2, 4-D (1.0 mg/ml) produced light brown and loose callus, which is beneficial for inducing faster callogenesis.

Addition of soil amenders like rice husk and sawdust along with soil, sand, and FYM in the ratio of 1:1:1:1 is effective for rooting 3 nodded leafy shoot cutting of *Ipomoea digitata* L. according to Raut et al., [87]. This can be planted in polythene bags containing a potting mixture and can be transplanted to the main field at a spacing of 50 x 50 cm, after about one month, when it generates sufficient root system [76]. In the main field, 60 x 60 cm spacing is also recommended, and the vines are to be trailed on support. The study carried at the Instructional Farm, College of Agriculture, Vellayani, Kerala Agricultural University, and Thrissur, India by Anilkumar et al., [3] indicated that organic manure addition and irrigation have a significant positive effect on tuber yield of milk yam. The plants give an economic yield of tubers by around 1.5 to 2 years after planting [57].

5.5 PHYTOCHEMICAL SCREENING

An ethnobotanical survey was conducted during January 2014 along the natural growing tracts of milk yam in Kerala, and twenty accessions (both tubers and vines) were collected [116]. During the survey, apart from relevant details on soil and climate from the place of collection, ethnobotanical information on milk yam was also collected. The morphological characters of *I. digitata* accessions were cataloged based on the descriptor (0–9 scale) developed for sweet potato by the International Board for Plant Genetic Resources (IPGRI) [19].

Collected tubers of each accession were powdered, and samples were subjected to systematic phytochemical screening by successive extraction (1:1 ratio) with different solvents (methanol, ethanol, chloroform, and hydroethanolic extract). Both qualitative and quantitative analyses of phytoconstituents were done following appropriate procedures. The data pertaining to the qualitative estimation of phytoconstituents (Table 5.1) showed the presence of alkaloids, carbohydrates, glycosides, saponins, phytosterols, fats, and oils, resins, flavonoids, and proteins, for all the accessions under different extraction methods.

TABLE 5.1 Qualitative Phytochemical Analysis of Different Extracts of *I. digitata* L

Phytochemical	Type of Extract			
	Methanol	Chloroform	Ethanol	Hydro-ethanolic
Alkaloid (Wagner's reagent)	+	–	+	+
Carbohydrate (Molisch's test)	–	–	+	+
Fat and Oil (copper sulfate test)	–	–	–	+
Flavonoids (Alkaline reagent test)	–	–	+	+
Glycoside (Keller Kelliani's test)	–	–	–	+
Phytosterols (Liebermann-Burchard test)	–	+	+	+
Protein (ninhydrin test)	+	–	+	+
Resins (CuSO$_4$)	+	–	+	+
Saponins (Foam test)	+	–	–	+

Note: + = Yes, the chemical is present; – = No, the chemical is not present

The quantitative estimation of alkaloids, carbohydrates, glycosides, saponins, phytosterols, fats, and oils, resins, flavonoids, and proteins was done following accepted procedures; and significant variation in all phyto-constituents was noticed among the accessions (Table 5.2).

5.5.1   DIVERGENCE ANALYSIS

About divergence of yield characteristics (fresh tuber yield plant^{-1}) recorded the highest percentage of divergence being 37% followed by dry tuber yield plant^{-1} (24%), length of tubers (18%), girth of tubers (12%) and number of tubers per stock (9%), respectively. Regarding contribution of phytoconstituents regarding divergence, highest divergence was observed for flavonoids (48%), followed by oils (16%), glycosides (14%), alkaloids (8%), carbohydrates (6%), fatty acids (3%), proteins (3%) and saponins (2%), respectively.

5.6   PHYTOCHEMISTRY

Plant cells produce primary metabolites like carbohydrates, lipids, and proteins, which are directly involved in growth and metabolism of plants; as well as secondary metabolites (such as: alkaloids, phenols, essential oils and terpenes, sterols, flavonoids, lignins, tannins, etc.) [73, 77]. All these phyto-chemicals are naturally occurring biochemicals that give color, flavor, taste, smell, and texture to the plant [16, 94]. According to Ayurvedic Pharmacopoea of India [106], phytoconstituents in milk yam tubers are: pterocarpan-tuber-osin, pterocarpanone-hydroxytuberosone, two pterocarpenes–(anhydrotu-berosin and 3 O-methylanhydrotuberosin) and a coumestantuberostan, an isoflavone-puerarone and a coumestan-puerarostan.

Mishra et al., [67] reported that milk yam contains 1.3% fixed oil and the components include oleic acid (60.1%), linolenic acid (19.38%), palmitic acid (8.15%), and linolenic acid (1.11%) in mixed acid fraction. Acetone extract can yield higher quantity of phenols (285.05 mg gallic acid equivalents (GAE)/g extract), tannins (190.02 mg GAE/g extract) and flavonoids (174.44 mg rutin equivalents (RE)/g extract) among different solvents studied (petroleum ether, chloroform, acetone, methanol, and hot water) [117]. Ethanol extract was also studied [6]. The presence of phenols [45], tannins [72], and flavonoids [10] in milk yam extracts might be the possible reason for its antioxidant activity [25].

TABLE 5.2 Phytochemical Constituents of Milk Yam (*I. digitata* L.) Accessions at the Time of Collection

Treatment ID	Alkaloids	Carbohydrates	Glycosides	Saponins	Proteins	Flavonoids	Fatty acid, KOH	Oils
			(%)			(mg g^{-1})		(%)
ID-1	31.12	61.21	0.62	9.23	5.12	0.61	0.19	1.09
ID-2	27.61	52.24	0.24	5.32	3.22	0.24	0.12	0.43
ID-3	30.99	59.99	0.55	9.21	4.11	0.31	0.16	0.94
ID-4	24.96	43.21	0.17	4.32	5.27	0.39	0.20	1.11
ID-5	24.20	48.92	0.21	4.76	3.14	0.27	0.17	0.72
ID-6	28.40	51.24	0.28	5.42	3.18	0.35	0.11	0.3
ID-7	32.33	62.31	0.61	9.11	5.32	0.67	0.22	1.08
ID-8	28.70	54.12	0.32	5.73	4.10	0.42	0.18	0.99
ID-9	30.33	56.21	0.34	8.00	5.02	0.60	0.24	1.13
ID-10	24.11	49.81	0.11	3.21	3.31	0.41	0.14	0.56
ID-11	28.12	46.14	0.31	5.24	3.93	0.39	0.13	0.45
ID-12	21.14	32.33	0.10	2.39	2.73	0.28	0.11	0.32
ID-13	28.63	49.11	0.23	5.37	4.41	0.43	0.16	0.89
ID-14	22.86	41.23	0.21	3.42	3.24	0.27	0.11	0.83
ID-15	27.65	43.21	0.24	5.32	4.09	0.47	0.17	0.86
ID-16	34.14	59.21	0.59	9.34	5.66	0.57	0.24	0.73
ID-17	25.14	50.00	0.32	5.53	5.21	0.56	0.19	0.14
ID-18	20.24	54.80	0.31	5.48	3.73	0.48	0.17	0.88
ID-19	18.13	47.71	0.24	3.77	3.91	0.36	0.16	0.43
ID-20	34.69	56.90	0.63	9.76	5.24	0.63	0.23	1.18
S.Em	1.67	3.49	0.02	0.29	0.26	0.02	0.01	0.05
C.D	4.633	9.674	0.066	0.823	0.743	0.081	0.032	0.142
CV (%)	10.65	11.87	12.47	8.58	11.07	11.70	11.41	11.82

Ono et al., [75] reported the presence of isobutyric, (S)-2-Methylbu-tyric, Tiglic, n-decanoic, n-dodecanoic cinnamic acids, and two glycosidic acids (Quamoclic acid-A and Operculinic acid-A). The resin glycoside 'digitajalapin' was also extracted from *I. digitata* by the same group of scientists. Their structures were also determined based on chemical and H-NMR and C-NMR spectral data. Investigation on phytochemical constituents present in *I. digitata* L. tubers by Monjur-Al-Hossain et al., [69]concluded the presence of phytoconstituents like alkaloids, tannins, steroids, gums, glycosides, carbohydrates, and saponins. The phytochemi-cals present in the plant make it a phytestrogen source since its activity is like estrogen present in human body, which justifies its use in curing ailments related to female reproductive system [5]. Ayurvedic Drug Company in Himalaya reported that erganovine, an alkaloid present in Ipomoea tubers, is used to stop menstrual bleeding. The tubers are rich in carbohydrate, starch, protein, vitamins etc. The leaves also contain caro-tene at the rate of 6.3 mg/100 g [71].

Collection of herbs at right maturity is one of the parameters, which affect the efficacy of medicinal plants. Maturity of the officinal part is also an important criterion for gaining maximum potency of the formulations developed out of the plant. Only mature (bigger size) tubers of *I. digitata* L. are used for preparing galactagogues and immunomodulatory herbal medicines by the traditional medical practitioners. Proximate composition of mature and immature tubers of *I. digitata* L. was studied by Khan et al., [46], who reported to have moisture (15%), total ash (17.35%), acid insoluble ash (1.4%), alcohol soluble extractive value (4.08%) and water-soluble extractive value (11.85%). Proximate composition of immature tubers was slightly lower compared to mature tubers that had 10% moisture, 16.9% total ash, 1.2% acid insoluble ash, 3.12% alcohol soluble extractive, and 12.25% water-soluble extractive value. Khan et al., [46] compared phytoconstituents and bioactivities of immature tubers with mature tubers using chromatographic methods. Mature tuber's chemical and biological profiles possess about twice phytoconstituents than that in immature tubers.

Qualitative evaluation of both mature and immature tubers revealed the presence of alkaloids, carbohydrates, glycosides, saponins, phytosterols, resins, flavonoids, and proteins. Immature tubers contain more proteins and saponins (6.60% and 7.28%) than mature tubers (4.4% and 1.65%), whereas mature tubers contain more starch (5.8%) than the immature ones

(2.10%). HPLC and HPTLC analysis of these tubers also supported the finding that there exists a variation in the phytoconstituents among immature and mature tubers, the mature ones having more phytoconstituents than the immature ones. The study at the Department of Plantation Crops and Spices, College of Agriculture, Vellayani [116] indicated variation in phytoconstituents of mature and immature tubers, and this was confirmed through phytochemical analysis of the tubers at different harvest periods (Tables 5. 3 and 5.4).

In the present study, the mature tubers were harvested at 2 years after plants that showed superior concentration of phytoconstituents than the immature source harvested at 1 year after planting. This enables us to conclude the authenticity of traditional recommendation.

5.7   CHROMATOGRAPHIC ANALYSIS

Thin layer chromatography (TLC) of *I. digitata's* methanolic extract showed seven different spots at Rf 0.19 (green), 0.34 (magenta), 0.45 (green), 0.48 (blue), 0.62 (blue), 0.67 (red) and 0.92 (dark pink) [106]. The *I. digitata* L. (Syn. *I. paniculata*) tubers reported to possess taraxerol, taraxerolacetate, β-sitosterol, scopoletin, and 7-O-β-D-glycopyranosylscopoletin (Scopolin) [94]. Triterpenid, coumarin, octadecyl Є-p-coumarate, t-cinnamic acid [undecyl Є-3-(4-hydroxyphenyl)-2-propeonate], an unknown coumarin (5-hydroxyl-7-ethoxy coumarin) and a lignin type resin glycoside were also present in the tubers [31, 54, 86]. Scopoletin and taraxerol can inhibit acetylcholinesterase activity; therefore, it is found beneficial for treating Alzheimer's disease [53].

Chromatographic techniques were used for estimating the amount of umbelliferone, another important phytoconstituent in the tuber. Percentage recovery of umbelliferone by HPLC and HPTLC were 97.90% and 98.90%, respectively [20]. Both HPLC and HPTLC have the same efficiency and sensitivity for determining umbelliferone from dried tuber powder. Umbelliferone is anti-coagulant and anti-HIV in action [62].

Scopoletin (7-hydroxy-6-methoxycoumarin) is hepatoprotective, spasmolytic, inhibits prostate cancer proliferation, antioxidant, anticoagulant, and anti-HIV [62]. Ethyl acetate: methanol: water: ammonia (13: 5: 1.8: 0.2) solvent system was used to isolate and quantify scopoletin in

TABLE 5.3 Phytochemicals in Milk Yam (*I. digitata* L.) Accessions Harvested after 365 Days of Planting

Treatment ID	Saponins	Glycosides	Carbohydrates	Alkaloids	Proteins	Flavonoids	Fatty acid (mg KOH)	Oils
			(%)			(mg/g)		(%)
ID-1	7.33	0.33	56.80	28.03	3.43	0.45	0.13	0.40
ID-2	6.26	0.39	58.10	28.36	3.41	0.18	0.15	0.76
ID-3	8.00	0.39	59.16	30.70	3.93	0.20	0.12	0.83
ID-4	9.20	0.36	57.53	28.00	4.29	0.32	0.12	0.90
ID-5	6.73	0.43	56.83	30.30	4.22	0.29	0.13	0.73
ID-6	7.46	0.33	55.16	27.63	2.95	0.23	0.11	0.93
ID-7	7.60	0.34	54.23	29.23	4.05	0.20	0.12	0.30
ID-8	7.73	0.38	55.50	27.96	4.10	0.18	0.16	0.53
ID-9	9.33	0.52	57.83	30.33	4.30	0.19	0.16	0.36
ID-10	7.06	0.40	55.16	27.80	4.00	0.47	0.15	0.43
ID-11	8.93	0.46	56.53	30.20	3.60	0.55	0.15	0.53
ID-12	8.53	0.39	56.03	28.33	3.71	0.35	0.13	0.96
ID-13	7.86	0.44	53.76	28.60	4.05	0.35	0.15	1.06
ID-14	8.60	0.36	55.90	27.26	4.22	0.44	0.12	0.73
ID-15	6.00	0.42	54.96	28.40	4.37	0.24	0.11	1.06
ID-16	9.13	0.43	56.26	27.16	4.30	0.22	0.12	0.83
ID-17	6.20	0.36	56.10	29.23	3.22	0.41	0.12	0.83
ID-18	7.06	0.42	53.43	30.46	3.41	0.46	0.12	0.83
ID-19	5.46	0.35	57.10	28.16	4.20	0.49	0.12	0.53
ID-20	6.46	0.45	55.43	29.60	4.16	0.19	0.12	0.36
S.Em	0.64	0.01	0.65	0.50	0.30	0.02	0.00	0.07
C.D	1.84	0.03	1.86	1.43	0.85	0.06	0.02	0.20
CV (%)	14.80	5.82	2.01	3.01	13.32	11.44	9.89	18.19

TABLE 5.4 Phytochemical Constituents of Milk Yam Accessions (*I. digitata* L.) Harvested after 2 Years After Planting

Treatment ID	Saponins	Glycosides	Carbohydrates	Alkaloids	Proteins	Flavonoids	Fatty acid (mg KOH)	Oils
	(%)					(mg/g)		(%)
ID-1	7.82	0.51	57.26	30.33	4.16	0.49	0.14	0.96
ID-2	7.23	0.54	54.43	29.16	4.14	0.23	0.18	0.89
ID-3	7.99	0.53	59.32	32.24	4.27	0.39	0.16	0.91
ID-4	9.34	0.53	58.56	30.07	4.39	0.38	0.17	0.93
ID-5	7.37	0.49	59.98	30.97	5.01	0.47	0.14	1.02
ID-6	7.93	0.39	56.63	31.36	3.90	0.33	0.13	0.98
ID-7	7.77	0.50	56.10	30.00	4.30	0.36	0.12	1.10
ID-8	8.13	0.42	57.80	27.47	4.67	0.32	0.18	1.03
ID-9	9.51	0.56	59.03	31.54	4.86	0.37	0.18	0.85
ID-10	7.63	0.51	56.00	29.99	3.99	0.51	0.16	0.92
ID-11	9.31	0.53	58.11	30.09	3.98	0.53	0.16	0.77
ID-12	9.23	0.52	59.23	30.53	4.41	0.48	0.18	0.87
ID-13	8.71	0.51	55.61	29.84	4.73	0.43	0.18	1.04
ID-14	7.00	0.56	56.01	29.61	4.42	0.51	0.16	0.82
ID-15	7.03	0.54	55.19	30.06	5.00	0.47	0.13	0.91
ID-16	9.19	0.53	58.33	30.04	5.42	0.43	0.15	1.02
ID-17	8.00	0.59	57.01	30.43	5.34	0.49	0.15	1.04
ID-18	8.23	0.57	54.99	33.10	4.33	0.48	0.17	0.98
ID-19	7.92	0.52	59.10	32.21	5.01	0.51	0.16	0.71
ID-20	7.40	0.54	56.97	34.69	5.09	0.34	0.16	0.65
S.Em	0.52	0.03	2.59	1.85	0.28	0.03	0.01	0.06
C.D	1.451	0.089	7.169	5.115	0.785	0.072	0.027	0.155
CV (%)	11.15	10.73	7.52	10.14	10.74	10.59	10.93	10.58

the methanolic extract of milk yam tubers by Karthik and Padma [43]. TLC fingerprint of the methanolic extract revealed the presence of Scopoletin (Rf value 0.56) and it was quantified as 0.029–0.034% using HPLC and HPTLC methods. Scopoletin and β-sitosterol glucoside, an antioxidant, were isolated from its ethanolic extracts and its structure was elucidated using HNMR and CNMR spectroscopic analysis by Khan and Hossain [47].

Viji and Paulsamy [117] conducted the GC-MS profiling of milk yam's acetone extract and revealed the presence of 27 compounds, namely:

- 1-Docosanol methylether;
- 1-Octadecene;
- 1-Octadecenol;
- 2-methyl-4,5-dihydroxybenzaldehye;
- 2, 2-Dideutro octadecanal;
- 4-acetylbutyricacid;
- 4-(3'4'(methylenedioxy) 6'formy phenyl) 6-fluro coumarin;
- 6, 8-dioxabicyclo (3, 2, 1) octan 3a-OL-2, 2, 4, 4-D4;
- 9, 12-Octadecadienoic acid (Z,Z);
- 9-Octadecene 1-ol(z);
- Chloroacetic acid, 4-hexadecylester;
- Dodecanoic acid;
- E-15-Heptadecanal;
- Ethyl-3-(trimethylsilyloxy) 8azabicyclo(4, 2, 1)oct-2-ene8 carboxylates;
- Hahnfett;
- Hexadecen-1-ol trans-9-Hexadecanoic acid;
- Hexatricontane;
- Isopropyl Myristate;
- n-Tetracosanol-1;
- Nonacosane;
- Octacosane;
- Octadecanoic acid;
- S-(2-aminoethylester);
- Tetradecanal;
- Tetradecanoic acid;
- Tetratetracontane; and
- Thiosulfuric acid ($H_2S_2O_3$).

The highest peak area (14.22%) was observed for hexadecanoic acid (palmitic acid), which is known for its selective cytotoxicity against human leukemia cells, antioxidant, anti-inflammatory, hypocholesterolemia, nematicide, pesticide, and lubricant activities [4, 32]. Research must be focused on isolating different potential chemical constituents from milk yam and to evaluate its therapeutic activities through *in vitro* and *in vivo* clinical trials [74].

5.7.1   THIN LAYER CHROMATOGRAPHY (TLC)

Chromatographic fingerprinting of alcoholic extract of *I. digitata* L. by TLC was done by Vidya [116]. The phyto-documentation of the plates showed numerous bands under UV 366 nm and visible light after derivatization. The Rf at 0.2 value of *I. digitata* extract is presented in Figure 5.2. The developed plates were photographed under UV 366 nm and under visible light, which confirmed the presence of alkaloids, saponins, and flavonoids, respectively.

5.7.2   HIGH PERFORMANCE LIQUID CHROMATOGRAPHY (HPLC)

High performance liquid chromatography (HPLC) was performed with the aqueous extracts of different samples of tubers of *I. digitata*. Samples collected from best-performing accessions were used for analysis. The HPLC profile of accessions in the aqueous extracts consistently exhibited a quantitative difference in the phytoconstituents particularly those eluting at 1.843 minute and 2.050 minute retention time (Rt) in the HPLC chromatograms. The results confirmed the presence of flavonoids and alkaloids in the sample. In the analysis, two new compounds were also identified, Rutin – a flavonoid (Figure 5.3), and nicotinic acid – an alkaloid (Figure 5.4).

The identity of the peak of rutin and nicotinic acid in sample solution was confirmed by comparing the retention time (Rt) or retardation factor (Rf) with that of standard rutin and nicotinic acid. The retention time (Rt) of rutin was 2.050 min, whereas the retention time of nicotinic acid was 1.843. The assay results indicate that amount of rutin (flavanoid) and nicotinic acid (alkaloid) estimated by HPLC method was 4.16 μg g^{-1} and 17.20 μg g^{-1}, respectively.

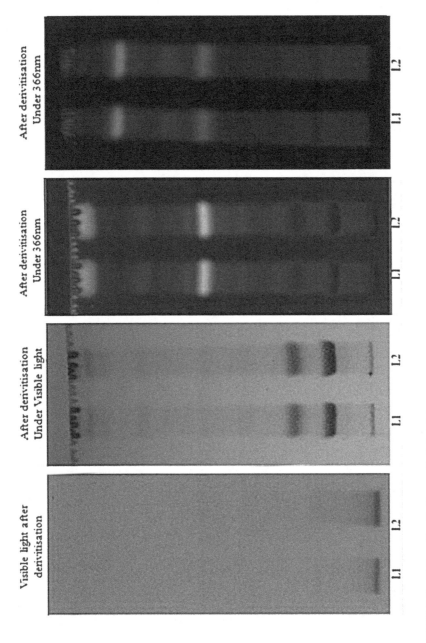

FIGURE 5.2 TLC profile of milk yam (*I. digitata* L.).

FIGURE 5.3   HPLC fingerprint of rutin fraction.

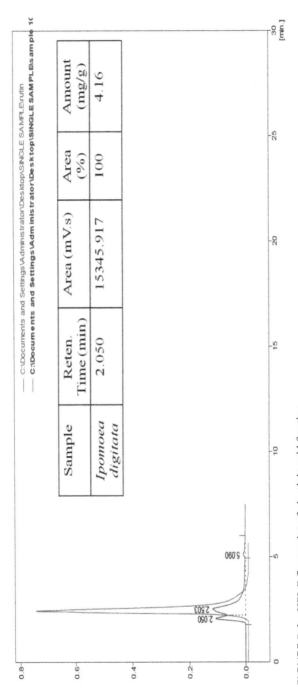

FIGURE 5.4 HPLC fingerprint of nicotinic acid fraction.

5.8 ETHNOBOTANY

The indigenous systems of medicine always depend on traditional knowledge for the drug development. In recent years, ethnobotanical studies have been used for the discovery of new drugs and new drug development. Milk yam plant and its medicinal, as well as nutraceutical properties, have been well mentioned in ancient literature. The '*Susruthasamhitha*' contains several prescriptions for using it as an aphrodisiac. In the Konkan region of India, tubers are washed, peeled, cut into small pieces and dried for using as an aphrodisiac [26]. Powdered root macerated in its own juice and administered with honey and butter is also an aphrodisiac [103]. Extract prepared by macerating powdered roots is used as an abortifacient in Senegal [82].

It is also used: as a galactagogue, restorative, carminative, expectorant, anthelmintic, stomachic, and appetizer [27]; and for treating leprosy, syphilis, gonorrhea, inflammations, burning sensation, vomiting, gastric ulcer, ulcerative colitis, hoarseness, etc. [18, 119]. It is a blood purifier as well as a voice and complexion improver, but it causes 'Kapha.' Its roots are used for applying on swellings of joints [23]. Its flowers are also sweet, cooling, aphrodisiac, used for curing biliousness but, it causes 'Vata' and 'Kapha' [49]. Khare [48] mentioned its use as a cholagogue, which increases bile flow to the intestine. In India, its seeds are used for coagulating milk [82].

Milk yam tubers are used in more than 45 formulations of Ayurveda or itself as a single drug [43]. In Midnapore district of West Bengal, its tubers are eaten raw and it is popular for its astringency for treating blood dysentery [96]. In Sreepur Upazilla-Magura district of Bangladesh, folk medicinal practitioners (*Kavirajees*) are using milk yam for curing diseases as well as a functional food for improving voice and general health of humans [84]. Kandha tribes of Orissa are using the tuber juice along with cow's milk continuously for seven days to increase lactation [9].

Rahmatullah et al., [83] documented use of milk yam by a folk medicinal practitioner (Kaviraj) of Pabna district, Bangladesh. Kaviraj recommended a formulation containing tuber of *Ipomoea digitata*, bark of *Pterocarpus santalinus*, fruits of *Phyllanthus emblica* and fruits of *Elettariacardamomum* mixed with ghee and honey to be taken orally @ of 3 g for 3–4 times a day for increasing appetite.

A decoction made of its roots and leaves are used to treat diarrhea, acting as a purgative and diuretic. Decoctions prepared out of its dried roots are used in Cote d Ivoire as enema to treat kidney pain and female sterility. The same decoction is used to ensure a good pregnancy and to avoid miscarriages. Leaves of milk yam along with leaves of *Croton lobatus* L. are used to treat colic infection in babies and decoction of leafy twigs is given to strengthen emaciated children in Benin (Africa). A root decoction is used as an alternative and aphrodisiac tonic in Nigeria [82]. Root decoction is also used against constipation [97].

Administration of 3–10 g of tuber powder (prepared by boiling the tuber with milk and drying under shade) dissolved in milk is a good galactagogue. Fresh tuber juice combined with cumin, coriander, fenugreek seeds and sugar can also be used as a lactagogue for curing spermatorrhoea [78]. Also, milk yam powder (prepared in the same way as said above) and barley powder in equal proportions (1: 1) can be mixed with ghee, milk, and sugar to prepare porridge for curing emaciation [76]. The root powder, equal parts of wheat flour and barley, milk, ghee, honey, and sugar can also be used as a restorative for emaciated and debilitated children [103].

According to literature, tuber flour of milk yam is being used for treating hepato-splenomegaly [96]. A preparation made from sun-dried milk yam tuber powder added with sugar and ghee is used for lowering menstrual discharge as well as to increase body weight [40]. *Yaogika-chikitsa* and *Dravyaguna* have mentioned its use for treating hypertension and heart diseases [89]. Also, milk yam tuber powder added with honey can be beneficial for reducing blood pressure and heart diseases. In Africa, the root mixed with palm wine is used as a galactagogue [82].

Yusuf et al., [120] described its use for treating sexual disabilities in Khagrachari district of Bangladesh. Generally, it is using as a *rasayana* drug, as a nutritional supplement [109]. *Louha Bhasma* is a Rasayana or immunomodulator prepared using iron for curing anemia. It is a practice of mixing milk yam juice with Louha Bhasma for the management of impotency during its unique method of preparation 'Putapaka,' as described by the ancient Indian chemist Nagarjuna around 100 AD [81].

In Gabon, tubers macerated with water are used as a wash against venereal diseases. Tuber macerated, along with *Ampelocissusbombycina*, is used to treat edema. Its young shoots are ground and applied to immature gangrenous and necrotic ulcers in Congo [82].

The ethnobotanical information gathered during a survey revealed that milk yam powder is generally used as a galactagogue/general tonic/aphrodisiac along with milk/honey in Kerala [116]. It acts as a rejuvenative herb for both mind and body and help to slow down the aging process of the body. In females, it helps in healthy menstruation and lactation; and in males, it helps in healthy production of sperms.

5.9 MEDICINAL USES

According to the Ayurvedic Pharmacopoeia of India [106], the therapeutic uses of *I. digitata* L. include: *Raktavikara, Vrana, Styanyavikara, Pittaja Sula, Mahavatavyadhi, Mutraroga, and Bhagna*. Studies on the pharmacological properties of the milk yam tubers revealed its potentiality against different types of lifestyle diseases or disorders like diabetes, hypertension, cardiovascular diseases, infertility, immune deficiency, etc. Besides it is proved to have galactogogue, antioxidant, antimicrobial, analgesic, revitalizing, anti-inflammatory, and spasmogenic activities. Even if all the plant parts contain different essential phytoconstituents, its tubers are said to be the official part. Tewati and Mishra [105] proved its hypotensive property and muscle relaxant activity in frogs, dogs, rats, and rabbits.

Administration of hydroalcoholic extract of milk yam tubers at a dose of 100–200 mg/kg bodyweight for 28 days showed significant antidiabetic activity and it may be due to the presence of active principles like flavonoids and β-sitosterol [79]. Anti-diabetic activity of hydro-ethanolic root extracts of *Ipomoea digitata* L. in alloxan-induced diabetic rats was studied by Minaz et al., [64], who reported that both alcoholic and aqueous extracts have antidiabetic activity as a result of significant reduction in fasting serum glucose, triglyceride, and cholesterol levels.

The effect of methanolic extract of *Ipomoea digitata* L. tuber was studied for reducing cholesterol levels in experimentally induced hyperlipidemic rats. A significant reduction in body weight with a simultaneous increase in HDL level was also achieved with the administration of tuber extract at the rate of 300 mg/kg [70].

Administration of antibiotics like gentamicin can cause kidney dysfunction which may result in increased level of urea, creatinine, sodium, and decreased level of protein, potassium, and non-enzymatic antioxidants such as vitamin C and vitamin E. Kalaiselvan et al., [42] reported that administration of ethanolic extract of *Ipomoea digitata* L. could restore the alteration of these parameters in gentamicin intoxicated rats.

Milk yam is a good candidate for managing male infertility since a significantly high sperm density was recorded in neem oil-induced infertile albino rats, according to Mahajan et al., [55]. A significant increase in sperm density, sperm motility, along with increase in testis and epididymis weight, serum levels of testosterone, follicle-stimulating hormone (FSH) and luteinizing hormone (LH) was induced in albino male rats at a rate of 250–500 mg/kg bodyweight for 40 days. The β-sitosterol may be the likely cause for the significant enhanced process of spermatogenesis.

Immunomodulatory activity of four different plant species traded under the common name 'Vidari' was tested by carbon clearance test on Wistar rats by Kuntal Das et al., [51], who reported that *Pueraraia tuberosa* followed by *Ipomoea digitata* showed close activity to *Withania somnifera* powder, an approved potent immunomodulatory agent used as positive control. Garodia et al., [28] recommended Vidari as an immune modulator for patients, who have undergone radiotherapy.

As the name 'Milk Yam' or 'Palmuthukku (Malayalam)' suggests, *I. digitata* L. is supposed to have the property of increasing breast milk production in cattle as well as human beings.

Vasagam et al., [113] studied the *in vitro* antioxidant activity of methanolic extract of *Ipomoea digitata* L. tubers and reported to have 230 µg/ml hydroxyl radical scavenging activity, 800 µg/ml by fluorescence recovery after photobleaching (FRAP) method and 7.51 mg/g total phenol content, which represents its significantly higher antioxidant activity. Alagumanivasagam et al., [2] also reported that methanolic extract of *Ipomoea digitata* L. is a significant source of natural antioxidant since its administration in high-fat diet rats showed an increased level of antioxidant enzymes like superoxide dismutase (SOD), catalase (CAT), glutathione peroxidase (GPx) and glutathione reductase. Monjur-Al-Hossain et al., [69] reported that *Ipomoea digitata* L. tuber extract had 164 µg/ml DPPH free radical scavenging activity. But, compared to methanolic extracts of *Withania somifera* (IC50–27.80 µg/ml) and *Oxalis corniculata* (IC50–19.98 µg/ml), *Ipomoea digitata* (IC50–86.48 µg/ml) had only moderate antioxidant activity in terms of DPPH free radical scavenging potential [93].

Ipomoea digitata flowers collected from Dhaka city of Bangladesh were evaluated for the presence of microorganisms by Sharmin et al., [95], who revealed that milk yam flowers had a total bacterial load and fungal load of 10^7–10^8 CFU/g and 10^5–10^7 CFU/g, respectively of which *Staphylococcus* sp. (~10^7 CFU/g) was the predominant microorganism. An unknown coumarin present in the milk yam showed antibacterial activity against *Pseudomonas aeruginosa* and *E. coli* [54]. Crude ethanolic extract of *I. digitata* L. tubers did not exhibit any significant anti-microbial activity as reported by Monjur-Al-Hossain et al., [69]. Drug-resistant attribute of *Staphylococcus* sp. was studied by Afrin et al., [1], who reported that it exhibits highest multi-drug resistance against 83% of drugs tested. Streptomycin and Tetracycline exhibited zone of inhibition against *Staphylococcus* sp., but ampicillin and cefalexim were found to be ineffective.

According to folklore and Indian systems of medicine, milk yam is used against skin infections like acne, dandruff, body malodor etc. [39]. A supporting study was done by Mahendra et al., [56] to find out the antimicrobial activity of milk yam extracts (petroleum ether, chloroform, ethyl acetate and methanol) against skin pathogens like *Malassezia furfur*, *Propionibacterium acnes* and *Corynebacterium diphtheria* (@ 1, 20 and 50 mg/ml). A mild antibacterial activity was exhibited by chloroform and ethyl acetate extracts against *C. diphtheria* (@ 20 and 50 mg/ml) both with an inhibition zone of 13–14 mm. Petroleum ether extract showed inhibition against *C. diphtheria* at 50 mg/ml. None of the extracts showed activity against *P. acnes* and *C. diphtheria*. Also, methanolic extracts showed no inhibition against any organisms studied.

Viji and Paulsamy [117] found out a compound named 'tetradecanal' having antimicrobial activity from acetone extract of milk yam. The E-15-Heptadecenal and Octadecanoic acid are the other specific compounds in the extract capable of killing bacteria. The 1-Docosanol methyl ether identified from the acetone extract of milk yam tubers has the ability to inhibit *in vitro* replication of many lipid-enveloped viruses, including HSV (Herpes Simplex Virus).

Crude ethanolic extract of *Ipomoea digitata* L. tubers has analgesic activity since it showed 71.15% and 80.77% writhing inhibition at a dose of 250 mg/kg and 500 mg/kg, respectively as reported by Monjur-Al-Hossain et al. [69].

Ancient literatures of Ayurveda mention that resin glycosides in milk yam are responsible for the anti-inflammatory activity. Hexadecanoic acid and Hexadecan-1-ol trans-9 are compounds with anti-inflammatory activities in milk yam tubers [117]. Paniculatin a glycoside isolated from *Ipomoea digitata* L. and when administered intraperitoneally in mice recorded an LD 50 value (48 hours) with 95% fiducial limits 867.4 (755.3–985.1) mg/kg. Administration of paniculatin resulted in elevated blood pressure, showed stimulant effect on myocardium and respiration, a vasoconstrictor and bronchoconstrictor effect with spasm-genic effect on smooth muscles of gut and also oxytocic activity [58].

5.10 PHARMACOGNOSY

Chandira and Jayakar [17] had formulated herbal tablets using aqueous extract of *Ipomoea digitata* L., since it contains most of the phytochemicals when compared to acetone, petroleum ether and alcohol extracts. Administration of tablet made from its aqueous extract at the dose of 300 mg/kg body weight in streptozotocin-induced diabetic rats could significantly decrease the blood glucose level indicates its potent antidiabetic activity.

Administration of *Ipomoea digitata* L. tuber powder (3 g) for a period of 12 weeks could reduce systolic, diastolic, mean blood pressure, serum total cholesterol, LDL cholesterol and atherogenic index with an increase of fibrinolytic activity and total antioxidant status in individuals having stage-I hypertension according to Jain et al., [40]. A decrease in systolic blood pressure by 40 points and diastolic pressure by 20 points can be achieved by milk yam administration [30].

A lactogenic polyherbal formulation containing milk yam named 'Lactovedic' was administered to rats having suckling pups to find out its galactagogue activity [61]. The study indicated that Lactovedic could increase milk yield, pups body weight, weight of mother rat, glycogen, and protein content of mammary gland tissue, serum prolactin and cortisol, which could prove its significant galactagogue activity. The herbo-mineral drug-containing different medicinal plants along with *Ipomoea digitata*, Zinc-ash complex, and high energy carbohydrate molecules were evaluated for its clinical efficacy by Rani et al., [85]. An overall improvement ranged between 69–77% in all the patients having symptomatology of general weakness, appetite, sleeplessness, etc., which could prove its potential revitalizing effect in humans.

5.11 NUTRACEUTICAL POTENTIAL

Ipomoea digitata L. can be regarded as a potential nutraceutical agent but remains as an underutilized plant [99]. In India, its tubers are considered as an alternate source of Vidari and are widely used to prepare popular Ayurvedic nutraceutical products like *Chavanaprasha* [115]. For preparing nutraceuticals, the tubers are subjected to several postharvest operations. Harvested fresh tuberous roots of the plant were cut into small pieces and sun-dried and then dried in an oven at reduced temperature ($\leq 50°C$) [47].

Yield and quality of the final product are a function of a dynamic relationship between maturity, yield, and quality parameters [21].

Harvested mature tubers must be washed, peeled, and shredded into small pieces, again washed for at least three times in good quality running water and excess water has to be drained-off by keeping in a bamboo basket for sufficient period of time before drying. Drying can be done in a hot air oven maintained at 60°C for 12 hours and then can be powdered to yield milk yam tuber powder (Figure 5.5) having high nutritional value per 100 g as: protein (12.58 g), total carbohydrate (24.36 g), soluble carbohydrate (4.78 g), fat (1.28 g), fiber (7.2 g), ash (5.23 g) and vitamin C (68.35 mg/g) [100].

 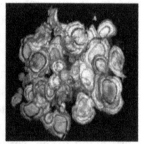 

FIGURE 5.5   Milk yam tuber slices (left); (B) Dried chips (center); (C) Milk yam powder (right).

Its tuber powder is grayish brown in color having no characteristic odor, bitter in taste, having parenchyma filled with starch, septate fibers in the form of crystals as well as bulb-like pipette (Figure 5.5). Vessels have simple and scalariform cross perforation plates, stone cells, and starch [106]. The tuber powder is coarse to moderately coarse in texture. The vessel elements are short having a terminal to sub-terminal, simple perforation plates and with lateral or circular or elliptical pitting. The pitting belongs to pseudo-scalariform, silt-like and circular to elliptical crowded types. Parenchymatous cells are rich in a minute to large-sized (10 μm to 50 μm) starch grains having spherical or sub-spherical shape. Fibers, resin ducts, sclereids are absent [43]. Nutraceutical products from milk yam are available in different types of formulations such as: powder, tablets, rasayana, linctus, confection, decoction, milk decoction, etc. [33].

According to the Ayurvedic Pharmacopoeia of India [106], an important formulation containing milk yam is *Sivagutika*, and the recommended dosage is 5–10 g. It is also an important constituent in Ayurvedic formulations like Vidaryadikvatha Churna, Vidaryadi Ghrita, Marma Gutika, Manmathabhra Rasa, Pugakhanda (Aparah), Dasamoolarishtam, and Chavanaprasam [76].

A herbo-mineral drug-containing seven different herbs along with milk yam, zinc-ash complex and high energy carbohydrate molecules namely Revivin has been evaluated for its clinical efficiency and safety by Rani et al., [85], who reported an overall improvement of 69–77% for patients having various symptomatology of general weakness, appetite, sleep mood and lack of concentration. They also observed that irritability was reduced by 69% and weakness was improved by 77% due to Revivin administration. In most of the patients, an improvement in terms of wellbeing, decreased lethargy, and capacity for working even in stressful conditions without any tiredness even at the end of the day was noticed.

National R & D Facility for Rasayana [27] has documented certain promotive and curative recipes using Vidari particularly for rejuvenative, strength-giving (Balya) and blood purifying (Raktadoshahara) properties as well as for using vidari as a galactagogue (Stanya Vriddhi), aphrodisiac (Vrishya), emaciation (Shosha) and hemoptysis (Raktavamana). Consumption of Vidari Kalka (paste) @10 g dissolved in half glass of milk twice daily can rejuvenate our body, purify blood, improve strength and stamina and can cure blood vomiting [80]. Vidari tuber juice (25ml) or Vidari powder (10 g) added to milk and consuming thrice daily after food could act as a galactagogue [11].

For increased sperm production, vidari juice/powder (10 g) mixed with 5 g ghee and 150ml milk can be consumed thrice daily after food for two months for beneficial effects. There is also a list of some Rasayana drugs containing Vidari (viz., Vidaryadi Gritha, Vidaryadi Kasaya, Vidaryadi Leha, and Chavanaprasha). Vasu Research Center, a division of Vasu Healthcare Pvt. Ltd., Vadodara, Gujarat has developed a product 'Mahamash oil' containing Kshervidari [114]. Phala Ghrita is also an Ayurvedic formulation containing milk yam [13].

Ipomoea digitata is a major ingredient in various mixtures used as nutritive, diuretic, and expectorants [38]. Milk yam is one of the ingredient of a popular rasayana drugs 'Aswagandharistam,' 'Balarishtam' and 'Dhanwanthararishtam,' which are principally made out of *Withania somnifera* (Aswagandha) having several health benefits like, epilepsy, faintness, fatigue, psychic problems, piles, indigestion, rheumatic complaints, improving memory power etc.[8]. Dhanwanthararishtam is used for curing all types of rheumatic complaints, hernia, and vaginal diseases [60]. Manmathabra rasa is also a rasayana drug containing milk yam, useful for curing general debility [102].

Satveda herbs Ltd. [91] is selling Vidarikanda powder, a micronized powder potentiated with Vidari's decoction as a dietary supplement with different therapeutic claims, such as: for men and it helps in healthy, semen production; and in females, it helps in healthy menstruation and lacta-tion. In both sexes, it supports fertility and vitality. Also, it is rejuvenative, immune booster, anti-aging, improves complexion, and so on. Hamdard Laboratories in Pakistan is also selling milk yam products under the brand names, Habb-e-Asgand and Sharbat Mufarreh Muqawwi-e-Qalb. They notified that excessive intake of milk yam is supposed to be harmful for people with warm temperament [22].

Milk yam capsules @ one tablet thrice daily along with milk, tender coconut water or turmeric juice is effective to cure leucorrhoea, urinary calculi, diabetes, burning micturition, etc. Sathavarigulam also contains milk yam and is recommended at a rate of 10 g daily twice with milk. Marmagulika is also a capsule recommended at a dose of ½ to 1 tablet thrice daily with milk or jeeraka (cumin) water for curing leucorrhoea, body pain due to trauma and burning micturition.

Ayu Muscle Tone is an ayurvedic muscle tablet containing milk yam along with Shatavari (*Asparagus racemosus*), Ashwagandha (*Withania somnifera*) and Amalaki (*Emblica officinalis*). It is recommended as a bone and muscular tonic consumed at a dose of two tablets twice a day. Lukowin tablets another product contain 80 mg of milk yam and is recommended at a dose of 2 tablets with water daily for curing menstrual disorders [35].

5.12 MISCELLANEOUS USES

Milk yam is often grown for ornamental purposes by training against trel-lises and pillars [7] for its beautiful pink or purple flowers and shade. The stem and leaves of the plants are used as fodder [71]. In India and Senegal (Africa), the plant is browsed or fed to the cattle for enhanced milk production [82].

5.13   FUTURE RESEARCH

An immense scope for research in *I. digitata* L. exists since its functional potential has not been exploited fully. Its tuberization morphology as well as anatomy, phytochemical characterization using sophisticated and accurate analytical tools like chromatography, spectroscopy, etc. can be beneficial for authenticating the crude drug to prevent its adulteration with Vidari (*Peuraria tuberosa*). Development of nutraceutical products from its tubers is also a potential area to evaluate.

5.14   SUMMARY

Ipomoea digitata L. (milk yam) is a type of morning glory plant naturalized in many parts of the world. The root is considered as tonic, alterative, aphrodisiac, demulcent, galactagogue, and cholagogue. In India, tubers of *Ipomoea digitata* (*Vidari*) and many of the Ayurvedic industries use Vidari in popular Ayurvedic nutraceutical products. This chapter focuses on the various ethnobotanical, medicinal, and nutraceutical properties of milk yam. Diversity analysis of milk yam accessions collected from different agro-ecological regions, detailed account of reproductive biology and phytochemical analysis of milk yam tubers have also been discussed.

KEYWORDS

- accessions
- callogenesis
- descriptor
- epipetalous
- fingerprinting
- laticifers
- nutraceutical
- phytochemical`

REFERENCES

1. Afrin, S., Basak, P., & Rahman, T. Demonstration of drug–resistant bacteria among commonly available flowers within Dhaka metropolis and assessment of their anti–bacterial properties. *Stamford J. Microbiol.,* **2015**, *5*(1), 5–8.

2. Alagumanivasagam, G., Muthu, K. A., Kumar, S. D., Suresh, K., & Manavalan, R. *In vivo* antioxidant and lipid peroxidation effect of methanolic extract of tuberous root of *Ipomoea digitata* L. in rat fed with high fat diet. *Int. J. Appl. Biol. Pharm. Technol.,* **2010**, *1*(2), 214–220.

3. Anilkumar, A. S., Suharban, M., Hajilal, M. S., Jayasree, P., & Rengini, P. R. Effect of cropping system, organic manure and soil moisture on yield and biochemical constituents of medicinal plants used for the preparation of 'Sathavarigulam.' *Proceedings of the 6ᵗʰ National Seminar on Medicinal Plants* (pp. 111–122). Department of Pharmacognosy, Govt. Ayurveda College, Thiruvananthapuram, Kerala, India, **2007**.

4. Aparna, V., Dileep, K. V., Mandal, P. K., Karthe, P., Sadasivam, C., & Haridas, M. Anti-inflammatory property of N-Hexadecanoicacid: Structural evidence and kinetic assessment. *Chem. Biol. Drug Des.,* **2012**, *80*, 434–439.

5. Ashajyothi, V., Pippalla, R. S., & Satyavati, D. *Ipomoea digitata*–update. *Int. J. Pharm. Res. Rev.,* **2013**, *1*(1), 13–21.

6. Ashajyothi, V., & Satyavati, D. The FTIR studies of ethanolic extract of *Ipomoea digitata. Int. J. Pharm. Biol. Sci.,* **2016**, *6*(1), 30–32.

7. *Australian Tropical Rainforest Plants* (6ᵗʰ edn., p. 312). (RFK6), CSIRO (Commonwealth Scientific and Industrial Research Organization), Canberra ACT, Australia, **2010**.

8. Ayurvedic Medicine Information, PGDPSM, Utah, US, **2015**. www.ayurmedinfo. com/2011/06/29/balarishta-uses-side-effects-dose-and-side-effects/ (Accessed on 30 September 2019).

9. Behera, S. K., Panda, A., Behera, S. K., & Misra, M. K. Medicinal plants used by the Kandhas of Kandhamal District of Orissa. *Indian J. Tradit. Know.,* **2006**, *5*, 519–528.

10. Benavente-Garcia, O., Castillo, J., Marin, F. R., Ortuno, A., & Del-Rio, J. A. Uses and properties of citrus flavonoids. *J. Agric. Food Chem.,* **1997**, *45*(12), 4505–4515.

11. Bhatt, N., Singh, M., & Ali, A. Effect of feeding herbal preparations on milk yield and rumen parameters in lactating crossbred cows. *Int. J. Agric. Biol.,* **2009**, *11*(6), 721–726.

12. Bihrmann's Caudiciforms (*Ipomoea digitate*). http://www.bihrmann.com/caudici-forms/subs/ipo-dig-sub.asp (Accessed on 30 September 2019).

13. Bimbima: Daily life experience of ayurvedic medicines, complementary therapies, Anupama Prabhakar, Sundervihar, New Delhi, India, **2014**. http://www.bimbima. com/ayurveda/medicinal-herb-vidarialligator-yam-ipomoea-digitata/363/ (Accessed on 30 September 2019).

14. Biodiversity of India. A Wiki Resource for Indian Biodiversity. http://www.biodiver-sityofindia.org/index.php?title=Ipomoea_digitata (Accessed on 30 September 2019).

15. Boopathi, C. A., & Sivakumar, R. Phytochemical screening on leaves and stem of *Andrographis necsiana* Wight–an endemic medicinal plant from India. *World Appl. Sci. J.,* **2011**, *12*(3), 307–311.

16. Brouns, F. Soya isoflavones: A new and promising ingredient for the health food sector. *Food Res. Int.,* **2002,** *35,* 187–193.

17. Chandira, M., & Jayakar, B. Formulation and evaluation of herbal tablets containing *Ipomoea digitata* Linn. extract. *Int. J. Pharm. Sci. Rev. Res.,* **2010,** *3*(1), 101–110.

18. Chaudhari, A. B. *Endangered Medicinal Plants* (p. 216). Daya Publishing House, New Delhi, **2007.**

19. CIP, AVRDC. *IBPGR: Descriptors for Sweet Potato* (p. 12). International Board for Plant Genetic Resources (IBPGR): Rome, Italy, **1991.**

20. Dighe, V., & Adhyapak, S. Comparison of HPLC and HPTLC techniques for determination of umbelliferone from dried tuber powder of *Ipomoea mauritiana* Jacq. *Int. J. Pharm. Sci. Res.,* **2011,** *2*(11), 2894–2900.

21. Douglas, J. A., Smallfield, B. M., Burgess, E. J., Perry, N. B., Anderson, R. E., Malcolm, H., & Douglas, I. A. G. V. Sesquiterpene lactones in *Arnica montana*: A rapid analytical method and the effects of flower maturity and stimulated mechanical harvesting on quality and yield. *Planta Med.,* **2004,** *70,* 166–170.

22. *Drug Information System.* Pharma Professional Services, Karachi, Pakistan, **2002.**

23. Edison, S., Unnikrishnan, M., Vimala, B., Pillai, V. S., Shella, M. N., Sreekumari, M. T., & Abraham, K. *Biodiversity of Tropical Tuber Crops in India* (p. 96). Scientific Bulletin Number–7, NBA (National Biodiversity Authority), Chennai, Tamil Nadu, India, **2006.**

24. Efloraofindia. Database of Indian Plants: Developed by the members of Efloraofindia Google group, **2007.** http://sites.google.com/site/efloraofindia (Accessed on 30 September 2019).

25. Fazal, H., Ahamed, N., & Khan, M. A. Physiochemical, phytochemical evaluation and DPPH scavenging antioxidant potential in medicinal plants used for herbal formulation in Pakistan. *Pak. J. Bot. (Spec. Issue),* **2011,** *43,* 63–67.

26. *Flowers of India.* Department of Physics, Jamia Millia Islamia, New Delhi, India, **2016.** www.flowersofindia.net/catalog/slides/Giant%20Potato.html (Accessed on 30 September 2019).

27. *FRLHT's Clinically Important Plants of Ayurveda.* CD-ROM, Version 1.0, (Foundation for Revitalization of Local Health Traditions (FRLHT), Bangalore, India, **2002,** p. 117.

28. Garodia, P., Ichikawa, H., Malani, N., Sethi, G., & Aggarwal, B. B. Ancient medicine to modern medicine: Ayurvedic concepts of health and their role in inflammation and cancer. *J. Soc. Integr. Oncol.* [Online], **2006,** *5*(1), E-article, www.metaphysicspirit.com/books/Ayurvedic%20Concepts%20of%20Health.pdf (Accessed on 30 September 2019).

29. *GBIF Backbone Taxonomy.* GBIF (Global Biodiversity Information Facility), Universitetsparken 15, Copenhagen, Denmark, **2013.** www.gbif.org/species/8249252.2013 (Accessed on 30 September 2019).

30. Haider, P. *Ipomoea Digitata for Lowering Blood Pressure, Cholesterol and Immune Enhancement.* OMTimes Team [Online], **2013.** http://www.omtimes.com/wbp-content/uploads/2013/08/ipomoea-digitata-for-lowering-blood-pressure-and-cholestrol (Accessed on 30 September 2019).

31. Hao-Fu, D., Jiang, X., Jun, Z., & Zhong-Tao, D. The chemical constituents from roots of *Ipomoea digitata. Acta Metall. Sin.,* **2000,** *22*(2), 1–3.

32. Harada, H., Yamashita, U., Kurihara, H., Fukushi, E., Kawabata, J., & Kamei, Y. Anti-tumor activity of palmitic acid found as a selective cytotoxic substance in a marine Red Alga. *Anticancer Res.,* **2002**, *22*, 2587–2590.

33. Herbal Remedies. *Vidhari Kand–Fights All Toxic Substances in the Body*, **2014**. www.herbalsatt.blogspot.in/2014/01/68-vidari-kand-fights-all-toxic.html (Accessed on 30 September 2019).

34. Himalaya Global Holdings Ltd. *Himalayawellness: Herb Finder*. http://www.himalayawellness.com/herbfinder/ipomoea-digitata.html (Accessed on 30 September 2019).

35. Indiamart. *Lukowin Tablets*. India MART, InterMESh Ltd., Noida, Utter Pradesh, India, **1996–2016**. www.indiamart.com/proddetail/lukowin-tablets-4716655430.html (Accessed on 30 September 2019).

36. Islam, M. S., & Bari, M. A. Rapid *in-vitro* multiplication, callogenesis and in-direct shoot regeneration in *Ipomoea mauritiana*–A rare medicinal plant in Bangladesh. *Med. Aromat. Plants*, **2013**, *2*(6), 138.

37. Islam, S., Chowdhury, M. H. R., Hossain, I., Sayeed, S. R., Rahman, S., & Azam, F. M. S. A study on callus induction of *Ipomoea mauritiana*: An ayurvedic medicinal plant. *Am. Eurasian J. Sustain. Agric.,* **2014**, *8*(5), 86–93.

38. Iyer, K. N. *Pharmacognosy of Ayurvedic Drugs* (Vol. 5, p. 117). University of Kerala, Trivandrum, India, **1962**.

39. Jain, S. K. *Dictionary of Indian Folk Medicine and Ethnobotany* (p. 522). Deep publications, New Delhi, **1991**.

40. Jain, V., Verma, S. K., & Katewa, S. S. Therapeutic validation of *Ipomoea digitata* tuber (*Ksheervidari*) for its effect on cardio–vascular risk parameters. *Indian J. Tradit. Know.,* **2011**, *10*(4), 617–623.

41. Joy, P. P., Thomas, J., Mathew, S., & Skaria, B. P. Medicinal plants. In: Bose, T. K., Kabir, J., Das, P., & Joy, P. P., (eds.), *Tropical Horticulture* (Vol. 2, pp. 449–632). NayaProkash, Calcutta, **2001**.

42. Kalaiselvan, A., Anand, T., & Soundararajan, M. Reno-productive activity of *Ipomoea digitata* in gentamicin induced kidney dysfunction. *J. Ecobiotechnol.,* **2010**, *2*(2), 57–62.

43. Karthik, S. C., & Padma, V. Phytochemical and microscopic analysis of *Ipomoea mauritiana* Jacq. (Convolvulaceae). *Phcog. Mag.,* **2009**, *5*, 272–278.

44. *Keralaplants*. Ratlin Computers, (Roy, M. M.), Thrissur, Kerala, **2015**. http://ww.keralaplants.in/keralaplantsdetails.aspx?id=Ipomoea_mauritiana (Accessed on 30 September 2019).

45. Kessler, M., Ubeaud, G., & Jung, L. Anti- and pro-oxidant activity of rutin and quercetin derivatives. *J. Pharm. Pharmacol.,* **2003**, *55*, 131–142.

46. Khan, M. S., Nema, N. M. D., & Khanam, S. Chromatographic estimation of maturity based phytochemical profiling of *Ipomoea mauritiana*. *Int. J. Phytomed.,* **2009**, *1*, 22–30.

47. Khan, U. M., & Hossain, S. Md. Scopoletin and β – sitosterol glucoside from roots of *Ipomoea digitata*. *J. Pharmacogn. Phytochem.,* **2015**, *4*(2), 5–7.

48. Khare, C. P. *Indian Medicinal Plants: An Illustrated Dictionary* (p. 213). Springer, Verlag Berlin/Heidelberg, **2007**.

49. Kirtikar and Basu's illustrated Indian medicinal plants: Their usage in Ayurveda and Unani. *Indian Medicinal Science Series* (Vol. II, p. 321), Sri Satguru Publications, A Division of Indian Books Center, New Delhi, India, **2000**.

50. Kirtikar, K. R., & Basu, B. D. *Indian Medicinal Plants* (2nd edn., Vol. III, pp. 1717–1718). International Book Distributors, Dehradun, India, **1987**.

51. Kuntal, D., Raman, R., & Shilpashree, V. K. Pharmacognostic authentication and constituent validation by HPLC for four different species of *Vidari* marketed in India. *South Pacific J. Pharma. Bio Sci.,* **2015**, *3*(1), 217–229.

52. La-Medica (India) Private Limited. *Herbal Medicines in India: Raw Herbs.* www.la-medicca.com/raw-herbs-Ipomoea-digitata.html (Accessed on 30 September 2019).

53. Lee, J. H., Lee, K. T., Yang, J. H., Baek, N. I., & Kim, D. K. Acetylcholinesterase inhibitors from the twigs of *Vaccinium oldhami* Miquel. *Arch. Pharm. Res.,* **2004**, *27*, 53–56.

54. Madhavi, D., Rao, B. R., Sreenivas, P., Krupadanam, D. G. L., Rao, P. M., Reddy, J. K., & Kishore, P. B. K. Isolation of secondary products from *Ipomoea digitata* a medicinally important plant. Online, **2010**, Pharmaceutical Information, Articles and Blogs. http://www.pharmainfo.net/articles/isolation-secondary-products-ipomoea-digitata-medicinally-important-plant (Accessed on 30 September 2019).

55. Mahajan, K. G., Mahajan, R. T., & Arun, M. Y. Improvement of sperm density in neem–oil induced infertile male Albino rats by *Ipomoea digitata* L. *J. Intercult. Ethnopharmacol.,* **2015**, *4*(2), 125–128.

56. Mahendra, C., Gowda, D. V., & Babu, U. V. Insignificant activity of *Gmelina asiatica* and *Ipomoea digitata* against skin pathogens. *J. Pharm. Negative Results,* **2015**, *6*(1), 27–32.

57. Mathew, G., Joy, P. P., Skaria, B. P., & Mathew, S. *Cultivation Prospects of Tuberous Medicinal Plants* (p. 8). ResearchGate: Conference paper (Online), **2005**. https://www.researchgate.net/publication/305995299 (Accessed on 30 September 2019).

58. Matin, M. A., Tewari, J. P., & Kalani, D. K. Pharmacological effects of Paniculatin – A glycoside isolated from *Ipomoea digitata* Linn. *J. Pharm. Sci.,* **1969**, *58*(6), 757–759.

59. *Medicinal Plants Database of Bangladesh.* Ethanobotany Lab, Department of Botany, Chittagong University, Bangladesh. www.mpbd.info/plants/ipomoea-mauritiana.php (Accessed on 30 September 2019).

60. *Medicinal Plants India,* **2016**, Shalu Francis, Athirampuzha, Kottayam, Kerala, India. www.medicinalplantsindia.com/dhanwantararishtam.html (Accessed on 30 September 2019).

61. Meera, S., & Narasimhulu, K. Phytochemical screening and the evaluation of the antioxidant, antimicrobial and analgesic properties of the plant *Ipomoea mauritiana*. *Int. J. Green Pharm.,* **2011**, *4*(2), 61–64.

62. Meira, M., Da Silva, E. P., David, J. M., & David, J. P. Review of the genus *Ipomoea*: Traditional uses, chemistry and biological activities. *Braz. J. Pharmacogn.,* **2012**, *22*(3), 682–713.

63. Menon, P. Conservation and consumption: a study on the crude drug trade inthreatened medicinal plants in Thiruvananthapuram District, Kerala. KRPLLD Research Report No. 85/99, Kerala Research Program on Local Level Development (KRPLLD), Center for Development Studies (CDS), Trivandrum, Kerala, India, **2003**, p. 87.

64. Minaz, N., Rao, V. N., Nazeer, A., Preeth, M., & Shobana, J. Anti-diabetic activity of hydro–ethanolic root extracts of *Ipomoea digitata* in alloxan induced diabetic rats. *Int. J. Res. Pharm. Biomed. Sci.,* **2010**, *1*(2), 76–81.

65. Mishra, B. *Maharishi University of Management Vedic Literature Collection. Bhava Praksaha* (p. 164). H. 4, Verse 180–181, Maharishi University of Management, Fairfield, Iowa: United States, **1969**.

66. Mishra, S. S., & Datta, K. C. A preliminary pharmacological study of *Ipomoea digitata* Linn. *Indian J. Med. Res.*, **1962**, *50*, 43–45.

67. Mishra, S. S., Tewari, J. P., & Matin, M. A. Investigation of the fixed oil from *Ipomoea digitata* tubers. *J. Pharm. Sci.*, **1964**, *3*, 471–472.

68. Mohanan, M., & Henry, A. N. Flora of Thiruvananthapuram. In: *Flora of India* (pp. 315–316). Botanical Survey of India, Calcutta, **1994**.

69. Monjur-Al-Hossain, A. S. M., & Hasan, M. Md., Shamsunnahar, K., Avijit, D., & Khan, R. Md. Phytochemical screening and the evaluation of the antioxidant, antimicrobial and analgesic properties of the plant *Ipomoea mauritiana*. *Int. Res. J. Pharm.*, **2013**, *4*(2), 60–63.

70. Muthu, K. A., Alagumanivasagam, G., Kumar, S. D., & Manavalan, R. Effect of methanolic extract of tuberous root of *Ipomoea digitata* L. on hyperlipidemia induced by rat fed with high fat diet. *Res. J. Pharm. Biol. Chem. Sci.*, **2011**, *2*(3), 183–191.

71. Nair, K. K. N. *Manual of Non–Wood Forest Produce Plants of Kerala* (p. 65). Research Report 185, Kerala Forest Research Institute (KFRI), KAU, Thrissur, Kerala, India, **2000**.

72. Narasimhan, S., Shobana, R., & Sathy, T. N. Antioxidants–Natural rejuvenators that heal, detoxify and provide nourishment. In: Rakesh, K. S., & Rajesh, A., (eds.), *Herbal Drugs: A Twenty First Century Prospective* (pp. 548–557). J.P. Brothers Medical Publishers, New Delhi, **2006**.

73. Ndam, L. M., Mih, A. M., Fongod, A. G. N., Tening, A. S., Tonjock, R. K., Enang, J. E., & Fujii, Y. Phytochemical screening of the bioactive compounds in twenty (20) Cameroonian medicinal plants. *Int. J. Curr. Microbiol. Appl. Sci.*, **2014**, *3*(12), 768–778.

74. Ojha, G., Mishra, K. N., & Mishra, A. Pharmacological uses and isolated chemical constituents of *Ipomoea digitata*: A review. *IOSR J. Pharm. Biol. Sci.*, **2016**, *11*(3), 1–4.

75. Ono, M., Fukuda, H., Muraya, H., & Miyahara, K. Resin glycosides from the leaves and stems of *Ipomoea digitata*. *Nat. Med. Aro.*, **2009**, *63*(2), 176–180.

76. *Oshadhasasyangal: Krishiyum Upayogavum* (p. 89). Kerala Agricultural University Press, Mannuthy, Thrissur, Kerala, India, **2013**.

77. Pal, A. *Biotechnology, Secondary Metabolites, Plants and Microbes* (p. 215). Scientific Publishers: Portland, US, **2007**.

78. Panda, H. *Herbs Cultivation and Medicinal Uses* (p. 98). National Institute of Industrial Research, New Delhi, India, **1999**.

79. Pandey, A., Gupta, P. P., & Lal, V. K. Preclinical evaluation of hypoglycemic activity of *Ipomoea digitata* tuber in streptozotocin–induced diabetic rats. *J. Basic Clin. Physiol. Pharmacol.*, **2013**, *24*(1), 35–39.

80. Pandey, N., & Tripathi, Y. B. Antioxidant activity of tubers in isolated from *Pueraria tuberosa* Linn. *J. Inflam.*, **2010**, *7*, 47–54.

81. Pandit, S., Biswas, P. K., Debnath, P. K., Saha, A. U., Chowdhury, U., Shaw, B. P., Sen, S., & Mukherjee, B. Chemical and pharmacological evaluation of different ayurvedic preparations of iron. *J. Ethnopharmacol.*, **1999**, *65*, 149–156.

82. *Plant Resources of Tropical Africa (PROTA): Ipomoea Mauritiana* (p. 78). Wageningen University, Netherlands–Europe, **2015**.

83. Rahmatullah, M., Anzumi, H., Rahman, S., & Islam, M. A. Uncommon medicinal plant formulations used by a folk medicinal practitioner in Nagaon District, Bangladesh. *World J. Pharm. Pharm. Sci.,* **2014**, *3*(12), 176–188.

84. Rahmatullah, M., Mollik, A. H. M., Islam, K. M., & Islam, R. M. A survey of medicinal and functional food plants used by the folk medicinal practitioners of three villages in SreepurUpazilla, Magura District, Bangladesh. *Am. Eurasian J. Sustain. Agric.,* **2010**, *4*(3), 363–373.

85. Rani, U. P., Naidu, M. U. R., Kumar, R. T., & Sobha, J. C. Evaluation of the efficacy and safety of a new herbal revitalizer revivin. *Ancient Sci. Life,* **1997**, *16*(3), 190–195.

86. Rao, B., Suseela, K., Rao, P. V. S., Krishna, P. G., & Raju, G. V. S. Chemical examination of some Indian medicinal plants. *Indian J. Chem.,* **1984**, *23B*, 780–787.

87. Raut, S. S., Rane, A. D., Wanage, S. S., & Bhave, S. G. Vegetative propagation through leafy shoot cuttings of *Ipomoea mauritiana* under different potting media. *J. Tree Sci.,* **2011**, *30*(1 & 2), 16–19.

88. Sagar, P. K. Adulteration and substitution in endangered asu herbal medicinal plants of India, their legal status, scientific screening of active phytochemical constituents. *Int. J. Pharm. Sci. Res.,* **2014**, *5*(9), 4023–4039.

89. Sarkar, P. R. *Yaogika Cikitsa and Dravya Guna (Compound Therapy and Fluid Properties)* (1ˢᵗ edn., p. 68). AMPS Publication, Tiljala, Calcutta, India, **1969**.

90. Sasidharan, N., & Sivarajan, V. V. Flowering plant of Thrissur forest: Western Ghats, Kerala, India. *Journal of Economic and Taxonomic Botany: Additional Series* (Scientific publishers: Jodhpur, India), **1996**, *4*, 570–579.

91. Satveda–By Herbs Forever (Since 1932). HerbsForever, Los Ángeles, California, USA. www.satveda.com/p/vidari-kanda-powder (Accessed on 30 September 2019).

92. Shahidullah, A. K. M., & Haque, C. E. Linking medicinal plant production with livelihood enhancement in Bangladesh: Implications of a vertically integrated value chain. *The Journal of Transdisciplinary Environmental Studies* (Online), **2010**, *9*(2), 1–18. http://www.journal-tes.dk/ (Accessed on 30 September 2019).

93. Shakhawat, H., Islam, J., Ahmed, F., Hossain, M. A., Siddiki, M. A. K., & Hossen, S. M. M. Free radical scavenging activity of six medicinal plants of Bangladesh: A potential source of natural antioxidant. *J. Appl. Pharm.,* **2015**, *7*(1), 96–104.

94. Sharma, S. C., Shukla, Y. N., & Tandon, J. S. Constituents of *Colocasia formicate, Sagittariasagittifolia, Arnebianobilis, Ipomoea paniculata, Rhododendron niveum, Paspalumscrobiculatum, Mundulea sericea* and *Duabangasonneratiodes. Phytochem. Rep.,* **1972**, *11*, 2621–2623.

95. Sharmin, M., Banya, P. D., Paul, L., Chowdhury, F. F. K., Afrin, S., Acharjee, M., Rahman, T., & Noor, R. Study of microbial proliferation and the *in vitro* antibacterial traits of commonly available flowers in Dhaka metropolis. *Asian Pac. J. Trop. Dis.,* **2015**, *5*(2), 91–97.

96. Singh, M. P., & Panda, H. *Medicinal Herbs and Their Formulations* (Vol. II, p. 217). Daya publishing House: New Delhi, India, **2005**.

97. Singh, V., Srivastava, V., & Sethi, R. *Ipomoea digitata* seed gum and the gum–g-polyacrylamide: Potential pharmaceutical gums. *Pharm. Biol.,* **2004**, *42*, 230–233.

98. Sivarajan, V. V., & Balachandran, I. *Ayurvedic Drugs and Their Plant Sources.* Oxford and IBH publishing Co. Pvt. Ltd., New Delhi, India, **1994**, p. 174.

99. Sonia, N. S., & Jessykutty, P. C. *Abstract Compendium, Pharmacia 2016* (p. 41). International Conference on Emerging Pharma Innovations–Challenges and Strategies, Srikrishna College of Pharmacy and Research Center, Parassala, Thiruvananthapuram, Kerala, **2016**.

100. Sonia, N. S., & Jessykutty, P. C. *Abstract Book. Convention for Multidisciplinary Healthcare Consortium: World Congress on Drug Discovery and Development* (pp. 250, 251). J.N. Tata Auditorium, Indian Institute of Science, Bengaluru, India, **2016**.

101. *Species 2000 & IT IS Catalogue of Life*. Naturalis, Leiden, the Netherlands, **2000**, p. 53. www.catallogueoflife.org/col/details/species/id/09d2f03c0e2ca56539c96da2a4a 15de1 (Accessed on 30 September 2019).

102. *Sree Dhanwantri Herbals*. Houston, US, **2015**. www.shreedhanwantri.com/shop/home/151-manmathabhar-rasa.html (Accessed on 30 September 2019).

103. Stuartxchange. *Philippine Medicinal Plants*. Aurorang–gubat. www.stuartxchange.com/AurorangGubat.html (Accessed on 30 September 2019).

104. T-Herb Directory, **1999–2016**. Oshims herbal online pharmacy, World of Herbal Remedies and Alternative Medicine, Anaheim, US. http://www.oshims.com/herb-directory/v/vidari-kanda (Accessed on 30 September 2019).

105. Tewari, J. P., & Mishra, S. S. Pharmacological investigations of *Ipomoea digitata*. *Vijnana Parishad Ausandhan Patrika*, **1965**, *7*, 85–88.

106. *The Ayurvedic Pharmacopoeia of India, Part–I* (1st edn., Vol. V, p. 119). AYUSH (Department of Ayurveda, Yoga and Naturopathy, Unani, Siddha and Homoeopathy), Ministry of Health and Family Welfare, Government of India: New Delhi, **2006**.

107. *The International Plant Names Index*. The Plant Names Project: Kew, Richmond, UK, **2012**, http://www.ipni.org (Accessed on 30 September 2019).

108. *The Plant List-A Working List of All Plant Species* [Online], Kew Science, London, Online, September **2013**, http://www.theplantlist.org (Accessed on 30 September 2019).

109. Trivedi, P., Verma, U., Singh, R., Joshi, P. K., & Rout, O. P. 'Rasayana' herbs used in Ayurveda–A review. *World J. Pharm. Pharm. Sci.*, **2015**, *4*(5), 1829–1837.

110. *U. S. National Plant Germplasm System*. United States Department of Agriculture: Agricultural Research Service, Washington D C, USA, **2015**.

111. Unnikrishnan, E. *Materia Medica of the Local Health Traditions of Payyanur* (p. 43). Discussion paper No. 80, Kerala Research Program on Local Level Development (KRPLLD), Center for Development Studies (CDS), Trivandrum, Kerala, India, **2004**.

112. Varier, P. S. *Indian Medicinal Plants: A Compendium of 500 Species* (Vol. III, p. 217). Orient Longman Publishing, Kottakkal, Kerala, India, **1997**.

113. Vasagam, G., Muthu, K. A., & Manavalan, R. *In – vitro* antioxidant potential of tuberous root of methanolic extract of *Ipomoea digitata* (Linn.). *Int. J. Pharm. Bio Sci.*, **2010**, *1*(2), 1–5.

114. Vasu Research Center: A Division of Vasu Health Care, Vadodara, Gujarat, India. www.vasuresearch.com/ipomoea-digitata/html (Accessed on 30 September 2019).

115. Venkatasubramanian, P., Kumar, S. K., & Venugopal, S. N. Use of 'Kshiravidari' as a substitute for 'Vidari' as per ayurvedic descriptions. *Indian J. Tradit. Knowl.*, **2009**, *8*, 310–318.

116. Vidya, K. M., *Diversity Analysis and Reproductive Biology of Milk Yam (Ipomoea digitata L.)* (p. 219). PhD thesis (unpublished), Kerala Agricultural University, Thrissur, Kerala, India, **2010**.

117. Viji, Z., & Paulsamy, S. Phytoconstituent analysis and GC – MS profiling of tubers of *Ipomoea mauritiana* Jacq. (Convolvulaceae). *Int. J. Recent Adv. Multidisciplinary Res.,* **2016**, *3*(3), 1345–1349.

118. Wiart, C. *Ethnopharmacology of Medicinal Plants: Asia and the Pacific* (p. 318). Springer Science and Business Media, New York, **2007**.

119. Williams, C. *Medicinal Plants in Australia*: *Plants, Potions and Poisons* (Vol. III, p. 118) (Online). EBL eBooks online: Rosenberg Publishing, 3 White Hall Road, Kenthurst NSW 2156, Australia, **2012**. https://books.google.co.in/books?id=ieVUA QAAQBAJ&printsec=frontcover&source=gbs_ge_summary_r&cad=0#v=onepage &q&f=false (Accessed on 30 September 2019).

120. Yousuf, M., Begum, J., Hoque, M. N., & Chowdhury, J. U. *Medicinal Plants of Bangladesh (Revised and Enlarged)* (p. 219). Paramount Printing Service: Andarkilla, Chittagong, Bangladesh, **2009**.

121. Zafar, M., Khan, M. A., Ahmad, M., Sultana, S., Qureshi, R., & Tareen, R. B. Authentication of misidentified crude herbal drugs marketed in Pakistan. *J. Med. Plants Res.,* **2010**, *4*(15), 1584–1593.

INDEX